Advances in
Planar Lipid Bilayers and Liposomes

EDITORIAL BOARD

Advances in Planar Lipid Bilayers and Liposomes

Volume 2

Editor

A. Ottova-Leitmannova

Department of Physiology
Michigan State University
East Lansing, Michigan
USA
and
Center for Interface Sciences
Microelectronics Department
Faculty of Electrical Engineering & Information
Slovak Technical University, Bratislava
Slovak Republic

Founding Editor

H.T. Tien

Department of Physiology
Michigan State University
East Lansing, Michigan
USA

2005

ELSEVIER
ACADEMIC
PRESS

Amsterdam • Boston • Heidelberg • London • New York • Oxford • Paris
San Diego • San Francisco • Singapore • Sydney • Tokyo

ELSEVIER B.V.
Radarweg 29
P.O. Box 211, 1000 AE Amsterdam
The Netherlands

ELSEVIER Inc.
525 B Street, Suite 1900
San Diego, CA 92101-4495
USA

ELSEVIER Ltd
The Boulevard, Langford Lane
Kidlington, Oxford OX5 1GB
UK

ELSEVIER Ltd
84 Theobalds Road
London WC1X 8RR
UK

First edition 2005

Library of Congress Cataloging in Publication Data
A catalog record is available from the Library of Congress.

British Library Cataloguing in Publication Data
A catalogue record is available from the British Library.

ISBN: 0 12 369453 1
ISSN: 1554-4516

♾ The paper used in this publication meets the requirements of ANSI/NISO Z39.48-1992 (Permanence of Paper).
Printed in The Netherlands.

CONTENTS

CONTRIBUTORS

PREFACE

The inspiration for lipid bilayer research, without question, comes from the biological world. Although the first report on self-assembled bilayer lipid membranes (BLMs) *in vitro* was reported in 1961, experimental scientists have been dealing with these interfacial phenomena since Robert Hooke's time (1672). In this connection, the conclusion that the fundamental structure of biomembranes is a lipid bilayer is based on three pivotal experimental findings:

- Firstly, the elegant and simple experiment of establishing the orientation of amphipathic molecules at interfaces by Langmuir in 1917,
- Secondly, using Langmuir's method Gorter and Grendel reported in 1925 that the extracted lipids from the plasma membrane of red blood cells (RBC) that occupied the area on the surface of a Langmuir trough which was twice that of the original membrane, and suggested that its structure is not unlike that of a soap bubble, and
- Thirdly, as deduced from the above indirect evidence, and from Fricke's electrical measurements of RBC suspension, as well as from the Unit Membrane Hypothesis of J.D. Robertson based on electron microscopy in the 1950s, the lipid bilayer structure was dramatically demonstrated after D.O. Rudin and his associates reconstituted a black lipid membrane (BLM) from the lipids extracted from cow's brain (*Nature*, 1962, **194**, 979).

Today, from all lines of research findings, all cell plasma membranes possess a lipid bilayer structure, thereby underlying the lipid bilayer principle of biomembranes. Thus, the lipid bilayer research over the past four decades has evolved as an interdisciplinary effort, benefited by a cross-fertilization of ideas. The lipid bilayer associated with life sciences and biotechnology is of current interest to a diversity of investigators, including biochemists, biologists, biophysicists, bioengineers and technologists, electrochemists, physiologists, pharmacologists, surface and colloid scientists, and others working on ultrathin films and membrane phenomena. In particular, BLMs have been used in a number of applications ranging from basic membrane biophysics studies to the conversion of solar energy *via* water photolysis, and to biosensor development using supported bilayer lipid membranes (s-BLMs). s-BLMs (supported planar lipid bilayers) provide the foundation for a variety of lipid bilayer-based molecular sensors that are sensitive, versatile, inexpensive (i.e., disposable) and open to all sorts of experimentation and development.

With the above background in mind, the present *Advances* series on BLMs (planar lipid bilayers and liposomes) include invited chapters on a broad range of topics, ranging from theoretical investigations, specific studies, experimental methods, to practical applications. The author(s) of each chapter solicited to focus mainly on the work of his/her laboratory, with minimal reviews of others. The goals of the chapters are for newcomers, but also for experienced scientists, and for others who are not familiar with the research areas dealt with. An aim of the series is to cover all aspects of lipid bilayer investigations, both fundamental and applied. These contributed chapters are entities to themselves, thereby strengthening the lipid bilayer principle of biomembranes, and are related to the overall lipid bilayer venture. We are grateful to each of the contributors for their expert knowledge and willingness to share generously their information.

H. Ti Tien
Angelica Ottova-Leitmannova

CHAPTER 1

Electrochemical Methods and Their Application

Sławomir Kalinowski*

Department of Chemistry, University of Warmia and Mazury, 10-957 Olsztyn, Poland

Contents

---

* Corresponding author. Tel.: +48-89-5233711; Fax: +48-89-5240408;
E-mail: kalinow@uwm.edu.pl

ADVANCES IN PLANAR LIPID BILAYERS AND LIPOSOMES, VOLUME 2
ISSN 1554-4516 DOI: 10.1016/S1554-4516(05)02001-6

Abstract

The chapter presents basic principles of the most common electrochemical techniques used for investigations of bilayer lipid membranes, and examples of electronic equipment: potential meters, 2-, 3-, and 4-electrode potentiostats and galvanostats. It also describes the origin of transmembrane potential in membrane systems, examples for application of potentiometry, amperometry, voltammetry, and chronopotentiometry. An important and characteristic parameter of membranes is their capacitance. Methods of capacitance measurement and their application are presented. Due to their parameters, lipid membrane systems require very high resistance, extremely low currents and high capacitance. They are sensitive to distortions and external noises. Noise-protection methods are also described in the chapter.

1. INTRODUCTION

Life processes are dependent on electric phenomena taking place in cell membranes. Transport of ions through the membrane and generation of the membrane potential are among basic life processes. The development of electrochemical research methods allows to understand many important processes observed in cells. These methods, known from classical electrochemistry, turned out to be very useful for analysis of cell membranes and artificial lipid membranes. The understanding of certain membrane phenomena enabled their practical application, mainly to analytical chemistry, medicine, and biochemistry. Bilayer lipid membranes were found to be a good matrix for receptor molecules separated from natural membranes and a wide variety of artificial transducers.

Electrochemical methods of analysis are relatively inexpensive and simple. Measurement devices can be miniaturized and applied as portable instruments. Rapid development of electronics makes it possible to employ numerous electrochemical measurement techniques. The application of computers for controlling measurement devices and data acquisition allows to use more and more advanced electrochemical techniques with simple electronic analog circuits, e.g., amperometry with programmable potential, electrochemical quartz crystal microbalance (EQCM), impedance analysis. Many companies, e.g., National Instruments, Advantech, CyberResearch, offer interfaces useful for data acquisition and controlling external devices as relays, solenoid valves, electrical motors, measurement of frequency, oscilloscope cards.

Measurement devices applied to membrane electrochemistry must be characterized by certain important parameters like high current sensitivity in a range of picoampers or lower, necessary for analysis of, e.g., single-channel proteins. Very high resistance of membranes disturbs the measurement of the transmembrane potential. Capacitance measurements performed using typical capacitance meters are impossible due to a too high measurement signal applied to the membrane by the meter, usually higher than 1 V. Bilayer lipid membranes are broken by a much lower potential, often below 200 mV.

This chapter describes the equipment useful for membrane studies and examples of its applications.

2. POTENTIOMETRY

2.1. Measurement equipment

Measurement of the transmembrane potential is one of the most commonly used and simplest techniques in membrane electrochemistry. Figure 1 shows examples of the application of this kind of measurements to membrane systems. This method is generally restricted to membranes separating two electrolyte solutions, where the Donnan or diffusion potential is generated. Systems with membranes covering solid support, e.g., metal, glassy carbon, conductive glass, do not generate this potential.

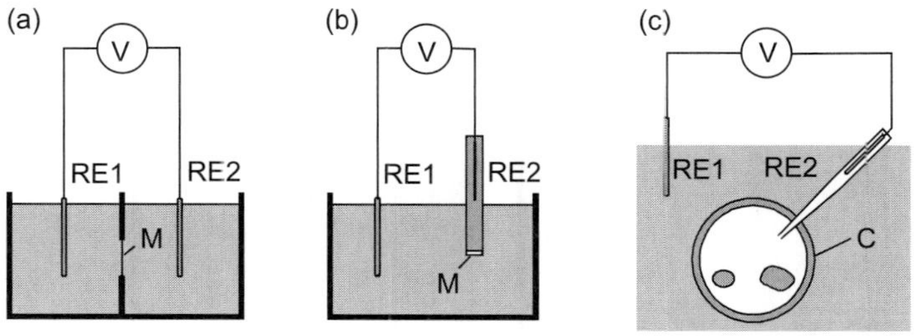

Fig. 1. Examples of potentiometric measurements in membrane systems: (a) membrane is separating two water solutions, (b) the membrane is supported on a gel electrode, (c) measurement of the membrane potential in a live cell using application of a microelectrode. RE1, RE2 are reference electrodes, M is membrane, C is cell.

Correct measurement of the transmembrane potential in lipid membranes depends on several conditions. Membranes sometimes exhibit resistance in a range of a few dozen gigaohms. In this case the inputs of the devices used for transmembrane potential measurements must show very high resistance, at least two ranges higher than the resistance of membranes. Another important parameter is input offset current, whose value should be as low as possible, lower than 1 pA. Otherwise the read value of the potential would be burdened with a gross error. Figure 2 presents the important parameters of the system measured and measurement device which have an influence on the measurement accuracy. Since the measurement was performed under static conditions, the capacitances of this model circuit could be ignored. The membrane, electrolytes and electrodes are shown as a source of the potential E_M composed of the membrane potential (Donnan or diffusion potential) and electrode potentials on both sides of the membrane. The membrane with electrodes and electrolytes exhibits the resistance shown in Fig. 2 as R_M. The membrane resistance is usually several ranges higher than the resistances of electrodes and electrolytes. In this case it can be assumed that R_M is equal to the membrane resistance.

Measurement devices do not have ideal input parameters. The main source of measurement errors can be the input resistance of the measurement device R_{inp} and the input offset current i_{off}. The input resistance R_{inp} causes a decrease in the read value of the potential. Additionally, the input of the measurement device is a source of low direct current. In the case of measurement of the potential generated by a signal source with high internal resistance an additional value appears, causing a measurement error. The input offset current i_{off} generates additional voltage added to the measured signal.

If the input offset current i_{off} is ignored, the read value of the potential can be described by the equation:

$$E_P = E_M \frac{R_{inp}}{R_M + R_{inp}} \tag{1}$$

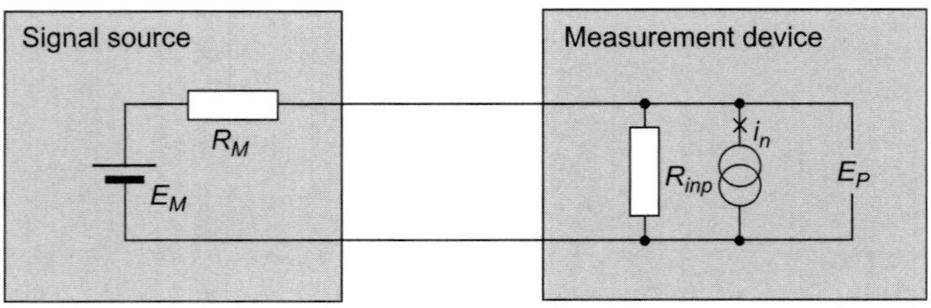

Fig. 2. Essential parameters of the potential source and the measurement device affecting the accuracy of the measurement. E_M is the membrane potential, R_M is the membrane resistance, R_{inp} is the input resistance of the measurement device, i_{off} is the input offset current, E_P is the displayed value of potential.

If the membrane resistance R_M is equal to 10^{10} Ω, the input resistance R_{inp} equal to 10^{12} Ω results in a measurement error of about 1%. In practice, the input resistance of the measurement device R_{inp} should be in the range of 10^{13}–10^{14} Ω. The input offset current i_{off} causes an additional error ΔE_{off} which can be calculated from Ohm's law:

$$\Delta E_{off} = i_{off} R_M \quad (2)$$

For the membrane resistance R_M equal to 10^{10} Ω and the current i_{off} equal to 1 pA the error ΔE_{off} is equal to 10 mV.

Many devices designed for voltage measurement are available on the market. Their suitability for membrane applications depends on the input resistance R_{inp} and the input offset current i_{off}. Simple voltage meters are useless due to their low input resistance, usually *ca* 1 MΩ. The best devices for membrane potential measurement are electrometers – instruments with extremely high input resistance, reaching 10^{15} Ω, and input offset current in the range of femtoampers. Also pH-meters and ion-meters can be used for the measurement of potential, as they have relatively good parameters. pH-meters are designed for co-working with glass electrodes with internal resistance in a range of hundreds of megaohms.

The input resistance of the device is usually given in the technical documentation attached to the device. The input parameters can be estimated by making several simple observations and measurements. If the device displays stability, the voltage equal to zero or close to zero usually indicates low input resistance of the device. Devices with high input resistance are sensitive to static electrical fields occurring, e.g., when clothes pick up static electricity. Keeping a hand close to the wire connected to the input causes changes in the displayed voltage. It is important to be careful because the inputs can be destroyed by this electrical charge.

The input resistance can be measured using a 1.5 V battery and a reference resistor of the same range as the input resistance. First it is necessary to measure the voltage E of the battery connected directly to the input (Fig. 3a). Next the resistor R should be connected in series with the battery (Fig. 3b) and the voltage E' should be measured again. The input resistance R_{inp} can be calculated

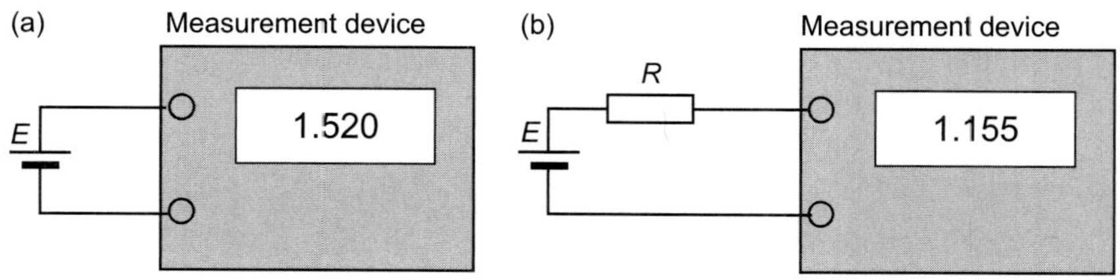

Fig. 3. Estimation of the input resistance of a potential meter: (a) battery voltage measurement, (b) voltage measurement with resistor R connected in series.

from the equation:

$$R_{\text{inp}} = R\frac{E'}{E - E'} \tag{3}$$

This method is useless when the measurement device is characterized by very high input resistance R_{inp} and relatively high input offset current i_{off}. If the input is not connected with any signal, this current generates such high voltage that the meter indicates the exceeding of the measuring range. In this case another method is useful. The capacitor C should be connected to the input of the device (Fig. 4) and changes in the displayed voltage should be recorded. After discharge caused by short circuit, the capacitor C (e.g., 10 nF) is charged with the current i_{off}. The voltage on the capacitor C increases and a part of the total current i_{off} flows through the resistance R_{inp}. When the displayed voltage E reaches the value E_{max}, the total current i_{off} is flowing only through the input resistance R_{inp}. The current charging the capacitance C is equal to zero and the displayed voltage E does not change.

The initial slope of the dE/dt curve (Fig. 4b) is a function of the current i_{off} and the capacitance C. The input offset current i_{off} can be calculated from the equation:

$$i_{\text{off}} = C\frac{\mathrm{d}E}{\mathrm{d}t} \tag{4}$$

The known value i_{off} makes it possible to estimate the input resistance R_{inp}:

$$R_{\text{inp}} = \frac{E_{\text{max}}}{i_{\text{off}}} \tag{5}$$

The input parameters of the measurement device can be improved by a simple method, i.e., the application of an operational amplifier with adequate parameters, configured as a voltage follower connected between the potential source and

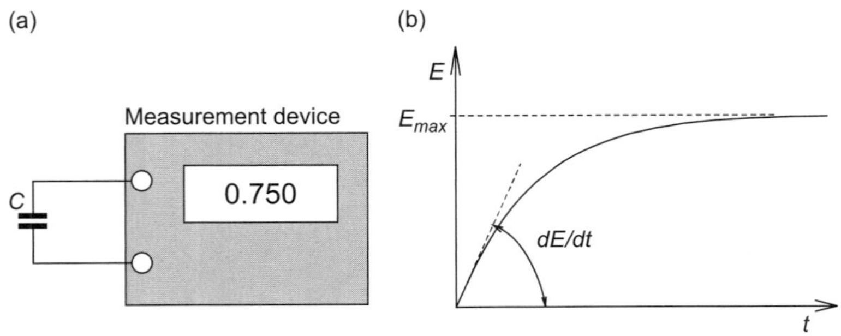

Fig. 4. Estimation of the input offset current and the input resistance of the device used for measurement of transmembrane potential: (a) connection of the capacitor to the input, (b) changes of displayed potential E in a time t.

the device input. Table 1 presents selected operational amplifiers for membrane potential measurements. This group of amplifiers includes both inexpensive and expensive high-performance ones. A simplified diagram of connections is shown in Fig. 5.

The operational amplifier should be placed next to the electrodes. The signal coming from the output of the amplifier is more resistant to external noises than the signal sent directly from the electrodes. This signal can be transmitted by a cable to the input of, e.g., a digital voltmeter, recorder, data acquisition card inserted to the computer. The operational amplifier should be supplied from two batteries placed close to it. This allows to protect the voltage follower against any noises from the power network. Detailed information concerning the application of amplifiers can be found in catalogues. In order to protect high input resistance and low input offset current the printed-circuit board should have additional conductive traces separating supply power lines and input lines. These traces protect inputs against currents flowing on the surface of the plate.

Table 1. Selected operational amplifiers applied as voltage followers to measurements of a membrane potential

Type	Producer	Input resistance (Ω)	Input offset current (pA)	Notices
AD 549K	AD	10^{15}	0.03	
OPA 128KM	BB	10^{15}	0.03	
OPA 129P	BB	10^{15}	0.03	
CA 3130	H	1.5×10^{12}	0.1	Supply max ±8 V
CA 3140	H	1.5×10^{12}	0.5	
CA 3160	H	1.5×10^{12}	0.5	Supply max ±8 V
CA 3240	H	1.5×10^{12}	0.5	Double
CA 3260	H	1.5×10^{12}	0.5	Double, supply max ±8 V
LMC 6041	NS	$>10^{13}$	0.001	Supply max ±8 V
LMC 6042	NS	$>10^{13}$	0.001	Double, supply max ±8 V
LMC 6044	NS	$>10^{13}$	0.001	Quadruple, supply max ±8 V
LMC 6061	NS	$>10^{13}$	0.005	Supply max ±8 V
LMC 6062	NS	$>10^{13}$	0.005	Double, supply max ±8 V
LMC 6064	NS	$>10^{13}$	0.005	Quadruple, supply max ±8 V

AD, Analog Devices; BB, Burr Brown; H, Harris; NS, National Semiconductor.

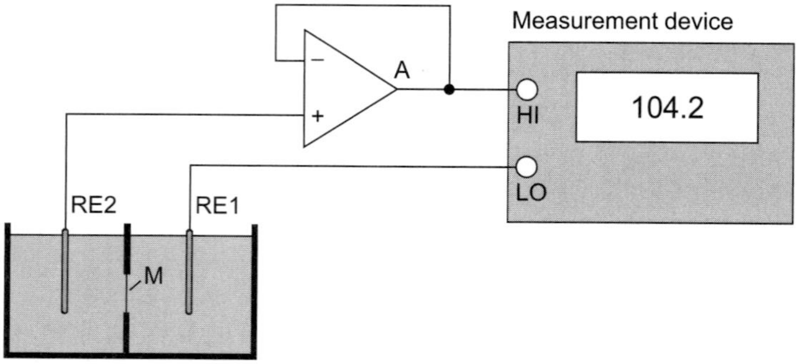

Fig. 5. A method for connecting of the operational amplifier A between the electrodes and the measurement device.

2.2. Origin of transmembrane potential

Selective transport of ions between electrolyte solutions separated by a membrane leads to a durable difference of potentials, dependent on the activity of the ions. The value of the potential is described by a thermodynamic balance in the system where the transport of electrical charges through the membrane is fixed and equal to both directions. When ions of one type alone are transported, the transmembrane potential is referred to as the Donnan potential. When ions of several types are transported, the transmembrane potential is a result of diffusion potential.

Life processes taking place in cells are dependent on the transmembrane potential, which is negative inside the cell, compared with the environment, and whose value varies from several to above 100 mV [1].

2.2.1. Donnan potential

The generation of the Donnan (Gibbs–Donnan) potential [1,2] is possible if the following conditions are fulfilled:

- two electrolyte solutions are separated by a semipermeable membrane,
- ions of one type alone are transported through the membrane,
- the ions are characterized by different activity on both sides of the membrane.

The origin of the Donnan potential can be explained by means of analysis of the kinetics of ion transport through the membrane, or thermodynamical equations. This potential is marked as $\Delta\phi$ in Fig. 6. It was assumed, for the purpose of the analysis, that solutions I and II contain cations *i* that can be transported through the membrane, and that the activity of the ions is higher in solution I ($a_i^{I} > a_i^{II}$), other ions in the solutions cannot be transported.

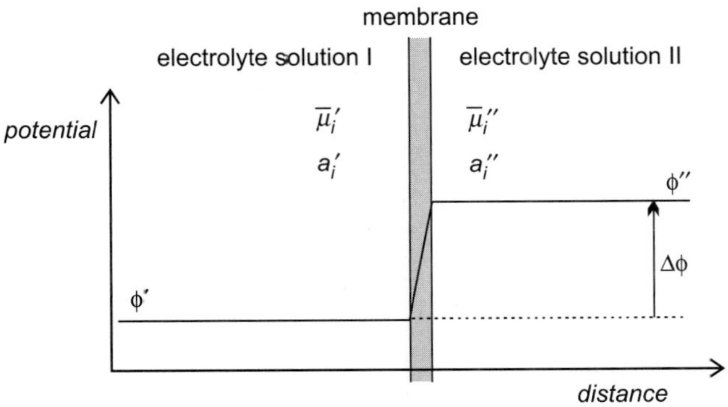

Fig. 6. Generation the Donnan potential in a system of two electrolytes separated by semipermeable membrane.

Connecting both solutions by a semipermeable membrane results in cation diffusion in both directions. At first the ion beam flows from solution I to solution II. This causes separation of the charges in the diffusion space and generation of the gradient of potential. Even a very small unbalanced electrical charge generates a high difference of potential. The potential difference $\Delta\phi$ reduces the diffusion speed of ions flowing from solution I to solution II, and increases the diffusion speed of those flowing from solution II to solution I. The potential assumes a fixed value when the beams of diffusing ions are the same. The system is in a thermodynamical balance. The electrochemical potentials of the ions *i* in both solutions are equal:

$$\bar{\mu}_i^{\mathrm{I}} = \bar{\mu}_i^{\mathrm{II}} \tag{6}$$

The electrochemical potentials $\bar{\mu}_i$ in both solutions are a function of chemical potentials dependent on the activity of the ions *i* and the electrical potential ϕ:

$$\bar{\mu}_i = \mu_i + z_i F \phi \tag{7}$$

where z_i is the charge of an ion and *F* is the Faraday constant. The chemical potentials μ_i are described by the following equation:

$$\mu_i = \mu_i^0 + RT \ln a_i \tag{8}$$

where μ_i^0 is standard chemical potential, *R* is the gas constant, and *T* is temperature. The combination of equations (6–8) gives the equation:

$$\mu_i^0 + RT \ln a_i^{\mathrm{I}} + z_i F \phi^{\mathrm{I}} = \mu_i^0 + RT \ln a_i^{\mathrm{II}} + z_i F \phi^{\mathrm{II}} \tag{9}$$

Its transformation and simplification allows to obtain an equation describing the Donnan potential:

$$\Delta\phi = \phi^{\mathrm{II}} - \phi^{\mathrm{I}} = \frac{RT}{z_i F}\ln\frac{a_i^{\mathrm{I}}}{a_i^{\mathrm{II}}} \tag{10}$$

When the system is at room temperature (25 °C), a simplified form of this equation can be applied:

$$\Delta\phi = \frac{0.059}{z_i}\log\frac{a_i^{\mathrm{I}}}{a_i^{\mathrm{II}}} \tag{11}$$

If the system analyzed remains in agreement with the Donnan conditions, each 10-fold difference in the activity of univalent ions generates a potential equal to 59 mV.

Selective transport of ions can be carried out by different methods. Figure 7 presents examples of systems where the Donnan potential is generated. Ions can diffuse through a selective ion channel, e.g., gramicidin or porins. They can be also

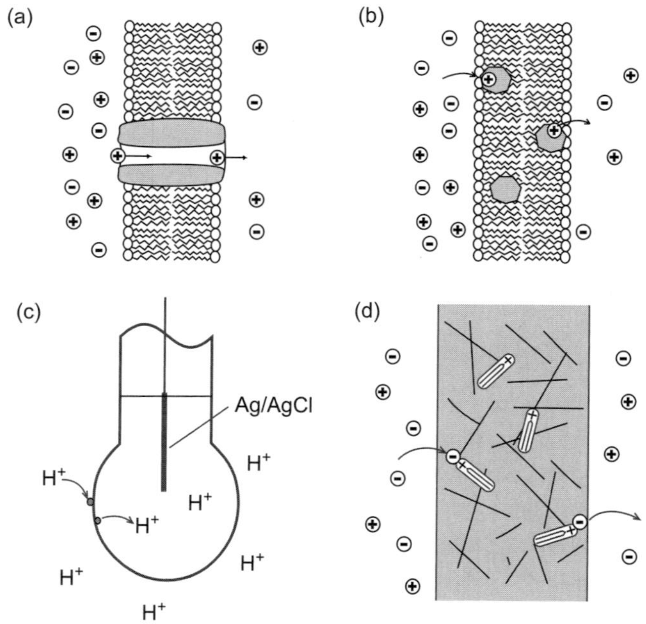

Fig. 7. Examples of systems where the Donnan potential is generated: (a) porine molecule in a bilayer lipid membrane forming a selective channel for cations, (b) valinomycin in a bilayer lipid membrane as a transporter for potassium cations, (c) glass membrane in an electrode for pH measurements enabling the transport of protons between the solution analyzed and the internal electrode solution, (d) lipophilic cations inside a liquid membrane enabling the transport of anions between electrolyte solutions.

carried by transporters binding them on one side of the membrane, and form a hydrophobic complex. After diffusion, the transporters liberate the ions on the opposite side of the membrane. This mechanism can be observed for valinomycin as a transporter of K^+.

One of the popular systems where the Donnan potential is generated is a glass electrode for pH measurements (Fig. 7c). Transport of H^+ ions does not take place by diffusion. The proton is built into the crystal structure of the glass on one side of the membrane and the other proton is released on the opposite side. The global effect is the same as in the case of diffusion transport. The potential response of glass electrodes is consistent with the Donnan equation in a wide range of concentration of H^+ ions.

Iiyama *et al.* [3,4] used in their investigations Millipore filters impregnated by dioctyl phosphate. These lipophilic anionic molecules caused membrane permeability to cations. The solutions at concentration of univalent cations 100 and 1 mM, separated by a membrane, generated the potential of 118 mV, which is in agreement with the Donnan equation.

2.2.2. Diffusion potential

Real systems, e.g., cell membranes, are more complex and are always permeable to more than one type of ion. The solutions on both sides of the membrane contain unequally distributed ions K^+, Na^+, H^+, Ca^{2+}; transport of anions also takes place. The artificial membrane systems used, e.g., in taste sensors are also characterized by complex transport of cations and anions. Analysis of the potential generated in these systems as the Donnan potential brings unsatisfactory results. Simultaneous transport of different ions leads to the generation of a diffusion potential. This potential appears when the following conditions are fulfilled:

- two electrolyte solutions are in contact, but without mixing,
- at least two types of ions are present in the solutions, and can be transported between the electrolytes,
- the ions transported between the electrolytes are characterized by different diffusion coefficients.

These conditions are fulfilled by many systems, e.g., two electrolytes separated by a porous ceramic membrane, a gel electrode placed in an electrolyte solution, a bilayer lipid membrane with a non-selective ion channel, an electroporated bilayer lipid membrane.

The simplest configuration for analysis of the diffusion potential consists of two solutions of the same electrolyte with different concentrations. The solutions contain cations and anions diffusing from the more concentrated solution to the less concentrated one. Ions with higher diffusion coefficients diffuse at a higher speed, which results in unequal spatial concentration of cations and anions. It causes a difference between the potentials of the solutions. This electrical field reduces the velocity of ions with higher diffusion coefficients, and increases the velocity of those with lower diffusion coefficients. The potential reaches the value at which the velocities of both types of ions are equal. This diffusion potential is higher for a higher difference in electrolyte concentration and diffusion coefficients. It can be calculated from the Henderson equation [1,2]:

$$\Delta\phi_D = -\frac{RT}{F}\frac{\sum_i z_i v_i (c_{i1} - c_{i2})}{\sum_i z_i^2 v_i (c_{i1} - c_{i2})} \ln \frac{\sum_i z_i^2 v_i c_{i1}}{\sum_i z_i^2 v_i c_{i2}} \qquad (12)$$

where z_i is the ion charge, v_i is the ion mobility. This equation must include all ions present in both solutions. When the solutions contain only one type of ions capable of diffusion between them, the Henderson equation assumes the form of the Donnan equation (equation (10)).

In many cases the occurrence of the diffusion potential is undesirable and should be eliminated. Real systems where a lipid membrane separates two electrolyte solutions contain additionally two electrodes (e.g., Ag/AgCl/Cl^- or calomel) which also are sources of diffusion potentials. The additional potentials can cause measurement errors. Complete elimination of these potentials is difficult, but it is often enough to keep their stable, low values. Table 2 shows the values of the diffusion potential observed in selected solutions.

Figure 8 presents a method for reducing the diffusion potential generated on reference electrodes. The diffusion potential on the contact surface of two solutions HCl at concentrations 0.1 and 0.01 M is equal to about 40 mV (Fig. 8a). When the solutions are separated by a saturated solution of KCl, two interfaces and two diffusion potentials are present (Fig. 8b). The more concentrated side generates a potential of 5 mV and the less concentrated one – of about 3 mV. In this case the resultant potential is equal to 2 mV. This confirms that accurate

Table 2. Experimental values of the diffusion potential for selected electrolyte solutions separated by a porous barrier [1,5]

Solution I	Solution II	Diffusion potential $\Delta\phi_D$ (mV)
0.1 mol/dm^3 HCl	0.1 mol/dm^3 KCl	−26.78
0.1 mol/dm^3 KCl	0.1 mol/dm^3 NaCl	−6.42
0.1 mol/dm^3 NaCl	0.1 mol/dm^3 NH_4Cl	+4.21
0.01 mol/dm^3 HCl	0.01 mol/dm^3 KCl	−25.73
0.01 mol/dm^3 KCl	0.01 mol/dm^3 NaCl	−5.65
0.01 mol/dm^3 HCl	0.1 mol/dm^3 KCl	−10
0.1 mol/dm^3 HCl	0.05 mol/dm^3 KCl	−53

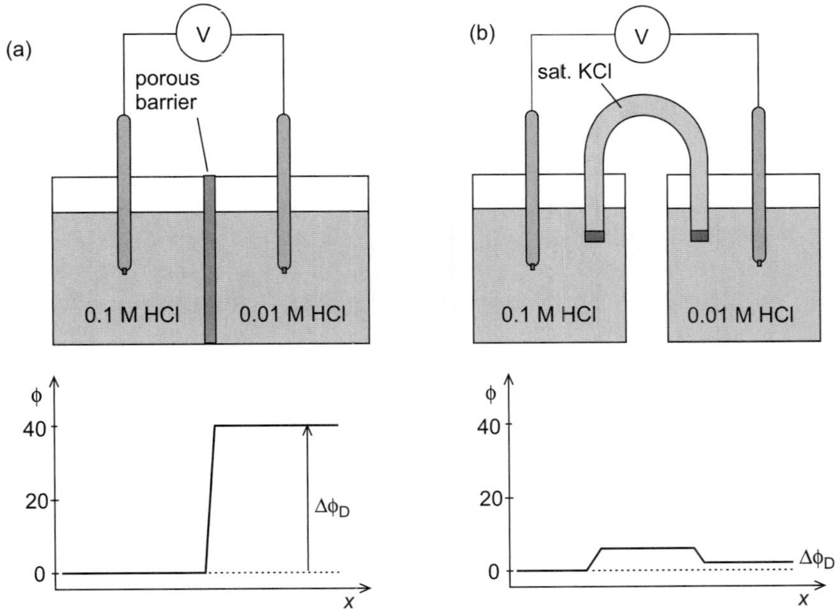

Fig. 8. Difference of potentials between two HCl solutions separated by a porous barrier (a) and separated by a salt bridge with saturated KCl (b).

measurement of the transmembrane potential requires the application of electrodes containing salt bridges separating the electrolytes adjacent to the electrode and to the membrane.

2.3. Application of potentiometry

2.3.1. Investigations of electrical processes in cells

Simple techniques, like potentiometry, allowed to explain many aspects of life processes taking place in cells, connected with different electrical processes which can be investigated by potentiometric measurements. Transport of neutral molecules through the cell membrane depends on ion motion and the membrane potential. Another group of phenomena includes signal conduction by the nervous system where the membrane potential is the most important parameter. The activity of many toxins against the living organisms also affects the cell membrane, as modifications of its permeability and depolarization result in energy exhaustion and death of cells which tried to maintain their characteristic membrane potential.

2.3.2. *Lipid membranes as ion selective electrodes*

Most applications of artificial lipid membranes concern analytical chemistry and construction of ion-selective electrodes. Potentiometric determination of ions is possible if the membrane enables their selective transport. This method consists in the generation of the Donnan potential by the membrane system. A classical example of this application is a detector for potassium cations where valinomycin is used as a selective transporter. Valinomycin forms hydrophobic complexes with K^+, able to diffuse by the hydrophobic core of the bilayer lipid membrane. Different crown ethers can be used as a transporter for cations, e.g., aza-18-crown-6 for K^+ [6].

2.3.3. *pH-sensitive electrodes*

Lipophilic molecules capable of protonation (e.g., 2,4-dinitrophenol, tridodecylamine) can be used for pH measurement. Lu *et al.* [7] described the application of bilayer lipid membranes supported on gel (sb-BLM) as pH-sensitive electrodes. They used tetrachloro-*o*-benzoquinone (TC-*o*-BQ) and tetrachloro-*p*-benzoquinone (TC-*p*-BQ) as a transporter for H^+ ions through a phosphatidylcholine membrane. These molecules undergo the protonation reaction:

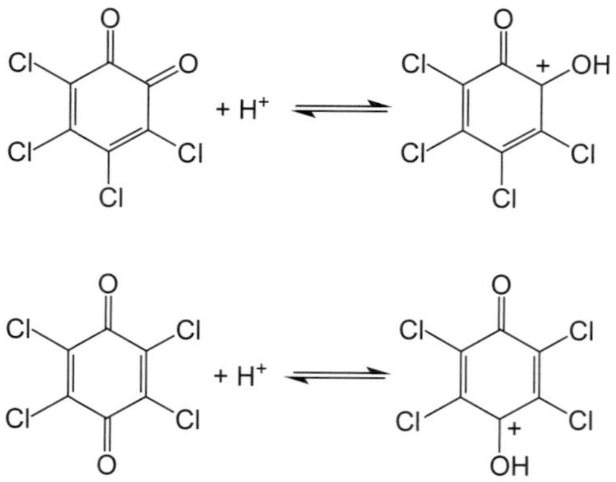

The response of the membranes containing TC-*o*-BQ corresponded with the Donnan equation to the highest degree.

2.3.4. *Potentiometric detection of molecules*

Molecules affecting ion transport can be detected using potentiometry. Usually the source of transmembrane potential is not the 'pure' Donnan potential, but the diffusion potential. Many mechanisms of signal transduction in cells are based on chemically controlled ion channel proteins. Opening of the channel by the trigger molecule causes membrane depolarization by flowing ions, and potential drop.

One of the interesting applications of potentiometry are taste detectors – electronic tongues. Such detectors are used in the food industry for controlling technological processes, e.g., during the production of wine, beer, juices, coffee. Detectors of this kind are generally based on PVC membranes modified by cationic and anionic lipids. Microporous membranes impregnated by lipids are also used. A matrix of several detectors with membranes showing different properties gives a set of potentials characterizing taste of the sample. In this kind of detectors the detailed composition of the solution analyzed is of secondary importance – the emphasis is placed on the reproducibility of the total signal, which provides the information that each part of the product has the same, appropriate taste. The signal can also inform about the completion of the technological process, e.g., fermentation.

2.3.5. Investigations of photo-effects

Photo-effects provide the basis for two processes important for life on Earth – photosynthesis in plants and vision processes in higher organisms. The processes are based on electron transfer through a membrane caused by light. The photo-effects in bilayer lipid membranes have been described in many papers. The generation of membrane potential was observed for many natural and artificial sensitizers, e.g., chloroplasts and their extracts [8–10], retinal [11], metalporphyrins [12–16], organic pigments [17], fullerens [18,19], CdS, In_2S_3, ZnS [20,21]. Changes in the membrane potential can be also observed in the case of light-induced changes in its permeability, when the diffusion potential is generated by the system. Lipophilic derivatives of azobenzene are often used as membrane sensitizers.

3. AMPEROMETRY AND VOLTAMMETRY

3.1. Potentiostats

Membrane processes are dependent on the difference of potentials on both sides of the membrane. Many electrochemical experiments require the application of a specified potential of a working electrode in relation to an electrolyte solution. In the case of membranes the potentiostat allows to maintain a specified potential between electrolytes separated by a membrane, or between the electrode and electrolyte solution. 2-, 3-, or 4-electrode systems are useful for membrane applications. Simplified diagrams with examples of potentiostat

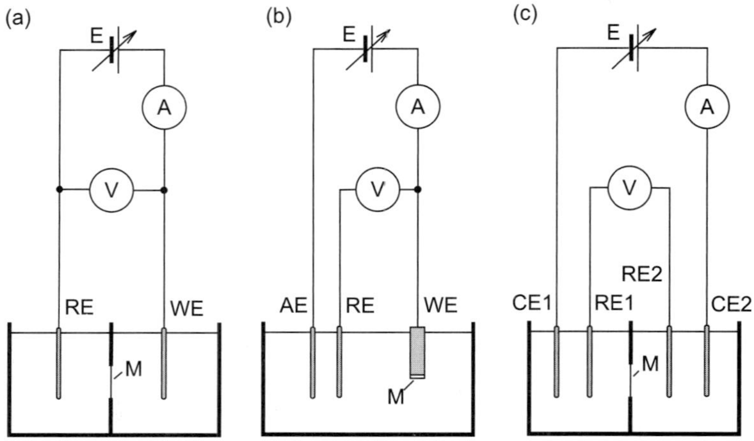

Fig. 9. Simplified diagrams illustrating the working principle of a 2-electrode (a), 3-electrode (b) and 4-electrode potentiostat. M is the membrane, WE is the working electrode, RE is the reference electrode, AE is the auxiliary electrode, CE1, CE2 is the current electrodes, RE1, RE2 is the reference electrodes, E is the regulated voltage source.

applications are presented in Fig. 9. Electrical models of these systems are shown in Fig. 10.

Experiments with a membrane separating two electrolyte solutions can be performed in 2- or 4-electrode arrangements. 3-electrode systems are generally used when a lipid membrane is supported on solid support, e.g., metal, glassy

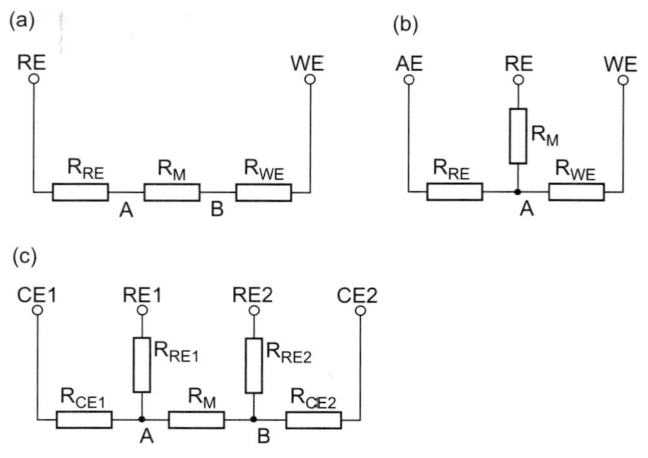

Fig. 10. Simplified electrical diagrams of the membrane–electrode–electrolyte systems studied in a 2-electrode (a), 3-electrode (b), and 4-electrode (c) arrangement. A detailed description can be found in the text.

carbon, conductive glass. The working principle of the potentiostat (Fig. 9) is based on a comparison of externally set voltage E with the potential difference of control electrodes (WE–RE, RE2–RE1). The potentiostat allows to apply such a high value of potential to current electrodes (WE–RE, WE–AE, CE2–CE1) that the potentials become equal. The potential E is regulated automatically by the potentiostat, according to the electrode, electrolyte, and membrane resistances.

The source of the external signal E_{inp} applied to the potentiostat (Figs. 11–14) can be, e.g., a resistor divider of the voltage, a generator of sinusoidal signal, a digital-to-analog converter. It is possible to employ different measurement techniques, depending on the signal applied to the input of the potentiostat and the method of analysis of the current flowing through electrodes, i.e., amperometry, voltammetry, capacitance measurements, impedance analysis, and variations of these techniques.

3.1.1. 2-electrode potentiostat

A 2-electrode potentiostat is used in direct current or low-frequency signal measurements. A minor measurement error is conditional upon the application of the electrodes RE and WE with low resistances R_{RE}, R_{WE}, much lower than the membrane resistance R_M. When the membrane resistance is equal to 1 MΩ and resistances of the electrodes 100 kΩ, the difference of potentials between the membrane surfaces (points A and B in Fig. 9a) is only 83% of the difference of potentials between the electrodes WE and RE. Usually the resistances of electrolytes are much lower than the resistances of electrodes. Their influence on the measurement accuracy is insignificant, so they were ignored in the electrical model. The electrolyte resistances should be considered when diluted electrolytes are used.

Figure 11 presents schemes of two 2-electrode potentiostats. The first one (Fig. 11a), simpler, consists of two operational amplifiers A_1 and A_2. The input voltage E_{inp} defines the potential of the working electrode WE in relation to the reference electrode RE. The working electrode WE has the same potential as the mass but is not galvanically connected with the ground. It is defined as virtual mass. In the circuit presented in Fig. 11b the working electrode WE is connected directly with the mass. This arrangement is more advantageous because it enables simpler and more convenient protection of the potentiostat against noises. Sometimes the system investigated can impose this solution, e.g., during experiments with neurons present in a water solution where one of the electrodes is inserted into the cell. Practically the best method is to connect the electrode in an external solution to the mass of the device. Another application of such potentiostats are investigations with a fragment of the cell membrane attached to the end of a very thin glass capillary (patch clamp method).

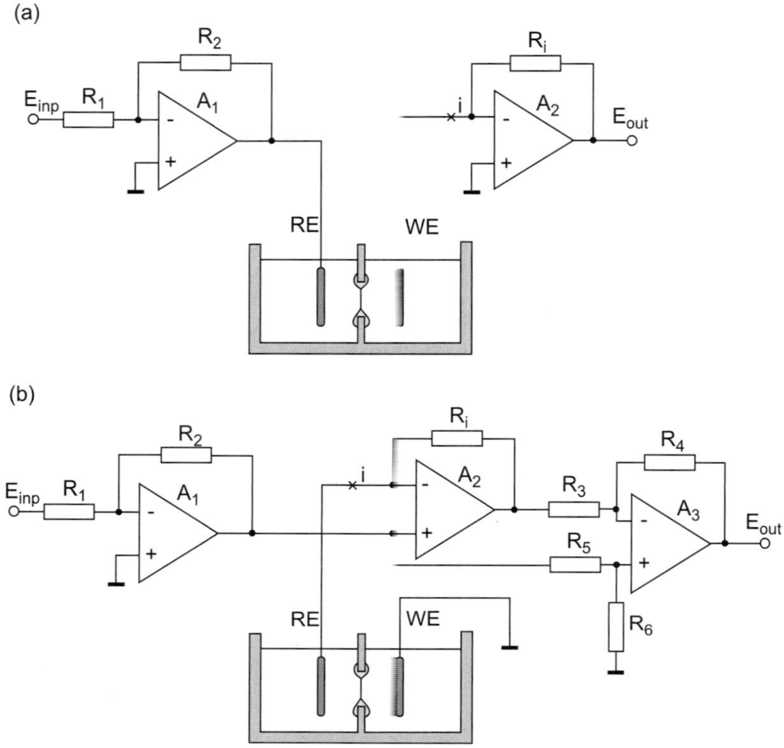

Fig. 11. Examples of 2-electrode potentiostats. RE is the reference electrode, WE is the working electrode, A_1–A_3 are operating amplifiers, R_1–R_6 are resistors 10 kΩ.

The amplifier A_1 can be any general-purpose amplifier, but better results are obtained for low-noise amplifiers (e.g., TL071, National Instruments). The membrane has high capacitance and even low-voltage noises can generate current noises comparable with the signal originating from membrane conductance. The amplifier A_2 works as a current-to-voltage converter. The output voltage E_{out} from the potentiostat is dependent on the current i flowing through the working electrode WE and the resistor R_i defined sensitivity of the potentiostat:

$$E_{out} = iR_i \tag{13}$$

In practice, for the lowest currents, the resistor R_i has a value up to 100 GΩ. The operational amplifier A_2 should be an amplifier with the lowest input currents and highest input resistance, selected, e.g., from Table 1. For the lowest currents in a range of several picoampers (range of currents for single-channel proteins) the most efficient solution is to apply 2-electrode potentiostats. Voltage drops on the electrode resistance are not recordable and simple electronic circuits have the best noise parameters and their dynamical properties are much better than those

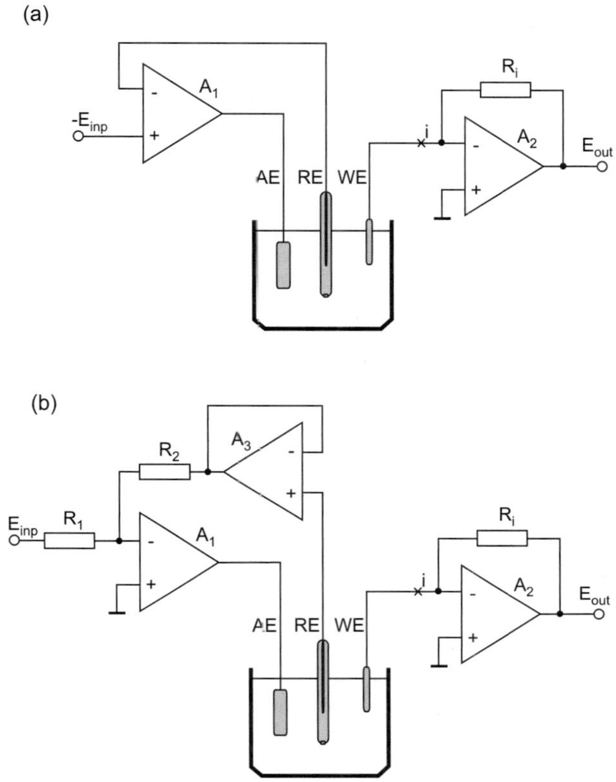

Fig. 12. Examples of 3-electrode potentiostats. AE is the auxiliary electrode, RE is the reference electrode, WE is the working electrode, A_1–A_3 are operating amplifiers, R_1, R_2 are resistors 10 kΩ.

of more complex 3- or 4-electrode systems. A simple potentiostat makes it possible to record fast processes of opening and closing ion channels, e.g., porin proteins. The amplifier A_3 in Fig. 11b can be a general-purpose amplifier like A_1 (e.g., low noise TL071).

3.1.2. 3-electrode potentiostat

3-electrode potentiostats are basic measurement equipment in classical electrochemical investigations [22]. Such potentiostats are also applied to electrochemical studies of membranes covering solid conductive support (metals, graphite, conducting glass, semiconductors), mercury or gel. The membrane covering the electrode performs the function of the working electrode WE.

Examples of simple potentiostats working in 3-electrode arrangements are shown in Fig. 12. The potentiostat sets potential of the working electrode WE

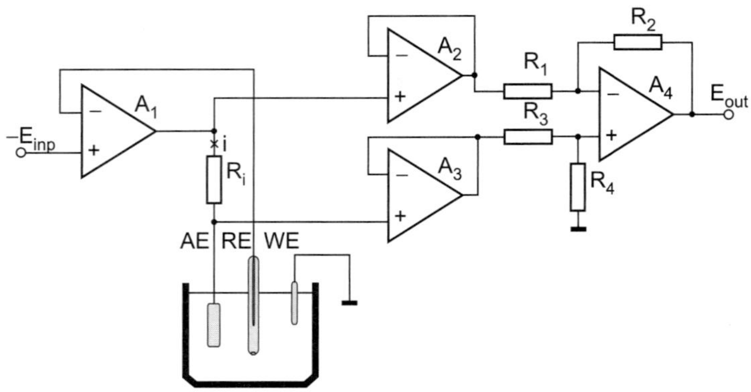

Fig. 13. An example of a Wenking type potentiostat, where the working electrode WE is connected to the mass of the device. AE is the auxiliary electrode, RE is the reference electrode, WE is the working electrode, A_1–A_4 are operating amplifiers, R_1–R_4 are resistors 10 kΩ.

in relation to the reference electrode RE equal to the input voltage E_{inp}. The operational amplifier A_1 is the main amplifier of the potentiostat. It compares the input voltage E_{inp} with the potential between the electrodes RE–WE and forces such high voltage on the auxiliary electrode AE that the potentials E_{inp} and RE–WE become equal. The output voltage E_{out} is a function of the current i flowing through the working electrode WE and the resistance R_i, regulating current sensitivity of the potentiostat (equation 13). The reference electrode RE is connected to the input of the operational amplifier with high input resistance (Fig. 12) and works practically currentless. Voltage drop on the electrode RE caused by this current is insignificant. Voltage drop on resistance of the electrolyte is also insignificant and no additional circuits are necessary for this compensation. 3-electrode potentiostats can be applied for direct current and alternating current experiments.

Figure 13 presents a potentiostat of the Wenking type. Its characteristic feature is that the working electrode WE is connected directly with the mass. The advantage of such potentiostats is the possibility of direct co-working with quartz crystal microbalances (QCM). A combination of a potentiostat with QCM is an EQCM. This instrument enables simultaneous recording of current and mass deposited on the electrode in a range of nanograms. The EQCM method is useful for analysis of mechanisms of electrode processes, adsorption, electropolymerization.

3.1.3. 4-electrode potentiostat

The working principle of a 4-electrode potentiostat–galvanostat is shown in Fig. 9c. Such a potentiostat is designed for keeping the right value of potential

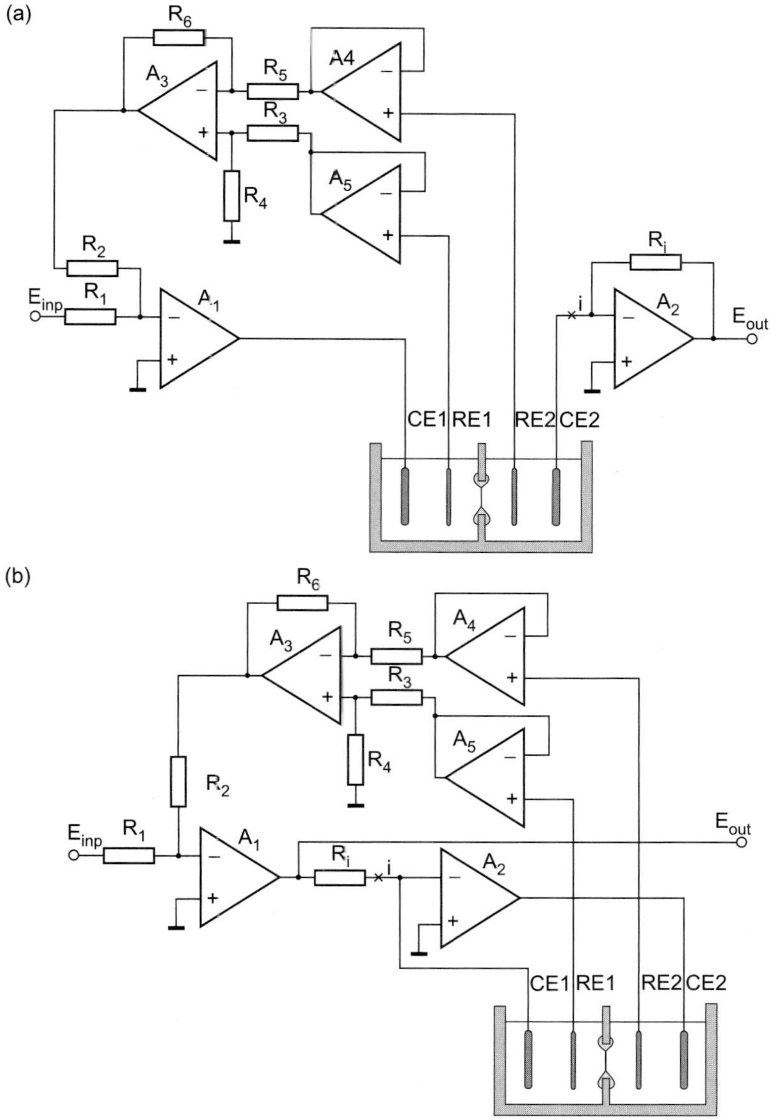

Fig. 14. Examples of a 4-electrode potentiostats applied for electrochemical investigations of the membranes. RE1, RE2 are reference electrodes, CE1, CE2 are current electrodes, A_1–A_5 are operating amplifiers, R_1–R_6 are resistors 10 kΩ.

between the control electrodes RE1 and RE2. The potentiostat allows to apply such high voltage to the current electrodes CE1 and CE2 that the potential difference between the electrodes RE1 and RE2 is equal to the input voltage E_{inp} (Fig. 14). This voltage defines the potential RE2 in relation to RE1. If the electrodes RE1 and RE2 are the same, the voltage E_{inp} describes the difference between the potentials of electrolytes on both sides of the membrane, i.e., the transmembrane potential. The electrodes RE1 and RE2 are connected to the inputs of the operation amplifiers A_3 and A_4 with high input resistance (amplifiers from Table 1). It makes it possible for the electrodes to work under currentless conditions. 4-electrode potentiostats are useful for impedance measurements. Automatic compensation of the electrode and, partially, electrolyte makes further analysis of the results simpler in the case of a smaller number of elements in the electrical model of the system. The results obtained are comparable with real parameters of the system. Compensation of the electrolyte resistance is optimal when the control electrodes RE1 and RE2 are placed as close to the membrane as possible.

Electrodes CE1 and CE2 can be of a different type, e.g., silver/silver chloride, platinum, stainless steel, graphite. In contrast to reference electrodes, their type and quality do not affect measurement accuracy. More information on 4-electrode potentiostats can be found in other papers [23–26].

The current sensitivity of the potentiostats presented in Fig. 14 depends on the resistor R_i. The amplifier A_2 with the resistor R_i works as a current-to-voltage converter. The amplifier A_2 should be selected from among high input resistance amplifiers (Table 1). The output voltage E_{out} is described by the equation:

$$E_{out} = -iR_i \tag{14}$$

3.2. Application of amperometry

Amperometry is a widely used analytical technique for investigations of membrane phenomena. Transport of ions or electrons across the membrane denotes flowing current. Usually this current is measured using a fixed electrode potential. Examples of application:

- measurement of membrane resistance and conductance,
- investigations of transport of lipophilic ions through the membrane,
- studies of ion channel proteins,
- investigations of photo-effects in membranes.

Measurements of current intensity are often used in biosensors and electrochemical sensors, e.g., DNA hybridization, electronic tongues and noses, enzymatic detectors.

3.3. Application of voltammetry

Voltammetry is an analytical technique where linearly sweep potential is applied to the working electrode and current flowing through this electrode is recorded.

A more common variant of this technique is cyclic voltammetry. The electrode potential is changed linearly in a fixed range of potentials. When the potential reaches the maximum or minimum value, the direction of sweep is changed.

Voltammetry is a popular technique in membrane studies. It can be applied to estimate membrane resistance and capacitance [27]. Cyclic voltammetry allows to determine the character of membrane conductance, i.e., whether it is connected with diffusion or redox reactions, and whether it is reversible, or not. Cyclic voltammetry is used for testing the quality of a mercaptan monolayer adsorbed on a gold electrode.

3.3.1. Measurements of membrane resistance and capacitance

The determination of membrane resistance and capacitance can be performed for membranes separating two electrolyte solutions or membranes supported on metal or graphite [28]. An example of a cyclic voltammetric curve is shown in Fig. 15. The slope di/dE of the curve is equal to membrane conductance and its inversion – to membrane resistance. Currents recorded during potential sweeping in opposite directions are shifted with the value Δi, dependent on the sweep speed V and capacitance C of the membrane. The current charging the membrane capacitance i_C is equal to half of Δi. The capacitance C can be calculated from the equation:

$$C = \frac{i_C}{V} \tag{15}$$

This estimated value C and the membrane area S allow to calculate the effective membrane thickness d:

$$d = \frac{\varepsilon_0 \varepsilon S}{C} \tag{16}$$

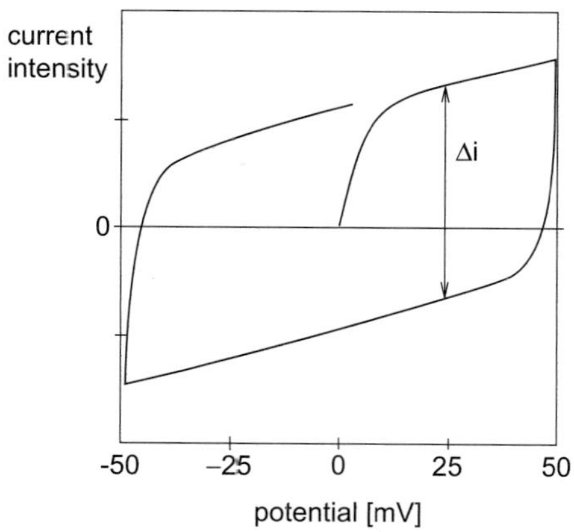

Fig. 15. A typical voltammetric curve obtained for a bilayer lipid membrane.

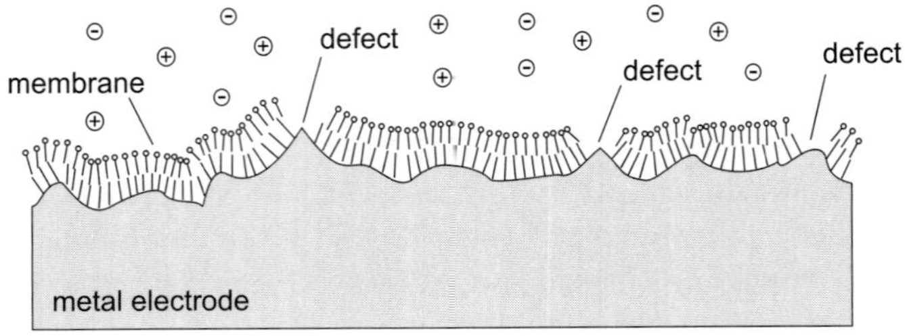

Fig. 16. Real surface of a polycrystalline gold electrode covered by a bilayer lipid membrane [92].

ε_0 is the electrical permittivity of vacuum (8.8542×10^{-12} F/m), ε is the relative permittivity of the membrane (it can be assumed as 2.1, like for hydrocarbons).

3.3.2. Estimation of quality of solid supported membranes

If a gold electrode is covered by a lipid membrane, the surface of the electrode is isolated from the solution. If the electrode is made of polycrystalline gold, its surface is not smooth but rough. In this case the membrane covering metal shows defects in the areas of direct metal–solution contact (Fig. 16).

Figure 17 presents two cyclic voltammetric curves recorded for a gold electrode without (a) and with a supported lipid membrane (b). The electrodes were immersed in a solution of sulfuric acid. The acid concentration is usually 0.1–0.5 M. The voltammetric curves have sharp peaks in the cathodic cycle. In the anodic cycle the surface of gold is oxidized and in the cathodic cycle this oxide is reduced [28]. The potential of the reduction peak is equal to about 900 mV in relation to the silver/silver chloride electrode. The peak of reduction is much lower for the electrode with a membrane (Fig. 16b). The ratio between the peak areas for

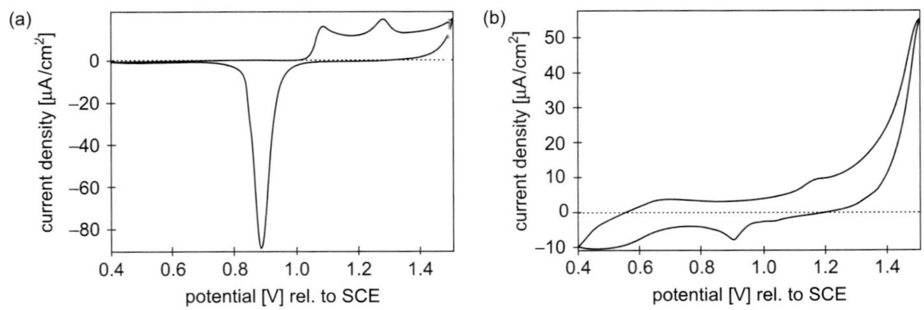

Fig. 17. Voltammetric curves obtained for a gold electrode without (a) and with (b) hexadecyl mercaptan (HDM). Electrolyte: 0.1 M H_2SO_4, sweep speed 10 mV/s. A peak recorded by potential about 900 mV during a cathodic cycle is used for estimating the area of defects in a membrane covering the electrode.

the electrodes with and without the membrane informs about the area occupied by defects and membrane quality.

Another method for estimating membrane quality is to record voltammograms in a solution containing molecules or ions capable of reversible redox reactions [28]. The most popular solution is $Fe(CN)_6^{3-/4-}$ at concentration 1–10 mM. Dopamine [29] or $Ru(NH_3)_6^{3+/2+}$ [30] can also be used. KCl at concentration 0.1 M is often used as a stock solution. The current of the reduction peak is three orders of magnitude lower for an electrode well covered by a membrane, compared with an electrode without this membrane.

4. CHRONOPOTENTIOMETRY

4.1. Galvanostats

Galvanostats are designed for controlling the intensity of current flowing through electrodes. The signal measured is the potential between electrodes forced by the flowing current. This technique is also known as a current-clamp method. Similarly as in the case of potentiostats, 2-, 3-, or 4-electrode galvanostats can be used. A scheme of the simplest 2-electrode galvanostat is presented in Fig. 18a. The operational amplifier A_1 is configured as a current source. The intensity of current flowing through the electrodes is described by the equation:

$$i = \frac{E_{inp}}{R_i} \qquad (17)$$

The output voltage E_{out} is equal to the potential of the reference electrode RE in relation to the working electrode WE.

A scheme of a 3-electrode galvanostat is shown in Fig. 18b. The current intensity is described by the same equation as for a 2-electrode galvanostat (equation (17)). The output voltage E_{out} is the difference between the potentials of the working electrode WE and the reference electrode RE.

A scheme of a 4-electrode galvanostat is presented in Fig. 18c. Current described by equation (17) is flowing from the current electrode CE2 to the electrode CE1. The output voltage E_{out} is equal to the potential of the electrode RE1 in relation to the electrode RE2. The operational amplifiers A_1, A_2, and A_3 should be selected from among amplifiers with the highest input resistance (Table 1).

4.2. Application of chronopotentiometry

Chronopotentiometry is used mainly for investigations of the electroporation phenomenon in bilayer lipid membranes [31–37]. Correctly selected current

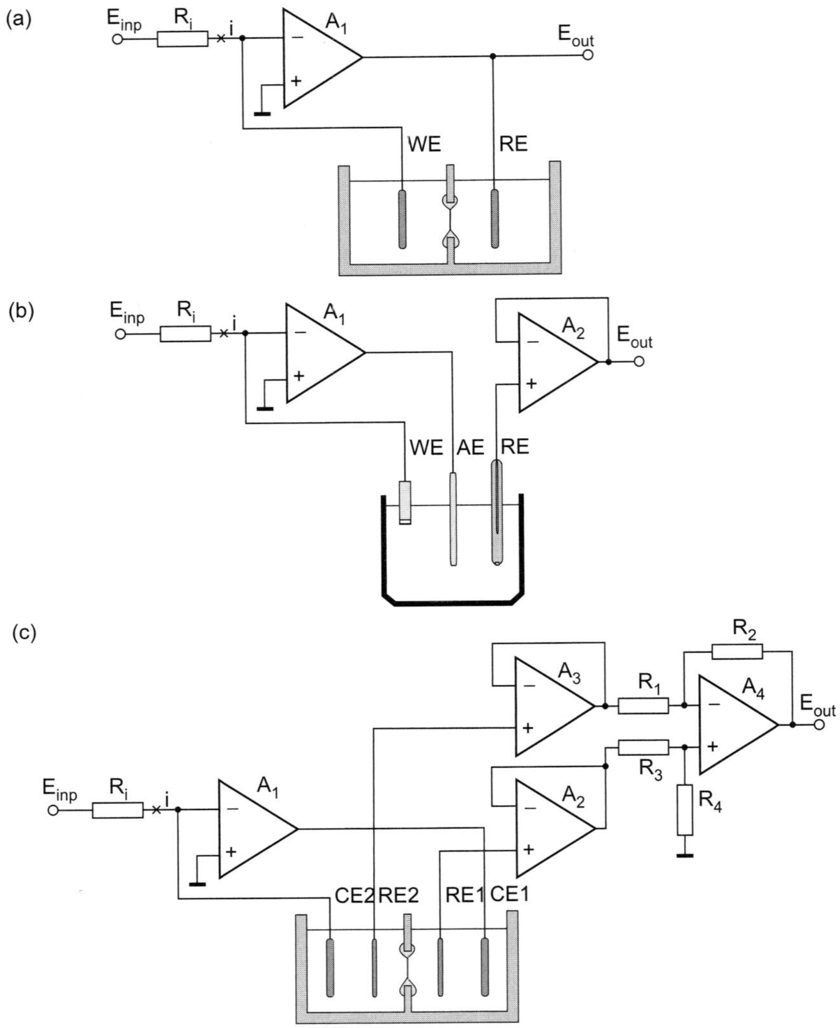

Fig. 18. Examples of 2-electrode (a), 3-electrode (b), and 4-electrode (c) galvanostats. AE is the auxiliary electrode, RE is the reference electrode, WE is the working electrode, RE1, RE2 are reference electrodes, CE1, CE2 are current electrodes, A_1–A_4 are operating amplifiers, R_1–R_4 are resistors 10 kΩ.

intensity allows to generate a single, stable, and long-life pore in the membrane, without its destruction [34,35,38]. A more advanced technique is programmable chronopotentiometry, where software controls the intensity and direction of current, and additionally enables short circuit and disconnection of current electrodes. This technique is useful for investigations of electroporation and pore resealing, and allows to analyze the recovery of membrane continuous structure [35].

Chronopotentiometry is also employed in experiments with filter membranes impregnated by lipids. These membranes show oscillations of the transmembrane potential under current-clamp conditions. The characteristics of these oscillations are dependent on the membrane parameters and solution composition. Substances adsorbing on the surface or influencing membrane fluidity have a significant effect on the oscillations. This property was used in taste and odor sensors [3,4,39–45].

5. CAPACITANCE MEASUREMENTS

5.1. Characteristics of the measurement methods

There are several methods for measurement of membrane capacitance. The measurement object is much more demanding than simple capacitors. The first restriction is the resistance of electrodes and membrane causing additional measurement errors. Another problem is low durability of membranes. They are often destroyed by a potential lower than 200 mV. Furthermore, the membrane potential changes its other parameters. Table 3 presents the parameters influencing capacitance measurements.

Measurements of membrane capacitance can be performed by several methods, differing in the way of capacitance determination and the type of a

Table 3. Parameters influencing measurements of membrane capacitance

Parameter	Characteristics
Amplitude of measurement signal	Capacitance of thin lipid membranes and membrane processes are dependent on the voltage applied to the membrane. The signal amplitude should be as low as possible. Usually it varies from several to a few dozen millivolts.
Electrode impedance	Electrodes with the lowest possible impedance should be used. Some methods enable compensation of the electrode impedance. Usually electrodes with resistance in a range of a few hundred ohms are used.
Membrane resistance	The total current flowing through the membrane has two components: resistance current and capacitance current. If the resistance current is comparable with the capacitance current, it is important to take its value into account during measurements.

Table 4. Characteristics of the methods applied to membrane capacitance measurements

Group of methods	Characteristics
Method of measurement	
Compensation methods	Compensation methods are based on a reference set of capacitors and resistors, modeling the system investigated. The measurement signal is induced both in the reference system and the system examined. The parameters of the reference system are regulated until the detector finds an identical response of both systems.
Parametric methods	In parametric methods the capacitance is calculated from the electrical signal dependent on the capacitance. The methods are based on, e.g., the measurement of capacitance current or analysis of curves of capacitance discharge.
Shape of the measurement signal	
Sinusoidal	Applied for impedance analysis and bridge methods.
Triangular	Applied for the measurement of capacitance current. For constant capacitance the current has a rectangular shape.
Rectangular	Applied mainly for pulse methods. The characteristics of membrane charging or discharging are analyzed.

measurement signal applied to the membrane. Table 4 presents the characteristics of the methods employed for capacitance measurements.

5.1.1. Measurements with a sinusoidal signal

The application of a sinusoidal voltage signal to the capacitor induces current flow. The voltage and intensity of the current are described by the equation:

$$U(t) = U_0 \sin \omega t \tag{18}$$

$$I(t) = C\frac{dU}{dt} = C\omega U_0 \cos \omega t \tag{19}$$

where U_0 is the amplitude of a voltage signal applied to the capacitor, ω is the angular frequency $= 2\pi f$.

Equations (18 and 19) show that the current flowing in the circuit is 90° ahead of the voltage. If only signal amplitudes are considered, the capacitance current is equal to:

$$I = U\omega C \tag{20}$$

The properties of the capacitor are the same as those of a resistor with resistance dependent on frequency:

$$R = \frac{1}{\omega C} \tag{21}$$

The application of algebra of complex numbers allows to consider the voltage, current, and phase shift in systems with capacitances. The reactance Z_C of the capacitor is given by the equation:

$$Z_C = \frac{-j}{\omega C} \tag{22}$$

where j is the imaginary number $\sqrt{-1}$.

A membrane with electrolytes and electrodes is a complex system of resistances and capacitances. Measurements of the system impedance at variable frequencies of the signal and mathematical analysis of the results allow to calculate the parameters of each element of the system: capacitance and resistance of the membrane, resistances of electrodes and electrolytes. More advanced analysis enables the determination of respective capacitances and resistances in a three-layer membrane model [46–50]. Measurements based on capacitance current have been described in many papers [2,51–55].

Stelzle and Sackmann [54] used a function generator HP8116A and a home-made analog-to-digital converter for measurements of the capacitance of membranes supported on chromium electrodes. The maximum output voltage of this meter is 100 mV. The voltage measured on the resistor connected in series with the capacitance examined was rectified and read by a computer using an analog-to-digital converter.

Hanke and Schlue [52] described a system for studies of membrane properties, which separates the capacitance and resistance current (Fig. 19). The sinusoidal signal is added to the direct current signal and both are applied to the electrode. The current-to-voltage converter generates the output voltage proportional to the current flowing through the electrode. Next this signal is separated giving two components. The low-pass filter passes only the direct current component of the signal, dependent on the membrane conductance. The alternating current signal, whose amplitude is a function of the membrane capacitance, is conducted along the other path, through a selective filter.

5.1.2. Measurements of capacitance with a triangular signal

A triangular signal is used for membrane excitation during measurements. Its advantage is that the capacitance current is a rectangular wave. This output signal can be read using an oscilloscope. After the rectification of the output signal direct current is obtained and no additional smoothing is necessary.

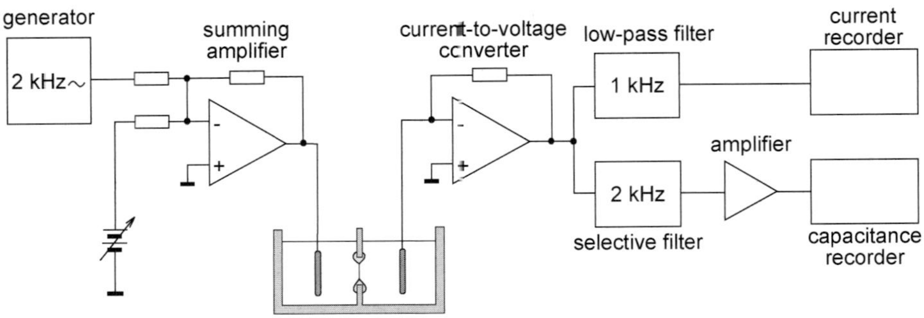

Fig. 19. A system for studies of BLMs based on the measurement of membrane capacitance current [52].

The application of linearly changed voltage to the capacitor causes the flow of capacitance current i with intensity:

$$i = VC \tag{23}$$

where C is the capacitance, V is the rate of voltage changes.

This method has already been described in literature [56–58]. A triangular measurement signal is also used for studying the effects of potential on membrane capacitance. The membrane is polarized by linearly changed potential in a wide range and the capacitance current is recorded as a potential function. A diagram of the measurement equipment is shown in Fig. 20. The triangular signal applied to the membrane had the amplitude ± 200 mV and changed at a rate of 100–500 V/s. The current-to-voltage converter gave the output signal, recorded by the oscilloscope.

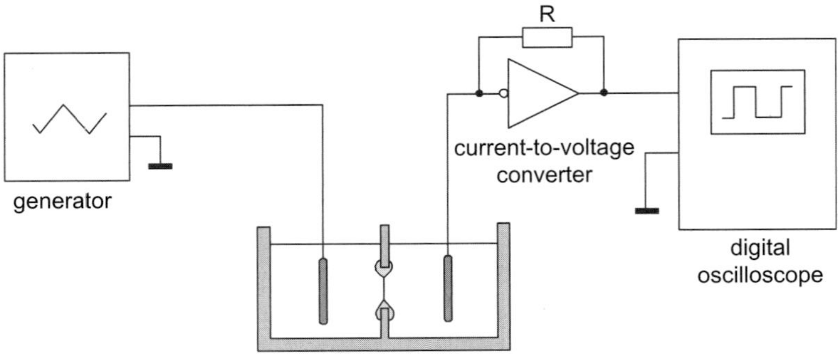

Fig. 20. Diagram of equipment used for measurements of membrane capacitance with using a triangular measurement signal [58].

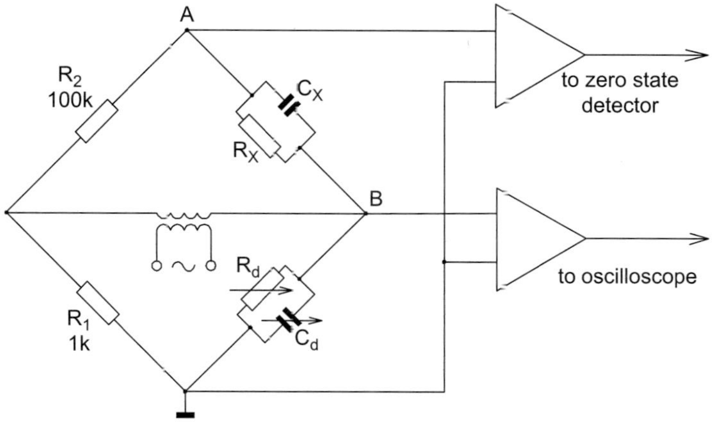

Fig. 21. Bridge used by White for measurements of membrane capacitance [59].

5.1.3. Bridge methods

Bridge methods belong to the group of compensation methods. They are often used for the determination of the membrane capacitance. These methods ensure high measurement accuracy. They are time-consuming, but enable direct reading of the value measured.

A precision bridge for capacitance measurements was elaborated by White [55, 59,60]. A simplified scheme of this bridge is shown in Fig. 21. There is a 100-fold difference between the resistances R_1 and R_2. Due to that the intensity of the current flowing through the bottom branch of the bridge is a hundred times higher than the intensity of the current flowing through the membrane. This circuit is less sensitive to disturbances and additional capacitances originate from the cables connecting decade capacitors and resistors with the bridge. The bridge is supplied from an external generator using an inductive or opto-electronic coupling. The bridge output (point A) is connected to the detector, which allows to find its balance. The oscilloscope is connected to point B, which enables the measurement of the signal amplitude.

This bridge was used by White in numerous experiments [60–67]. The bridge method was also employed by many other researchers [51,68–75]. The bridges were usually supplied by a sinusoidal signal with an amplitude from several to a few dozen millivolts and frequency of 500 Hz–5 kHz.

5.1.4. Pulse methods

Pulse methods applied for capacitance measurements have several varieties. They are simple and have already been described in many papers [2,51,52,68, 76–82]. Authors apply three general kinds of measurement systems: integration

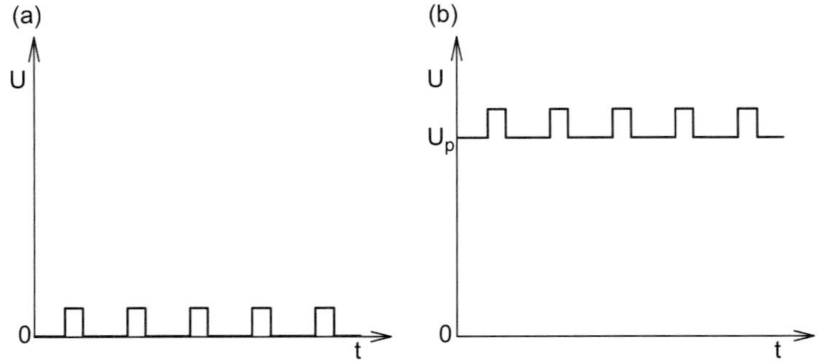

Fig. 22. Potential course during pulse measurements of membrane capacitance, (a) with polarization potential $U_p = 0$, (b) with imposed polarization potential U_p.

circuit, differentiation circuit, and the charge pulse method. The pulse methods are useful for measurements of capacitance of membranes with a defined polarization potential. In this case excited pulses are combined with a constant level of the potential (Fig. 22).

5.1.4.1. Integration circuits

A scheme of an integration circuit and voltage courses are shown in Fig. 23a, b. A jump of the excited potential on the input results in the flow of current through the resistor and capacitor with intensity:

$$i = C\frac{\mathrm{d}U}{\mathrm{d}t} = \frac{U_{\mathrm{inp}} - U_{\mathrm{out}}}{R} \tag{24}$$

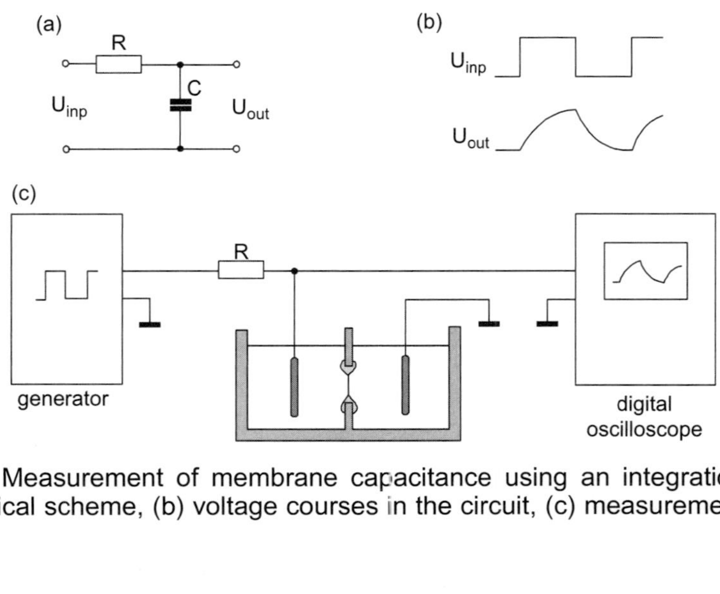

Fig. 23. Measurement of membrane capacitance using an integration circuit, (a) electrical scheme, (b) voltage courses in the circuit, (c) measurement setup.

After solving, this equation assumes the following form:

$$U_{out} = U_{inp} + Ae^{-t/RC} \tag{25}$$

Considering the boundary conditions $A = -U_{inp}$, the output signal is described by the equation:

$$U_{out} = U_{inp}(1 - e^{-t/rc}) \tag{26}$$

An example of practical application of the measurement setup is shown in Fig. 23c. The source of the excited signal is a generator with a rectangular voltage wave.

5.1.4.2. Differentiation circuit

A scheme of a differentiation circuit and voltage courses are shown in Fig. 24. When a voltage jump U_{inp} appears on the input, current i starts to flow through the capacitance C and resistance R:

$$i = C\frac{d}{dt}(U_{inp} - U_{out}) = \frac{U_{out}}{R} \tag{27}$$

After solving and taking into account the boundary conditions, the equation describes the output voltage U_{out} of the circuit:

$$U_{out} = U_{inp}e^{-t/RC} \tag{28}$$

A voltage jump on the input U_{inp} results in a jump of the output voltage U_{out}, which then falls exponentially to zero.

An example of practical application of the measurement setup is shown in Fig. 24c. A generator with a rectangular signal is the source of the measurement signal.

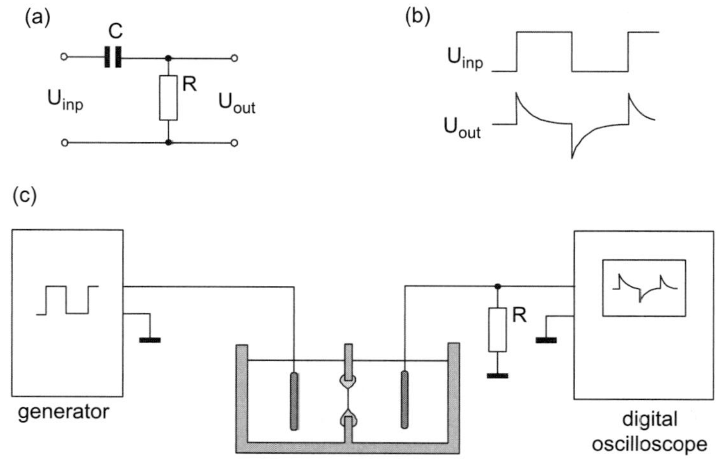

Fig. 24. Measurement of capacitance using a differentiation circuit, (a) scheme of the circuit, (b) voltage courses in the circuit, (c) measurement setup.

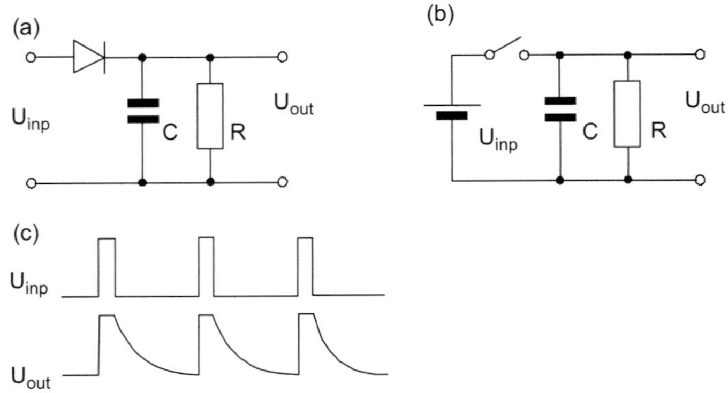

Fig. 25. Schemes of measurement circuits with pulse charging of the membrane, (a) circuit with a diode, (b) circuit with an electronic switch, (c) voltage courses in the circuits.

5.1.4.3. Charge pulse methods

Schemes of circuits for capacitance measurements are shown in Fig. 25. There are two methods of charging the membrane capacitance *C*. The membrane is charged by a short voltage pulse from a pulse generator through the fast switching diode D (Fig. 25a) or from a direct voltage source by an electronic switch (Fig. 25b). The voltage courses in these circuits are shown in Fig. 25c. The membrane is charged within a short time to the voltage U_{inp}. When the charging signal U_{inp} disappears, the membrane capacitance discharges by the resistance R. The membrane voltage falls down exponentially to zero.

$$U_{out} = U_{inp} e^{-t/RC} \tag{29}$$

5.1.5. Compensation methods

Alvarez and Latorre [76] applied a combination of the compensation method with pulse excitation of the membrane and reference circuit. A diagram of this system is shown in Fig. 26.

The rectangular voltage pulse is applied both to the membrane and the reference set of resistor R'_S and capacitor C'_M. The i/U converters convert the currents flowing through the membrane and the reference to voltage. The signal generated by the reference system is subtracted from the membrane signal and the difference is amplified. The values C'_M and R'_S are manually fitted during the measurement. When the output signal is equal to zero, it means that the parameters of the membrane C_M with solutions R_S and the reference values C'_M, R'_S are identical.

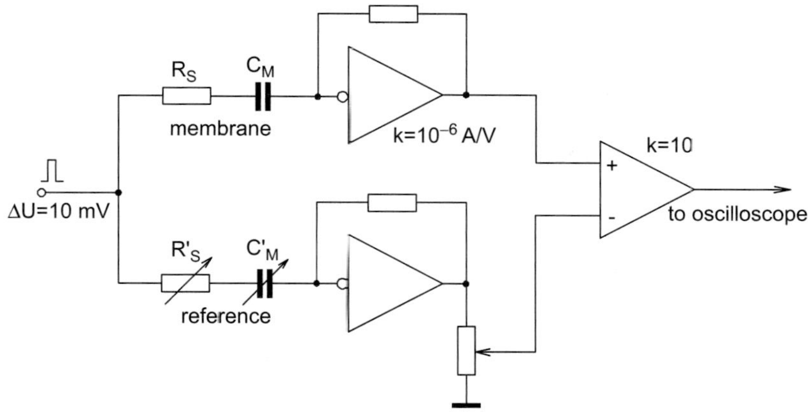

Fig. 26. Measurement system applied by Alvarez and Latorre [76].

5.1.6. Conversion of capacitance to frequency

The method of capacitance conversion to frequency described in Refs. [83,84] consists in cyclical capacitance charging and discharging by constant intensity current in a fixed range of potential (Fig. 27). The cycle period is proportional to capacitance and potential amplitude. Usually the potential amplitude is in the range of 10 mV. Measurements with the imposed membrane potential E_M consist in changing the measurement signal around this value (Fig. 27b).

A scheme of a simple, 2-electrode capacitance-to-frequency converter is shown in Fig. 28. Figure 29 presents a more complex, 4-electrode converter. The period of the output signal can be measured using a frequency/period meter, an oscilloscope or a computer interface for frequency measurements.

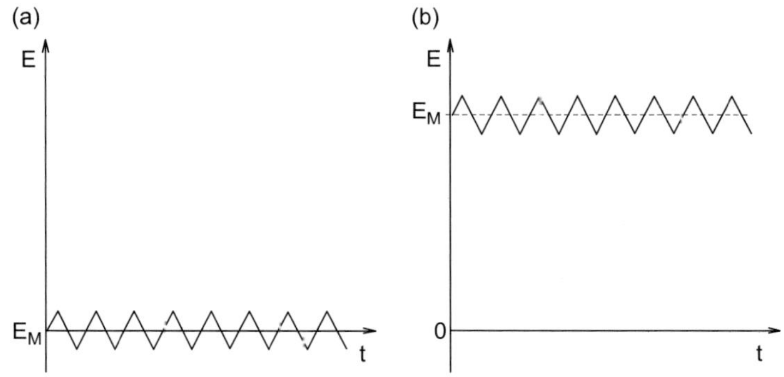

Fig. 27. Changes in the potential during capacitance measurement by the capacitance-to-frequency method, (a) with membrane potential E_M equal to zero, (b) with membrane potential E_M.

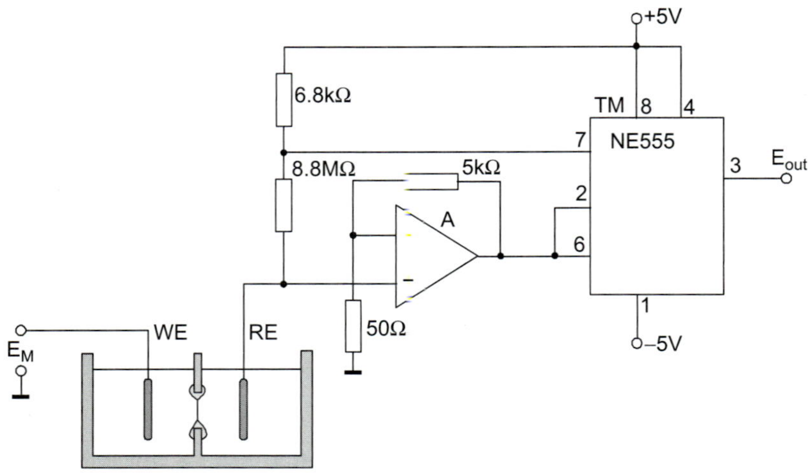

Fig. 28. Scheme of a 2-electrode system for measuring membrane capacitance based on the method of capacitance-to-frequency conversion [83].

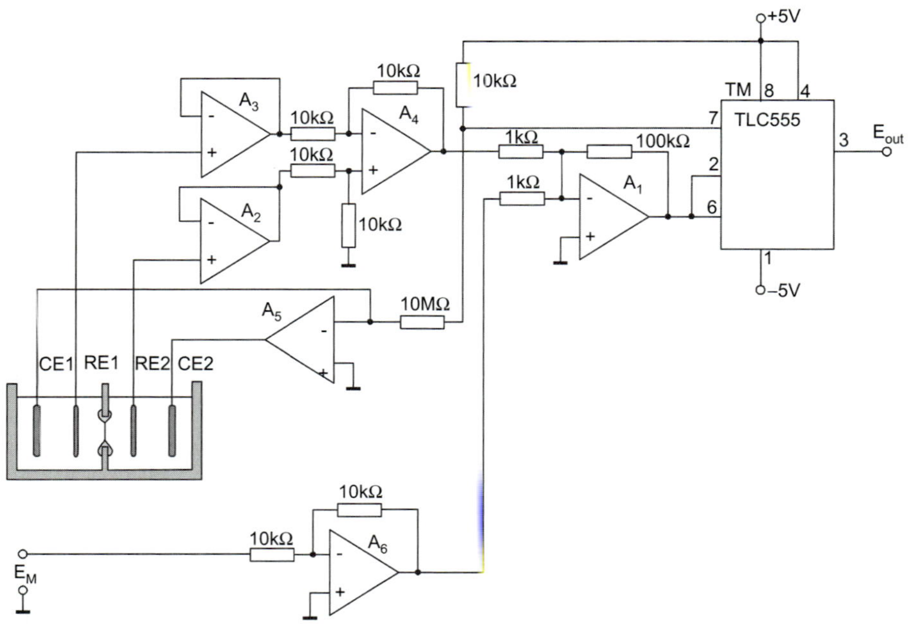

Fig. 29. Scheme of a 4-electrode system for measuring membrane capacitance based on the method of capacitance-to-frequency conversion [84].

Both converters are based on an universal integrated timer NE555, produced by many companies. This circuit is supplied by symmetrical voltage ± 5 V. This method of supplying makes the measurement signal oscillate around zero. The operating amplifier A (Fig. 28), connected as a non-inverting amplifier, ensures an appropriately low amplitude of the measuring signal on the membrane. Proper operation of the above 2-electrode converter requires the application of electrodes with low resistance, not higher than a few hundred ohms.

A 4-electrode converter is less demanding (Fig. 29). It is more complex, but less affected by measurement errors caused by the electrode resistance.

The advantage of capacitance-to-frequency conversion is that capacitance is directly converted to a digital signal, and more resistant to external disturbances and distortions.

5.1.7. Measurement of the minimum capacitance potential

The capacitance of lipid membranes depends on the membrane potential. The capacitance–potential characteristics have the shape of a parabola. The minimum value on this parabola is equal to the difference between the surface potentials of the membrane. These characteristics provide the basis for analysis of changes in the membrane surface potentials, and give information about many surface phenomena, e.g., adsorption processes, dissociation, formation of complexes.

Devices based on the measurement of the capacitance current are used for automatic recording of the minimum capacitance potential [58,85–87]. Simplified diagrams of these circuits are shown in Figs. 30 and 31. The signal from the current-to-voltage converter is conducted to the phase-sensitive detector through a selective amplifier. The phase-sensitive detector isolates a component, phase-shifted by 90° (capacitance current), from the signal. This signal is fed to the integration circuit. The output voltage from the integrator is added to the sinusoidal low signal from the generator and both are applied to the electrode. The circuit automatically compensates for the minimum capacitance potential.

5.2. Applications of membrane capacitance measurements

Measurements of membrane capacitance provide useful information. Due to their specific properties, thin lipid membranes require the application of modified methods. The basic requirement is the level of the signal exciting the membrane during the measurement, which must be as low as possible. The restrictions concerning measurement methods make it necessary to use non-commercial equipment. Several phenomena can cause changes in the membrane

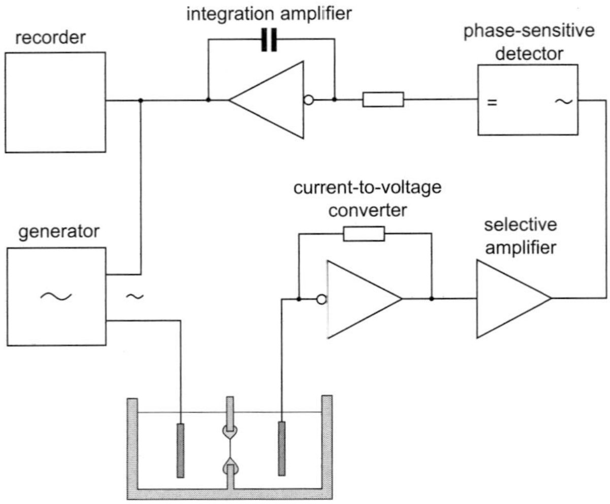

Fig. 30. Diagram of a circuit for the measurement of the minimum capacitance potential [58,85].

capacitance. One of them is a change in the permittivity of the hydrocarbon core of lipid bilayers, which is low – about 2.2 for pure lipids. Absorption of molecules inside the membrane makes this value increase, together with an increase in capacitance. Adsorption of molecules on the membrane surface usually increases its thickness and changes the permittivity of this layer, which is relatively high. This phenomenon can result in an increase or decrease in the capacitance, depending

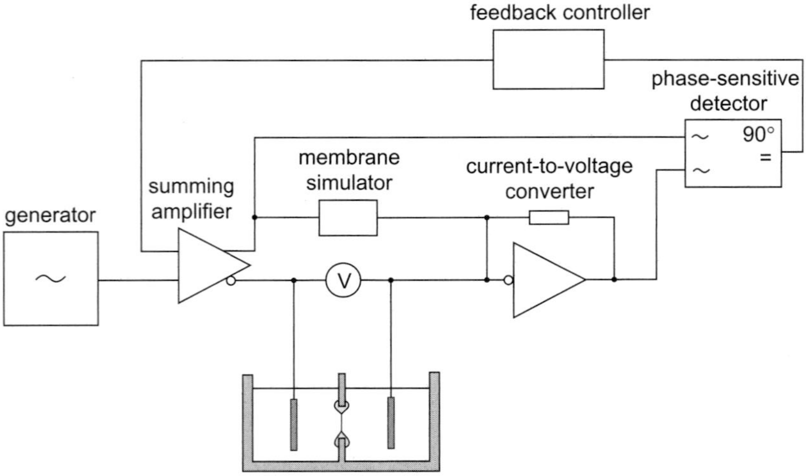

Fig. 31. Diagram of a circuit for the measurement of the minimum capacitance potential [85].

on the properties of adsorbed molecules. An additional effect is a change in the surface potential. This phenomenon leads to changes in electrostriction and, as a result, to changes in capacitance. Due to the fact that various phenomena cause changes in membrane capacitance, its measurements can be used for different purposes.

5.2.1. Measurement of membrane thickness

Membrane capacitance is a direct function of membrane thickness. The thickness d can be calculated from the equation:

$$d = \frac{\varepsilon_0 \varepsilon S}{C} \tag{30}$$

where ε_0 is the electrical permittivity of the vacuum, ε is the relative permittivity of the membrane, S is the area of the membrane, C is the capacitance of the membrane. The thickness d is referred to as an effective thickness because this is first of all the thickness of the hydrophobic core, with a small contribution of the thickness of the membrane hydrophilic region. More detailed results can be obtained by impedance analysis [46,49,71]. Some applications of capacitance measurements are described below.

5.2.2. Analysis of stability of bilayer lipid membranes

Lipid bilayers separating two electrolyte solutions from a system characterized by low stability. The equilibrium of the bilayer with the Plateau–Gibbs border surrounding it is labile and mechanical disturbance can cause changes in the bilayer area, followed by changes in membrane capacitance. The application of such membranes as cell membrane models requires stable and repeatable parameters. Capacitance measurements allow to determine the quality and usability of the membranes formed. They also make it possible to estimate the quality of membranes supported on metal, used for the construction of biosensors and electrochemical sensors.

5.2.3. Process of membrane formation

The self-assembling process of bilayer lipid membrane formation can be observed while recording membrane capacitance. The bilayer formed can separate two electrolyte solutions (Mueller–Rudin method) [88] or be supported on a metal electrode (Tien–Salamon method) [9]. The first stage of the process is characterized by low capacitance. It indicates that the membrane is thick. The second stage, when the capacitance increases relatively fast, is a stage of bilayer formation. When capacitance is stable and its value ranges from 0.3 to

1.0 μF/cm^2, the bilayer membrane is already formed and can be used for further applications.

5.2.4. Toxicological studies

Venom toxins are enzymes destroying membrane lipids. A bilayer lipid membrane supported on a solid electrode can be used as a sensor for these enzymes [89]. Hydrolysis of lipids caused by enzymes affects membrane thickness. Hydrolysis and removal of lipid molecules from the membrane reduce its thickness and, as a result, increase its capacitance.

Specific reactions of toxins with molecules supported on lipid membranes cause changes in the membrane properties. Adsorption of these molecules results in changes in the surface potential of membranes and electrostriction forces, reflected by capacitance changes. The effect of toxins can be also observed while measuring the minimum capacitance potential.

5.2.5. Investigation of photo-effects

Photo-effects in bilayer lipid membranes can be observed not only as photopotential or photocurrent. The capacitance of lipid membranes can be also changed under the influence of light. The effect is visible if a lipid membrane contains molecules changing their conformation under the influence of light, e.g., lipophilic azobenzene derivatives [90,91]. Double bonds of some molecules can change their conformation *cis–trans* and in this way affect changes in the membrane structure, leading to changes in its capacitance.

6. REFERENCE AND AUXILIARY ELECTRODES

6.1. Reference electrodes

When a membrane separates two electrolyte solutions, the difference of potentials between the electrolytes is of primary importance. Membrane processes depend on the transmembrane potential. If two identical reversible electrodes are immersed into electrolytes, the potential between them is equal to the transmembrane potential. An additional condition is that the current flowing through the electrodes should be so low as not to evoke recordable voltage drops on their resistance. It is recommended to use a 4-electrode arrangement. The measurement of the potential is performed using the reference electrodes RE1 and RE2 (Fig. 9c), under currentless conditions.

Electrodes of the second type are used most often as reference electrodes. Their potential depends on the concentration of anions forming sparingly soluble salt of the metal of which these electrodes are made.

6.1.1. Silver/silver chloride electrode

Ag/AgCl electrodes are the most widely used reference electrodes. The electrode redox process is as follows:

$$AgCl - e \Leftrightarrow Ag + Cl^-$$

The potential E of this electrode in a thermodynamical equilibrium (without current flow) is dependent on the concentration of Cl^- ions present in the solution. The Nernst equation for this electrode takes the form:

$$E = E^0_{AgCl/Ag} - \frac{RT}{F}\ln a_{Cl^-} \tag{31}$$

where $E^0_{AgCl/Ag}$ is the standard potential (+0.2224 V), R is the gas constant, T is the temperature (K), F is Faraday constant, a_{Cl^-} is the activity of Cl^- ions.

Silver/silver chloride electrodes are characterized by a stable and reversible potential, low-temperature hysteresis, and a wide range of temperatures at which they can be applied, up to 275 °C. One of the advantages of such electrodes is the possibility of their miniaturization and supporting on the same substrate as the working electrode.

Ag/AgCl electrodes in the form of a wire placed directly in the solution containing Cl^- ions, contacting with the membrane, are characterized by the best properties (Fig. 32a). They show low resistance and high capacitance, which results in good dynamical properties and makes them suitable for impedance analysis. This solution can be applied only for electrolytes not containing any strong oxidizers or reducers, forming coordination bonds with silver, e.g., CN^-, SCN^-, or ions forming salts more insoluble than AgCl, e.g., S^{2-}, I^-, Br^-. It should be also kept in mind that silver ions Ag^+ present in the electrolyte can be dangerous to some compounds, e.g., proteins. The solubility product constant of AgCl at room temperature is equal to 1.6×10^{-10}, which at KCl concentration of 0.1 M results in the presence of Ag^+ ions at the concentration of 1.6×10^{-9} M.

Direct immersing of Ag/AgCl electrodes in both solutions separated by a membrane can be restricted if the solutions differ in composition and concentration. Their potentials are different, and the difference between them is not equal to the transmembrane potential. In this case it is necessary to use electrodes placed in separate solutions, contacting with the membrane solution through porous contact, e.g., ceramic (Fig. 32b). A still better method is to apply electrodes with an additional salt bridge separating both solutions – adjacent to the membrane and electrode solution (Fig. 32c, d). It is more advantageous because the diffusion

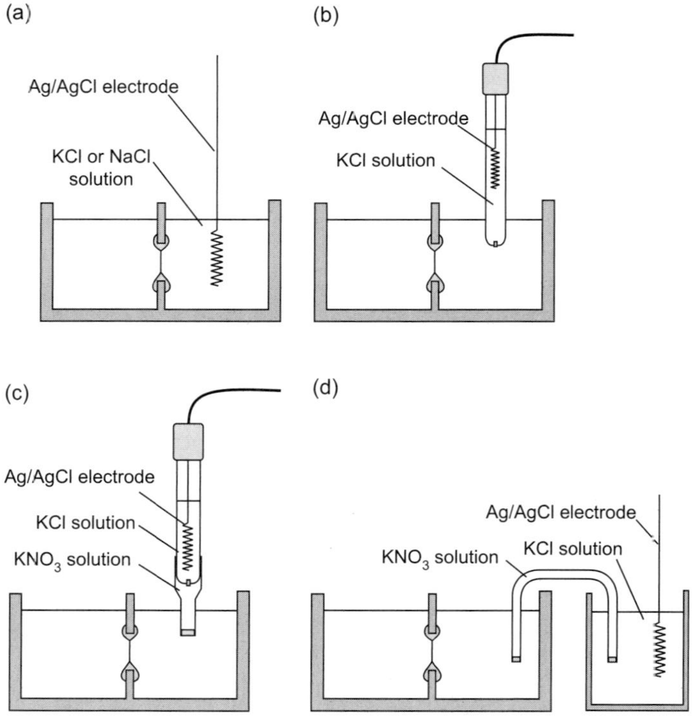

Fig. 32. Silver/silver chloride reference electrodes: (a) wire covered by AgCl immersed directly in the solution adjacent to the membrane, (b) Ag/AgCl electrode in solution of KCl (e.g., saturated), separated from the solution in the measurement vessel by a porous stopper, (c) electrode as in Figure b, but additionally separated by KNO_3 solution, usually saturated, (d) with an additional vessel with KCl solution and a Ag/AgCl electrode, connected with the membrane solution by a salt bridge filled with a saturated solution of KNO_3.

potentials of the electrodes are reduced. The salt bridge should be filled with saturated KNO_3.

The preparation of a silver/silver chloride electrode is relatively simple. A silver wire or plate can be covered by AgCl through anodic oxidation of the metal in a solution of HCl. An example of the equipment for the preparation of Ag/AgCl electrodes is presented in Fig. 33. A solution of hydrochloric acid should be stirred using, e.g., a magnetic stirrer. Electrolysis can be performed for 30 min with current density 10–30 mA/cm^2.

6.1.2. Calomel electrode

Calomel electrodes are also popular reference electrodes, but they are not used as widely as Ag/AgCl ones. Their parameters are not as good as those of Ag/AgCl

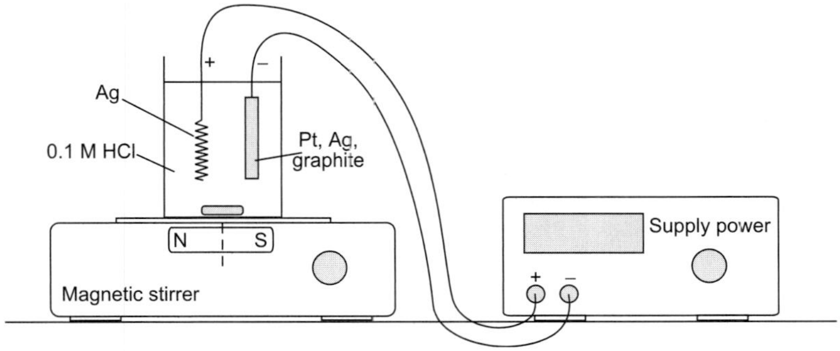

Fig. 33. A method for the preparation of a silver/silver chloride electrode. Silver is anodic oxidized in a solution of hydrochloric acid with current density 10–30 mA/cm^2.

electrodes, and their preparation is more complex. However, they are commercially available and considered useful for membrane investigations. The following redox process is observed in calomel electrodes:

$$Hg_2Cl_2 + 2e \Leftrightarrow 2Hg + 2Cl^-$$

The Nernst equation for this reaction assumes the form:

$$E = E^0_{Hg_2Cl_2/Hg} - \frac{RT}{F} \ln a_{Cl^-} \tag{32}$$

where $E^0_{Hg_2Cl_2/Hg}$ is the standard potential (+0.268 V), R is the gas constant, T is the temperature (K), F is Faraday constant, a_{Cl^-} is the activity of Cl^- ions. Examples of practical applications of calomel electrodes are shown in Fig. 34.

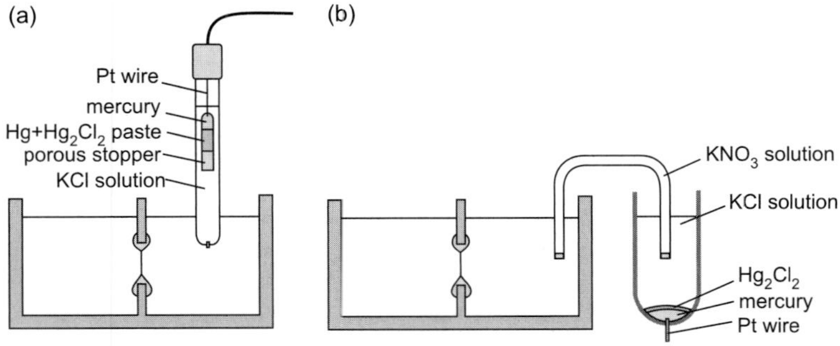

Fig. 34. Examples of practical applications of calomel electrodes, e.g., for membrane studies: (a) an easy-to-use calomel electrode, where the electrode internal solution is in contact with the electrolyte through a porous stopper, (b) a calomel electrode placed in a separate vessel, contacting with the electrolyte through a salt bridge.

A disadvantage of calomel electrodes is temperature hysteresis. The potential of these electrodes cannot reach a stable value for a few hours after exposure to temperature changes higher than 10 °C.

6.2. Auxiliary electrodes

Reference electrodes under currentless conditions are used in 3- and 4-electrode systems. In this case it is necessary to use additional current electrodes. These electrodes should have low impedance, but their reversibility is of secondary importance. Platinum electrodes are used most often, sometimes also stainless steel ones. Calomel or silver/silver chloride electrodes can be used as auxiliary electrodes.

7. PROTECTION AGAINST NOISES

During electrochemical investigations of membranes the signal measured is often in the range of picoampers. Such low signals need very effective protection against noises. Both measurement vessels and electronic equipment must be carefully shielded and connected. The accuracy of potential measurement is usually satisfactory at a level of 0.1 mV. and the measurement equipment does not cause any problems in this respect. But extremely low currents are difficult to protect. Alternating current with an amplitude 0.1 mV and frequency 50 Hz, induced in a cable connected to electrodes, can generate capacitance current of about 3 pA, with membrane capacitance of 1 nF. Such voltage can be undetectable by measurement devices, but its influence on the current signal can be totally destructive. Higher frequencies generate capacitance current with higher intensity. These frequencies can be generated especially by high-power devices and digital equipment.

The most common sources of noises:

1. Thyristor phase regulators of power, e.g., light controllers, electrical motor controllers.
2. Cables feeding high-power heaters. They generate strong magnetic fields.
3. Pulse power supply in hardware.
4. Computer monitors.
5. Magnetic stirrers.
6. Power supply of high-power illuminators, e.g., xenon lamps.

Recommendations for noise reduction:

1. Put measurement vessels to the Faraday metal cage.
2. Use the shortest possible cables for connecting electrodes to measurement devices.

3. Use signal cables with double shielding.
4. Do not put into the Faraday cage any equipment supplied from an electrical power network, e.g., magnetic stirrers, lamps.
5. Remember that each cable placed in the Faraday cage and sticking out of it, not connected with the mass, is a source of noises, because it acts as an antenna.
6. Using a thick copper wire, connect the cases of all devices to the common bus, connected to the earthing system.
7. Connect the shields of all signal cables together, using a thick, short wire. Make sure that the connection of the signal mass with the earth reduces the noises.
8. Choose the best point of connection of the Faraday cage with the signal mass or with the case of measurement devices.
9. If possible, use equipment supplied from batteries, e.g., pH-meters, interfaces for data acquisition, with opto-electronic isolation between the analog and digital parts.

REFERENCES

[1] K. Dołowy, Prog. Surf. Sci. 15 (1984) 245–368.
[2] P. Krysiński, H.T. Tien, Prog. Surf. Sci. 23 (1986) 317–412.
[3] S. Iiyama, Y. Suezaki, K. Toko, T. Murata, H. Kamaya, I. Ueda, K. Yamafuji, Biophys. Chem. 36 (1990) 141–147.
[4] S. Iiyama, K. Toko, T. Murata, Y. Suezaki, H. Kamaya, I. Ueda, K. Yamafuji, Biophys. Chem. 36 (1990) 149–154.
[5] J. Koryta, J. Dvořak, V. Boháčková, Elektrochemia, Państwowe Wydawnictwo Naukowe, Warszawa, 1980.
[6] H.T. Tien, A.L. Ottova, Colloids Surf. A 149 (1999) 217–233.
[7] X. Lu, A. Leitmannova-Ottova, H.T. Tien, Bioelectrochem. Bioeng. 39 (1996) 285–289.
[8] J.R. Lopez, H.T. Tien, Biochim. Biophys. Acta 597 (1980) 433–444.
[9] H.T. Tien, Z. Salamon, Bioelectrochem. Bioeng. 22 (1989) 211–218.
[10] H.P. Ting, W.A. Huemoeller, S. Lalitha, A.L. Diana, H.T. Tien, Biochim. Biophys. Acta 163 (1968) 439–450.
[11] M. Schadt, Biochim. Biophys. Acta 323 (1973) 351–356.
[12] E. Bienvenue, P. Seta, A. Hofmanová, C. Gavach, M. Momenteau, J. Electroanal. Chem. 162 (1984) 275–284.
[13] C.M. Drain, D.C. Mauzerall, Biophys. J. 63 (1992) 1556–1563.
[14] D.C. Mauzerall, C.M. Drain, Biophys. J. 63 (1992) 1544–1555.
[15] H.T. Tien, N.B. Joshi, Photobiochem. Photobiophys. 10 (1986) 241–251.
[16] X.D. Wang, B.W. Zhang, J.W. Bai, Y. Cao, X.R. Xiao, J.M. Xu, J. Phys. Chem. 96 (1992) 2886–2891.
[17] H.T. Tien, J. Kutnik, Photobiochem. Photobiophys. 7 (1984) 319–329.
[18] J. Bentz, S. Nir, Bull. Math. Biol. 42 (1980) 191–220.
[19] K.C. Hwang, D. Mauzerall, Nature 361 (1993) 138–140.
[20] J.H. Fendler, J. Macromol. Sci.-Chem. A27 (1990) 1167–1185.
[21] R. Rolandi, D. Ricci, O. Brandt, J. Phys. Chem. 96 (1992) 6783–6790.
[22] A.J. Bard, L.R. Faulkner, Electrochemical Methods, Wiley, New York, 1980.
[23] Z. Figaszewski, J. Electroanal. Chem. 139 (1982) 309–315.

[24] Z. Figaszewski, *Urzed Patentowy PRL*, Polish patent no. 135048.
[25] Z. Figaszewski, Z. Koczorowski, G. Geblewicz, J. Electroanal. Chem. 139 (1982) 317–322.
[26] S. Kalinowski, Z. Figaszewski, Meas. Sci. Technol. 6 (1995) 1050–1055.
[27] H.T. Tien, A.L. Ottova, Electrochim. Acta 23 (1998) 3587–3610.
[28] L. Ding, J. Li, S. Dong, E. Wang, J. Electroanal. Chem. 416 (1996) 105–112.
[29] A. Zebrowska, P. Krysiński, Z. otowski, Bioelectrochemistry 56 (2002) 179–184.
[30] N. Wilke, M. Baruzzi, J. Electroanal. Chem. 537 (2002) 67–76.
[31] L. Fittabile, M. Robello, A. Relini, M. De Rosa, A. Gliozzi, Thin Solid Films 284–285 (1996) 735–738.
[32] I. Genco, A. Gliozzi, A. Relini, M. Robello, E. Scalas, Biochim. Biophys. Acta 1149 (1993) 10–18.
[33] S. Kalinowski, G. Ibron, K. Bryl, Z. Figaszewski, Biochim. Biophys. Acta 1369 (1998) 204–212.
[34] S. Koronkiewicz, S. Kalinowski, K. Bryl, Biochim. Biophys. Acta 1510 (2001) 300–306.
[35] S. Koronkiewicz, S. Kalinowski, K. Bryl, Biochim. Biophys. Acta 1561 (2002) 222–229.
[36] A. Ridi, E. Scalas, M. Robello, A. Gliozzi, Thin Solid Films 327–329 (1998) 796–799.
[37] M. Robello, A. Gliozzi, Biochim. Biophys. Acta 982 (1989) 173–176.
[38] M. Kotulska, S. Koronkiewicz, S. Kalinowski, Acta Phys. Pol. B 33 (2002) 1115–1129.
[39] S. Iiyama, Y. Azuma, M. Nagaishi, K. Toko, Biophys. Chem. 61 (1996) 23–27.
[40] S. Iiyama, S. Ezaki, K. Toko, T. Matsuno, K. Yamafuji, Sens. Actuators B 24–25 (1995) 75–79.
[41] S. Iiyama, Y. Suzuki, S. Ezaki, Y. Arikawa, K. Toko, Mater. Sci. Eng. C 4 (1996) 45–49.
[42] S. Iiyama, K. Toko, T. Murata, H. Ichinose, Y. Suezaki, H. Kamaya, I. Ueda, K. Yamafuji, Biophys. Chem. 45 (1992) 91–100.
[43] S. Iiyama, K. Toko, K. Yamafuji, Sens. Mater. 3 (1991) 1–7.
[44] S. Iiyama, M. Yahiro, K. Toko, Sens. Mater. 7 (1995) 191–201.
[45] S. Iiyama, K. Toko, K. Yamafuji, Trends Phys. Chem. 1 (1991) 121–132.
[46] R.G. Ashcroft, H.G.L. Coster, J.R. Smith, Biochim. Biophys. Acta 469 (1977) 13–22.
[47] R.G. Ashcroft, K.R. Thulborn, J.R. Smith, H.G.L. Coster, W.H. Sawyer, Biochim. Biophys. Acta 602 (1980) 299–308.
[48] H.G.L. Coster, R. Simons, Biochim. Biophys. Acta 203 (1970) 17–27.
[49] D.R. Laver, J.R. Smith, H.G.L. Coster, Biochim. Biophys. Acta 772 (1984) 1–9.
[50] U. Zimmermann, R.G. Ashcroft, H.G.L. Coster, J.R. Smith, Biochim. Biophys. Acta 469 (1977) 23–32.
[51] R. Fettiplace, L.G.M. Gordon, S.B. Hladky, J. Requena, H.P. Zingsheim, D.A. Haydon, in: E.D. Korn (Ed.), Methods in Membrane Biology Vol. 4, Plenum Press, New York, 1975.
[52] W. Hanke, W.R. Schlue, Planar Lipid Bilayers, Methods and Applications, Academic Press, London, 1993.
[53] P. Schoch, D.F. Sargent, R. Schwyzer, J. Membr. Biol. 46 (1979) 71–89.
[54] M. Stelzle, E. Sackmann, Biochim. Biophys. Acta 981 (1989) 135–142.
[55] H.T. Tien, Bilayer Lipid Membranes: Theory and Practice, Marcel Deccer Inc., New York, 1974.
[56] A.N. Chanturiya, Biochim. Biophys. Acta 1026 (1990) 248–250.
[57] A.N. Chanturiya, H.V. Nicoloshina, J. Membr. Biol. 137 (1994) 71–77.
[58] V.V. Cherny, V.S. Sokolov, I.G. Abidor, Bioelectrochem. Bioeng. 7 (1980) 413–420.
[59] S.H. White, D.N. Blessum, Rev. Sci. Instrum. 46 (1975) 1462–1466.
[60] S.H. White, Biophys. J. 10 (1970) 1127–1148.
[61] S.H. White, Biochim. Biophys. Acta 196 (1970) 354–357.
[62] S.H. White, Biophys. J. 12 (1972) 432–445.
[63] S.H. White, Biochim. Biophys. Acta 323 (1973) 7–22.
[64] S.H. White, Biochim. Biophys. Acta 356 (1974) 8–16.

[65] S.H. White, in: C. Miller (Ed.), Ion Channel Reconstitution, Plenum Publishing Corporation, New York, 1986.
[66] S.H. White, W. Chang, Biophys. J. 36 (1981) 449–453.
[67] S.H. White, T.E. Thompson, Biochim. Biophys. Acta 323 (1973) 7–22.
[68] R. Benz, O. Fröhlich, P. Läuger, M. Montal, Biochim. Biophys. Acta 394 (1975) 323–334.
[69] H.G.L. Coster, R. Simons, Biochim. Biophys. Acta 163 (1968) 234–239.
[70] H.G.L. Coster, R. Simons, Biochim. Biophys. Acta 203 (1970) 17–27.
[71] H.G.L. Coster, J.R. Smith, Biochim. Biophys. Acta 373 (1974) 151–164.
[72] J.P. Dilger, R. Benz, J. Membr. Biol. 85 (1985) 181–189.
[73] M. Miyake, I. Inoue, Y. Kobatake, Biochim. Biophys. Acta 323 (1973) 367–377.
[74] A.D. Pickar, W.D. Amos, Biochim. Biophys. Acta 455 (1976) 36–55.
[75] R.C. Waldbilling, G. Szabo, Biochim. Biophys. Acta 557 (1979) 295–305.
[76] O. Alvarez, M. Brodwick, R. Latorre, A. McLaughlin, S. McLaughlin, G. Szabo, Biophys. J. 44 (1983) 333–342.
[77] R. Benz, K. Janko, Biochim. Biophys. Acta 455 (1976) 721–738.
[78] J. Anzai, T. Hoshi, T. Osa, Trends Anal. Chem. 13 (1994) 205–220.
[79] B. Fuks, F. Homblé, Biophys. J. 66 (1994) 1404–1414.
[80] P. Läuger, W. Lesslauer, E. Marti, J. Richter, Biochim. Biophys. Acta 135 (1967) 20–32.
[81] S. Takashima, R. Yantorno, R. Novack, Biochim. Biophys. Acta 469 (1977) 74–88.
[82] S. Toyama, A. Nakamura, F. Toda, Biophys. J. 59 (1991) 939–944.
[83] S. Kalinowski, Z. Figaszewski, Biochim. Biophys. Acta 1112 (1992) 57–66.
[84] S. Kalinowski, Z. Figaszewski, Meas. Sci. Technol. 6 (1995) 1043–1049.
[85] D.F. Sargent, T. Hianik, Bioelectrochem. Bioeng. 33 (1994) 11–18.
[86] P. Schoch, D. Sargent, Membrane proteins, in: A. Azzi, U. Brodbeck, P. Zahler (Eds.), A Laboratory Manual, Springer, Berlin, 1981.
[87] D. Sužnjević, P.A. Grigorijev, D. Vučelić, Bioelectrochem. Bioeng. 34 (1994) 19–24.
[88] P. Mueller, D.O. Rudin, H.T. Tien, W.C. Wescott, Nature (London) 194 (1962) 979.
[89] J. Stern, H.J. Freisleben, S. Janku, K. Ring, Biochim. Biophys. Acta 1128 (1992) 227–236.
[90] J.I. Anzai, T. Osa, Tetrahedron 50 (1994) 4039–4070.
[91] H. Yamaguchi, H. Nakanishi, Biochim. Biophys. Acta 1148 (1993) 179–184.
[92] M. Stelzle, G. Weismüller, E. Sackmann, J. Phys. Chem. 97 (1993) 2974–2981.

CHAPTER 2

Tethered Bilayer Membrane Sensors with Small Transmembrane Peptide Ion Channels – Recent Developments, Future Research and Potential Applications☆

Ping Yin*

Department of Chemistry, Macquarie University, North Ryde, NSW 2113, Australia

Contents

Abstract

This chapter covers the chemistry and experimental approach for studying tethered bilayer lipid membranes (tBLM) for incorporating the small peptide transmembrane ion channel alamethicin. It describes the chemical components developed by AMBRI and methods for

---

☆The experimental work described here was conducted at AMBRI limited during 1997–1999 and results were published in 2003, see Ref. [6].

* Corresponding author. E-mail: pyin@chem.mq.edu.au

ADVANCES IN PLANAR LIPID BILAYERS AND LIPOSOMES, VOLUME 2
ISSN 1554-4516 DOI: 10.1016/S1554-4516(05)02002-8

optimizing the tBLM containing alamethicin in order to construct stable and sensitive sensors. The sensor containing alamethicin channels was then probed with amiloride and its analogues in the μM concentration range. The potency of amiloride and analogues as channel blocker is rated against their molecular structure. Current research activities in related areas are also reviewed. The potential applications are studying "signature sequences" of large menbrane channel proteins, small molecule sensing and biodiscovery.

1. INTRODUCTION

There is an increasing interest in using tethered bilayer lipid membrane (tBLM) sensors [1–4] containing transmembrane ion channels for the screening of pharmacological compounds [5]. The tBLM sensors containing transmembrane channels can respond quickly and directly to small molecules with μM sensitivity [6]. These sensors can also be sufficiently stable for many applications under optimised conditions [17].

It is well established that medical conditions such as hypertension, arrhythmias, seizures and diabetes are associated with membrane proteins including ion channels, pumps and receptors [7]. One of the major challenges of using tethered membranes to study membrane proteins is to accommodate large proteins into the tethered membrane without denaturation or loss of functionality [8]. It is believed that the ion transport properties of small peptides may be relevant to that of some physiological channel proteins, which typically consist of 2000 or more amino acids [9]. It is, therefore, important to study small peptide channels as a model system.

Another potential application of tBLM sensor containing transmembrane ion channels is to detect molecules such as neurotoxins. For example, neurotoxins block the entry of the sodium ions required by cells to make 'action potentials', rendering the cells and tissue non-functional [10,11]. Ion channel-based sensors have an advantage over immunosensors as they are able to detect any compound that blocks the ion channel whereas immunosensors can only detect a specific compound, the antigen of the anchoring antibody [16].

tBLM is considered a useful cell mimic [12]. AMBRI established tBLM sensors with the ion-carrier valinomycin [13–15], or with gramicidin-based ion channel switches [16–18]. We chose the small peptide alamethicin as a model system to study its suitability for its transmembrane applications. We hope to extend our understanding from small peptide channels to channel-forming membrane proteins.

After the first observation of tBLM conductance induced by alamethicin, it took 2 years of laboratory work to show beyond doubt that the conductance was due to small peptide ion channels and to optimise the sensor. In this chapter, I shall discuss the experimental work as well as the overall optimisation process. There is also a brief review of related work in the field and suggestions for future work and potential applications in each section.

2. THE SENSOR PLATFORM – A TETHERED BILAYER LIPID MEMBRANE

The tBLM platform can be constructed by chemically attaching or physisorbing lipid components onto a solid support; for example, thiolated lipid components can be chemically attached onto gold surface forming a tethered lipid monolayer. Several recent review articles have summarised advances in the formation of the tethered lipid monolayer that consist of the first layer of the tBLM [19–21]. tBLM use the hydrophilic and hydrophobic properties of different parts of the lipid molecules to form a stable, self-assembled bilayer lipid membrane. Ion channels, or other conductance-inducing components, are incorporated to achieve controlled conductivity.

2.1. Chemical components of sealed tethered bilayer lipid membrane

A sealed tBLM is a tBLM in absence of any component that can induce a conductance. The compounds, MSLOH, DLP, MAAD, GDPE and DPEPC, used to construct a sealed tBLM for the alamethicin study are shown in Fig. 1 [1,22–26]. The tethered components are disulphides including MSLOH, DLP and MAAD. There are three important portions for MSLOH and DLP, namely lipid portion, reservoir portion and disulphide portion. MAAD is a surface coverage molecule without lipid and reservoir portions and acts as a spacer. Disulphide portions are responsible for chemically attaching the molecules onto the gold surface. The non-tethered components are GDPE and DPEPC. The lipid portions from all components including non-tethered components line up in parallel by hydrophobic force. The reservoir portions and the spacer molecules together create the reservoir compartment to hold the conducting ions. MSLOH is of special importance for membrane stability as it has a membrane-spanning portion that spans cross the bilayer (Fig. 2). Note that the reservoir portion of MSLOH and DLP consist of several ester bonds, the significance of these ester linkages will be discussed in Section 4.3.

2.2. Construction of the sealed tethered bilayer lipid membrane

An illustration of a sealed tBLM is shown in Fig. 2 [13,16–18]. The chemical components for each graphic symbol are shown in Fig. 1. The tethered layer, consisting of thiolated lipid components, was prepared by immediate immersion of a freshly sputtered or evaporated gold into an ethanol solution of thiolated components. The non-tethered lipid components were assembled by mixing of ethanol solution of these components by phosphate buffered saline followed by

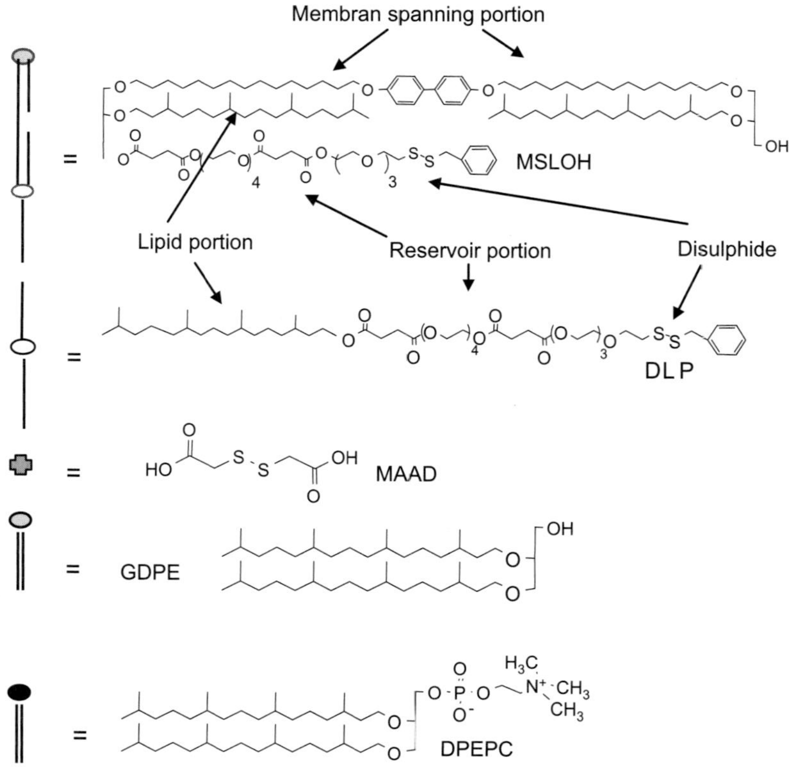

Fig. 1. Chemical structures and graphic presentation of the tethered bilayer membrane (tBLM) forming components.

several washing steps to remove excess lipids. The experimental procedures and methods for constructing such a membrane are detailed in a previous publication [6]. There are alternative methods to form assemblies of tethered membranes, such as vesicle fusion and deposition [27–29].

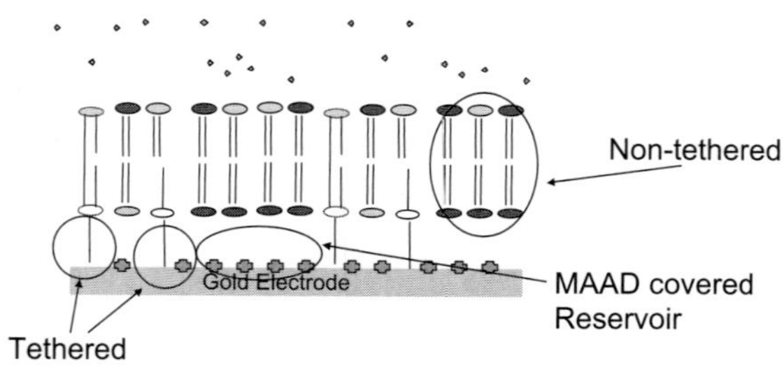

Fig. 2. A graphic illustration of a sealed tethered lipid bilayer membrane.

2.3. Stability of the sealed tethered bilayer lipid membrane

The disintegration of the membrane can be caused by the fragility of sulphur–gold bond, the chemical instability of the membrane components, changes in the balance of the hydrophobic force that holds membrane components in an organised arrangement, and possible mechanical damage.

To achieve a stronger and more uniform sulphur–gold layer that gave better stability, one of the strategies was to add a lipid section to all the tethered components to create a sealed layer directly above sulphur–gold bond [17]. Higher stability has also been achieved by tethering membrane lipids from bacteria that survive extreme environments. These bacteria contain lipid structures that span the entire membrane and have added chemical modifications such as replacing ester linkages with ether linkages. Since ester bonds can be hydrolysed under both acidic and alkaline conditions chemical modifications of MSLOH, Fig. 1, with all ether links were found to improve the stability of sensors [30–32].

The tethered lipid bilayer membrane can exhibit remarkable stability. It withstands drying and rehydrating. The sensors can be stable for months when using preservatives to stabilise freeze-dried sensor containing proteins [17]. Since crystallization of buffer salts upon drying may represent a source of mechanical damage, the method we used for drying sealed tethered lipid bilayer membranes for alamethicin study was to replace the buffer solution with distilled water, before drying under vacuum and storing in sealed bags under dry nitrogen. Rehydrating with distilled water then replaced by buffer restored the membranes successfully to its original state in 2 days [6].

2.4. Measuring configuration

The construction and measuring configuration for alamethicin containing tBLMs is illustrated in Fig. 3 [13]. An electrochemical cell is composed of the gold sensor electrode (supporting the tBLM), a counter and a reference electrode in a conventional three-terminal configuration. Electrical impedance spectroscopy measurements were conducted on a platform containing 16 independent sample cells, which permitted the simultaneous conduction of a range of experiments (e.g. a range of concentrations) and controls under nominally common preparation and experimental conditions. All 16 cells were concurrently connected to a purpose built multi-channel impedance analyser and the applied frequency swept between 1 Hz and 1 kHz (the applied peak-to-peak voltage was $\pm$50 mV). A d.c. bias voltage could also be independently applied and varied. Examples of impedance data are shown in Figs 4 and 5 in which the left-most four columns display the impedance spectra for the 16 cells (there are many sequential spectra overlaid in each graph, giving a graphical indication of the time evolution

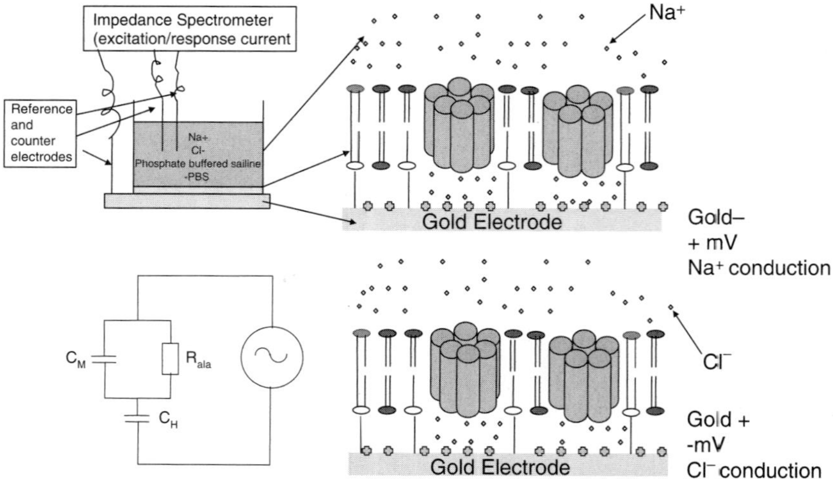

Fig. 3. Measuring configuration.

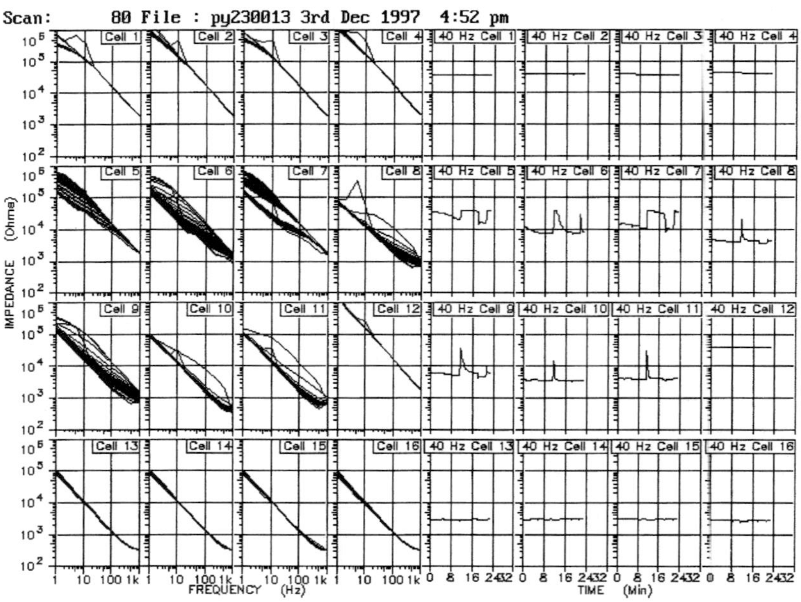

Fig. 4. Impedance spectra of the tethered bilayer lipid membrane with alamethicin in equilibrium in phosphate buffered saline at the concentration of 2.5 μM (second row), 6.25 μM (third row), and 25 μM (fourth row, loss of non-tethered membrane component at high alamethicin concentration) [50].

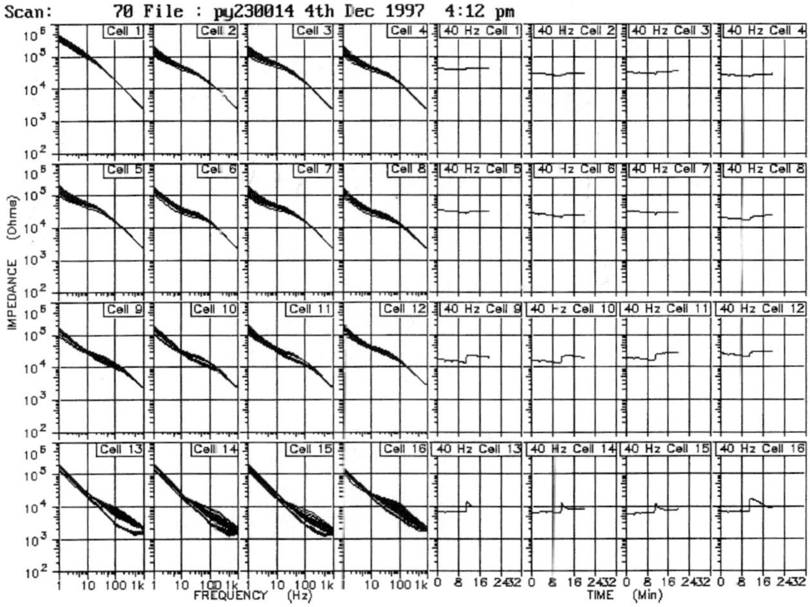

Fig. 5. Impedance spectra of the tethered bilayer lipid membrane with alamethicin in equilibrium in phosphate buffered saline at the concentration of 0.39 μM (first row), 0.78 μM (second row), 1.56 μM (third row), and 3.15 μM (fourth row) [50].

of the impedance spectrum). The right-most four columns show the time evolution of each cell's impedance as measured at 40 Hz. The equivalent circuit shown in Fig. 3 depicts a simplified electrical model of the tethered membrane where C_H is a combination of the Helmholtz capacitance of ions tightly bound to the electrode and the diffuse reservoir (the trapped volume beneath the tBLM) capacitance, C_M is the membrane capacitance and the impedance R_{ala} represents the amount of ionic conduction through transmembrane channels. Impedance spectroscopy is a convenient and powerful technique for determining the values and relative changes in the values of C_H, C_M and R_{ala}. However, in experiments where the tBLM is formed and channels are then incorporated, C_H and C_M should remain relatively constant and only R_{ala} should decrease in value as more channels incorporate into the tBLM and increase the overall conductance. For well-sealed membranes (i.e., those with a continuous and unbreached tBLM), the impedance at low frequencies is dominated by C_M and the impedance at high frequencies by C_H. The crossover between these two regimes gives a measure of the impedance of the channels R_{ala} as indicated by the inflections in the impedance spectra in Fig. 5 occurring between 10 and 100 Hz. This is the reason that 40 Hz is chosen for the time course graphs in the four right-most columns, in that it gives the largest response to ion channel incorporation into the tBLMs.

3. INDUCING ALAMETHICIN IN TETHERED BILAYER LIPID MEMBRANE

Alamethicin is an antibiotic alpha-helix peptide of 20 amino acids with acetylated N-terminus, characterised by X-ray crystallography [33,34].

Ac-Aib-Pro-Aib-Ala-Aib-Aib(Ala)-Gln-Aib-Val-Aib-Gly-Leu-Aib-Pro-Val-Aib-Aib-Gln(Glu)-Gln-Phe-OH(Pheol).

Firstly, we observed the conductance change of a sealed tethered bilayer membrane when we added alamethicin between 2.5 and 25 μM into PBS buffer adjacent to the membrane (Fig. 4). We did not remove excess alamethicin in PBS; therefore, the conductance of the membrane we observed was the effect of equilibrium of alamethicin in membrane and in solution. The sealed tBLM was made from a solution with 1:2 molar ratio of DLP:MAAD and 100:1 molar ratio of (DLP+MAAD):MSLOH (Fig. 1) for the tethered layer. The non-tethered membrane was prepared from a mixture of GDPE and DPEPC (Fig. 1). This initial experiment showed channel-like conductance that was similar to gramicidin channels in the tBLM. This experiment became the beginning of a process to understand the behaviour of alamethicin in tBLMs and the starting point to find optimal conditions under which to observe alamethicin channel formation, its conductance of anions (Cl^-) and cations (Na^+), as well as the direct blocking of the channel by amiloride and its analogues. Since this was the first time that channel formation of alamethicin in tBLMs was observed, we sought evidence to prove that the tBLM impedance changes were indeed induced by alamethicin channels formed in the lipid bilayers. Apart from gramicidin, the introduction of mellitin (1 μM) into tethered bilayers was also observed to change the impermeable bilayer to higher conductance by cyclic voltammetry [35]. The channel behaviour of alamethicin in suspended BLMs have been extensively studied [36–38]. The evidence of alamethicin insertion into tethered lipid bilayers and the formation of the ion channels is similar to those in suspended BLMs although there was no record of channel blocking by small molecules of pharmacological modulators. Although alamethicin is a small channel-forming peptide [33], its channel modulation by pharmacological molecules could be of particular significance. If the blocking were not electrostatic, this modulation would be a direct observation of the molecular interaction of a bioactive small molecule with a peptide. The aggregation of several peptide molecules (the average aggregates of alamethicin ion channels is believed to range from 5 to 8 molecules [39] with the hexamer being the most stable form [40]) could be playing an important role in the amplification of this interaction which otherwise may not be observable. These characteristics may lead to some exciting opportunities that shall be discussed in Section 6.

4. OPTIMISATION OF THE SENSOR AND MEASURING CONDITIONS

Compared with suspended bilayer lipid membranes, tBLMs provide a far more stable and well-sealed platform especially when tethered membrane-spanning lipids and lipids with phytanyl side chains are incorporated [17,41]. Of many tethering options such as polymer and gel supported bilayers [42–44], sulphur–gold tethering is by far the most extensively investigated method [45,46,20,21]. However, experimental results indicated little or no effective control over the relative numbers of different component species in a mixed film on the gold [47]. Apart from stringent control of the purity of all membrane components [48], good control over the assembly process seemed to be the key to achieving a reproducible self-assembled monolayer. While these understandings can guide the initial selection of a platform, the search for a suitable tBLM for specific applications could be achieved only by a matrix of optimisation experiments. Environmental conditions, such as clean room temperature and humidity, were kept constant. In this section, I shall describe the optimisation of the tethered bilayer membrane with alamethicin channels as well as the selection and optimisation experiments of inhibitors for alamethicin channels. This is essentially a process of finding the appropriate window in which alamethicin channel behaviour and its interaction with the channel blockers can be observed.

4.1. Concentrations and application methods of alamethicin

tBLMs prepared from a formulation of DLP and MAAD in a molar ratio of 1:2 (Fig. 2) was best for alamethicin-induced conductance when in equilibrium with alamethicin in PBS in the concentration range of 0.3–3 μM (Fig. 5). Within these concentrations, however, removal of alamethicin from PBS buffer by washing can remove the alamethicin from the tethered lipid bilayer membrane, recovering the sealed lipid bilayer membrane. When alamethicin concentration exceeded 25 μM, the non-tethered part of membrane eventually collapsed.

To prepare the assembly of the tBLM with alamethicin that could not be removed by washing a formulation of DLP and MAAD in a molar ratio of 2:1 (Fig. 2) was used. Although this ratio does not represent the coverage of DLP on the gold surface, as discussed in Section 4, the tBLM made with higher DLP ratio tolerated much higher alamethicin concentrations to enable permanent alamethicin insertion into the membranes. The alamethicin concentration range of 2.5 μM to 60 μM was tested and 32.2 μM was used to produce all tBLMs for various experiments [6].

Unlike gramicidin, the attempt to assemble alamethicin into the lipid bilayer with pre-mixed alamethicin and membrane components was unsuccessful.

Alamethicin was found to disrupt the formation of uniform membranes and it was necessary to add it after the formation of sealed tBLM. To obtain consistent and reproducible tBLM assemblies with alamethicin channels it is important to have consistent and reproducible sealed tBLMs. The amount of alamethicin channels in the assemblies is not only dependent on the concentration of alamethicin but also critically on the relative ratio of all lipid components.

The use of detergent such as SDS does not improve the consistency of the assembly but it does, however, improve the homogeneity of the alamethicin when dispersed in an aqueous phase.

In the context of assembly and application methods, alamethicin and gramicidin behave fundamentally differently. While gramicidin can easily be incorporated into a tBLM, we were not able to assemble alamethicin in the same manner. This may be because the channel for gramicidin is defined as a molecular unit, whereas the alamethicin channels are loosely bound aggregates that require voltage application to form. Use of preformed tBLM is a versatile approach for applications of many different ion transporters; for example, a preformed peptide-supported tBLM was used by Naumann *et al.* [49] to achieve incorporation of cytochrome *c* oxidase from bovine heart by fusion of liposomes. The preformed tBLM must have the fluidity and stability as well as be sufficiently spaced structurally to accommodate the non-transmembrane part of the protein without altering the 3D structure if protein functionality is to be preserved. Designer molecules of the tethered lipid components that exhibit these properties will be the foundation to generate such platforms.

4.2. Alamethicin isomers

Commercial alamethicin is a mixture of up to six components (Fig. 6) that vary according to supplier. These isomers can be separated by reverse phase HPLC using a solvent system of water, MeOH, acetonitrile and trifluoroacetic acid [50]. Alamethicin isomers were found to have different conductance properties in BLM [51,52]. Although the tBLM assemblies containing different isomers did have different conducting properties, use of a pure isomer was not advantageous for the experiments we were aiming to do. However, to fully explore the channel behaviour of alamethicin isomers, it would be worthwhile to look at the different behaviour of the isomers in tBLM, especially synthetically stabilised or mutated alamethicin isomers [53–55].

Three isomers from a mixture supplied by Sigma-Aldrich were separated by RP-HPLC and were used in parallel insertion experiments. At the same concentration, the conductivity induced by the three isomers A ($Rf=9.58$), B ($Rf=10.95$) and C ($Rf=12.90$) were in the order of $A \geq C > B$ although the difference was small [50].

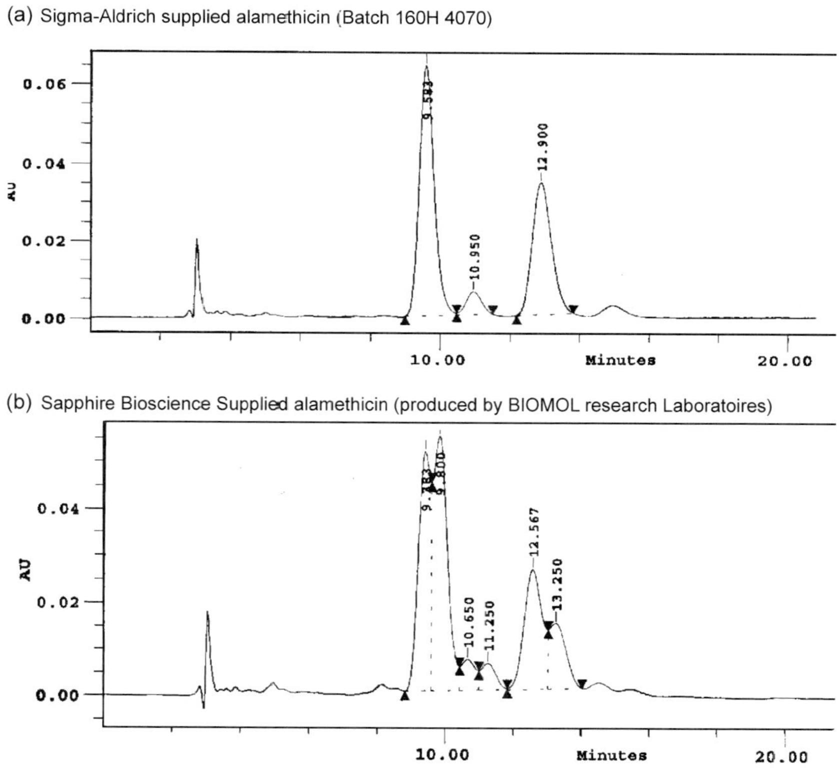

Fig. 6. Reverse phase HPLC (RP-HPLC) of commercial alamethicin (a) Sigma-Aldrich alamethicin (b) Sapphire Bioscience alamethicin, HPLC solvent: H_2O: MeOH:Acetonitrile = 20:20:60, with 0.05% trifluoroacetic acid at 1 ml/min, column Alltech Vydac218TP, UV detector at 214 nm [50].

4.3. Tethered lipid bilayer membrane components

The optimisation of tBLM for alamethicin application was based on the previous work of tBLM suitable for the applications of ion carrier valinomycin and gramicidin as an ion channel switch [13,16,17,22,56]. It was, however, of particular importance to reproduce the membranes consistently for alamethicin studies. Unlike the application of gramicidin, alamethicin was added after the formation of tBLM. The amount of alamethicin inserted into the membranes was extremely sensitive to the lipid constituents. Non-uniform membranes gave poor reproducibility so it is important to ensure the purity of each component and to adopt a consistent procedure to produce the membrane and to control very accurately the ratios of the components for the preparation of the membranes. These precautions proved to be critical for the optimisation of the platform to be used to study alamethicin.

Table 1. tBLM suitability for alamethicin study

	Concentration (μM)				Comments
	DLP	MAAD	MSLOH	DLP:MAAD:MSLOH	
tBLM 1	370	185	5.55	200:100:1	Suitable [6]
tBLM 2	277.5	277.5	5.55	150:150:1	Unable to maintain alamethicin in the tBLM
tBLM 3	185	370	5.55	100:200:1	At low alamethicin concentration, increased alamethicin concentration resulted in collapsed membranes

The optimisation of the sealed tBLM (a membrane in absence of any conductance inducing components) suitable for alamethicin study included a matrix of experiments varying the ratio of tethered components, DLP, MSLOH and MAAD and non-tethered components, GDPE and DPEPC (Fig. 1). The tBLM prepared from various ratios were then tested with a single alamethicin concentration to determine the most suitable platform for alamethicin studies. Optimisations were also conducted for GDPE and DPEPC and the optimal concentration for alamethicin studies was DPEPC (2.1 mM) and GDPE (0.9 mM) [6]. Membranes prepared from low-DLP ratio collapsed while when the DLP or MSLOH ratio was too high, insufficient alamethicin channels were produced. Table 1 summarises our findings for three different tBLMs we investigated.

4.4. Voltage application and measuring conditions

In a computer simulation study, Biggin *et al.* [57] clearly demonstrated the process and the time required for surface-absorbed alamethicin to insert into bilayers when the applied transmembrane voltage was 0, 100 and 200 mV. For the tBLM, the application of transmembrane voltage is offset by a Helmholtz capacitance C_H (Fig. 3 and Ref. [13]) so the voltage applied to tBLMs was higher than that to suspended BLM.

Optimisation of applied voltage 'formation' and measuring conditions were carried out between −500 and +500 mV. The optimal conditions for equilibrating alamethicin channels in the tBLMs was switching the voltage between +300 and −300 mV every 10 min until the conductance level became constant, reaching an equilibrium state (Fig. 7). This biosensor was made from a preformed sealed bilayer membrane. Alamethicin was added at a concentration of 38.5 μM (4 μl of 1 mM alamethicin ethanol solution into 100 μl of PBS). Excess alamethicin was washed off after 10 min incubation and before application of voltage (±300 mV,

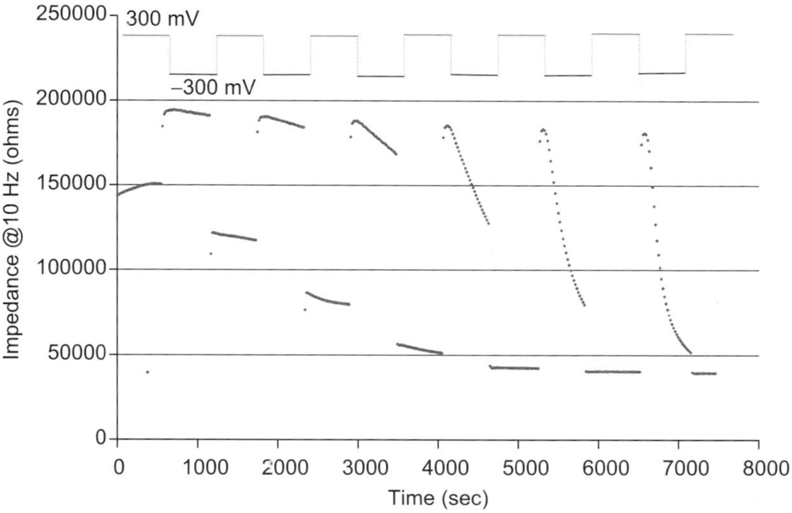

Fig. 7. Alamethicin insertion/ion channel formation assisted by the application of voltage to reach equilibrium state [6].

$t=0$). Higher voltages may result in ruptured membranes and the system took much longer to reach equilibrium at lower voltage.

At the equilibrium state, the conductance for Na^+ (at −300 mV) became constant and the conductance for Cl^- (at +300 mV) was limited by the diffusion of the Cl^- through the channel pore. It was known that the permeability of a single alamethicin channel for Na^+/Cl^- was approximately 1.6 [58].

4.5. Cation and anion conductance, polarity dependency and comparison with other channels

The central pore of small peptide ion channels such as gramicidin and alamethicin gave the conductance to the channels. Unlike the selective K^+ ion channels crystallised and structurally characterised by MacKinnon's group where the interaction of the channel with the ion is crucial for the ion transportation through the channel [59], gramicidin and alamethicin ion channels achieve conductance by creating a hole in the membrane through which ions can pass. This conductivity is similar to that of a leak caused by imperfect lipid bilayers except that latter may or may not be large enough to let certain ions through. In comparison with gramicidin ion channels that are completely non-conducting when the solid support is positive (anion conducting, Fig. 8) [60–63] channels formed by alamethicin (Fig. 7) can let Cl^- through. While both alamethicin and gramicidin conductance is polarity dependent, gramicidin does not have the ability of alamethicin to conduct by Cl^- anion.

(a)

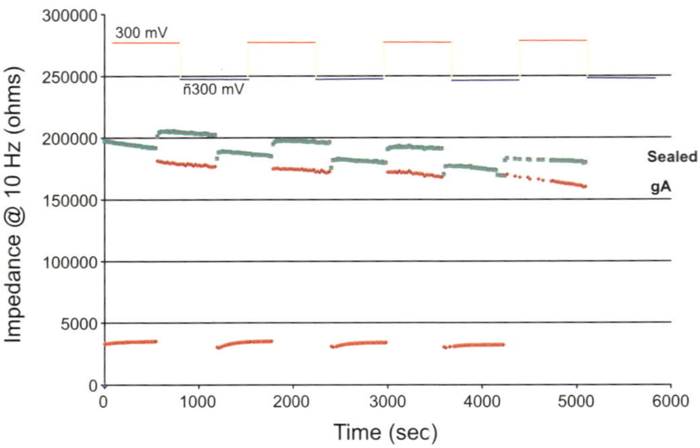

(b)

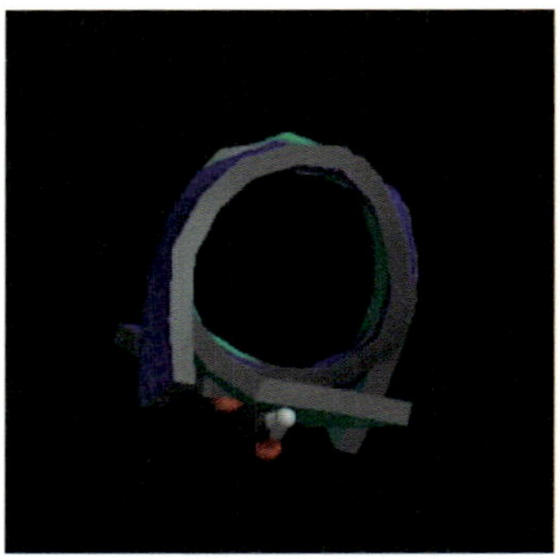

Crystal structure of gramicidin A (gA) dimmer (PBD ID 1av2, [81])

gAYYSSBn

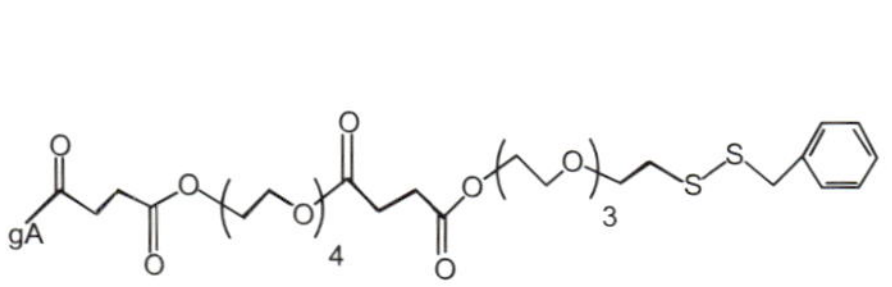

gA5XB

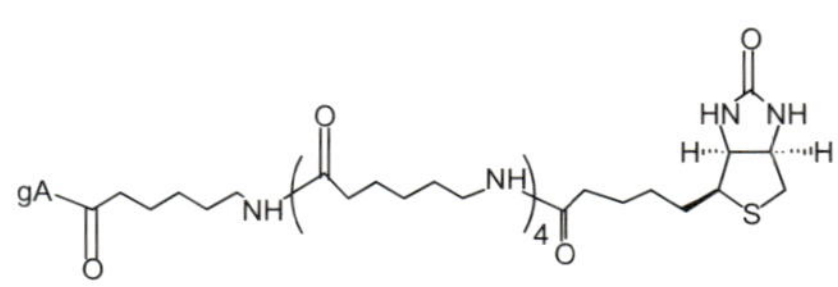

gA

HCO-X-Gly-L-Ala-D-Leu-L-Ala-D-Val-L-Val-D-Val-L-Trp-D-Leu-L-Tyr-D-Leu-L-Trp-D-Le u-L-Trp-$NHCH_2CH_2OH$

Fig. 8. (a) Application of voltage on tethered bilayers with Gramicidin ion channel (gAYYSSBn as tethered component and gA5XB as non-tethered component, red line) or without ion channel (Sealed, green line) [6] (b) gA dimmer crystal structure from Protein Databank (PDB ID 1av2, [81], Copyright 1998, National Academy of Sciences, USA), gA sequence, gAYYSSBn and gA5XB structure [13,16,82].

4.6. Channel conductivity modulating

After establishing the formation of ion channels in tethered lipid bilayers, it was important to further probe the channels with molecules that could regulate the ion transport. There was, however, no precedent for the interaction of alamethicin and amiloride, although amiloride and analogues are widely used as probes for ion transport studies. There are over a thousand amiloride analogues and many are commercially available. We, therefore, decided to investigate whether amiloride and its analogues were able to regulate the ion transportation of alamethicin channels [64]. We found they were excellent modulators for alamethicin channels and can easily form a convenient series of compounds of different potency from commercially available compounds.

Computer modelling of aggregated alamethicin channels (Fig. 9) [39] illustrated parallel helix bundles of alamethicin channels with the C-terminus mouth slightly larger than the N-terminus mouth. The orientation of the bundle helix of alamethicin after insertion into the tethered lipid bilayer is not known. The more hydrophilic C-terminus (Gln-Pheol) rather than acetylated N-terminus (Ac-Aib), however, is more likely to be exposed to the lipid–water interface. Amiloride may stop the conductance of alamethicin ion channels by electrostatic interactions or a direct molecular interaction between amiloride and the aggregated alamethicin

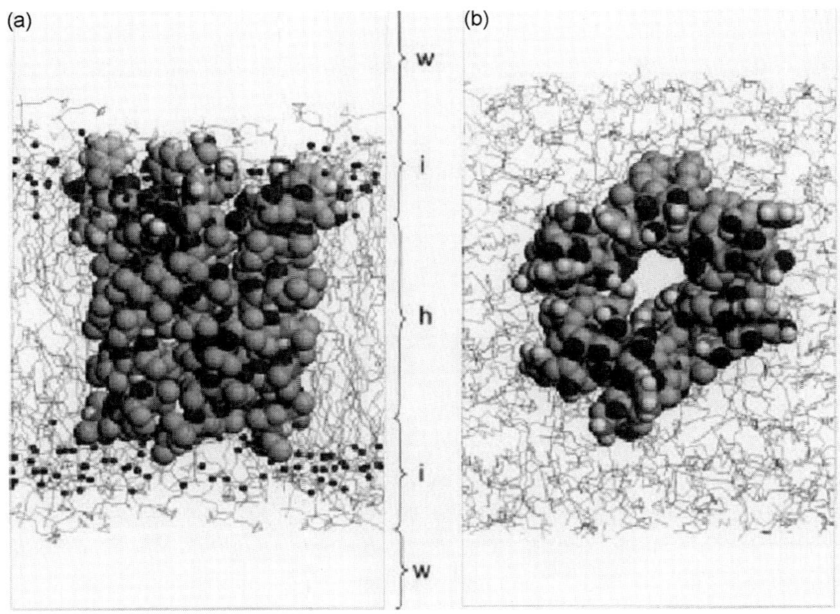

Fig. 9. Computer simulated alamethicin hexamer channel aggregates in lipid bilayer (With the permission of Elsevier) [39].

molecules. Direct molecular interaction at the C-terminus was shown to stop ion flow through alamethicin channels. Futaki *et al.* [65] biotinylated the C-terminus of alamethicin and demonstrated the channel blockage by streptavidin (Mr= 40,000), while the electrostatic interaction was also observed to cause channel blocking. Rink *et al.* [66] reported that peptide polycations salmon protamine with molecular weight of 4000 and poly-L-lysine with molecular weight of 100,000 were found to block ion permeation through the channels formed by synthetic alamethicin, natural alamethicin and phosphorylated alamethicin. In both these instances, the channel blockers are large molecules. The direct response of small molecular analytes are often more difficult to detect. For example, in the immunodiagnostics area, small molecule analysis often uses conventional competitive assays. If there are 1000 units of binding sites, when there is no analyte, the answer is 1000 bound. When there is one analyte, the answer is 999 bound. This method, being an indirect assay measuring difference, can be more complicated and much less sensitive than a direct assay when there is no amplification mechanism [67]. If the interaction of amiloride and alamethicin is not electrostatic, modulation of alamethicin ion channels by small molecules is of significance because this is a direct observation of the interaction between two small molecules. The aggregation of multiple alamethicin molecules probably enhances the interaction of the alamethicin molecules with amiloride by making several interaction points in very close proximity. This observation presents some future opportunities in sensing of small bioactive molecules. The C-terminus of alamethicin with COOH group can be accessed by a number of chemical reactions to attach receptor molecules, alamethicin analogues can be purified by HPLC, and the aggregations of channels place 5–10 molecules in close proximity to amplify the interaction, and this interaction can be measured directly by the conductance of the channels. All these factors combined will have the potential to create an ideal platform to observe the interactions of small molecules.

One of the other aspects of bilayer membranes containing a functional target ion channel is its application in bioactive molecule discovery. Bioactive molecules are often found to be active toward more than one biological target. For example, the natural product psammaplin A, discovered in 1987 [68–70], along with a number of derivatives [71–73] has interesting biological activities including antibacterial and anti-tumour activity [71–73]. In the earlier stages of biodiscovery, it is important that the molecules of interest are tested against a portfolio of targets instead of a specific target to form a bioactivity profile of the small molecules. By studying the binding patterns to these targets, combined with structural information, a majority of compounds can be eliminated from further investigation. Depending on the size and chemical diversity of the library and the available resources, the lead compounds can be 1–5% of the total. The lengthy screening process involves scientists from many disciplines and the demand for quantities of these compounds can be very high, especially in screening agro chemicals when plant assays are required. Computational chemistry can eliminate the 'unlikely active' compounds prior to screening thus saving time and money. However, the structural data for the target proteins in computational screening are generally taken from crystal structure data. Unless the protein of interest has a crystal structure, the computational screening may not be reliable. If considering the interaction of small molecules with proteins at the molecular level, these interactions are interactions of small molecules with certain amino acid sequences in a defined 3D geometry. The ability of a bioactive molecule to fit into this specific 3D geometry and to interact with the amino acids in this space *via* hydrogen bonds or hydrophobic force defines the bioactivity of the small molecule. The capacity of a small molecule to interact with different amino acid sequences is, therefore, a very important parameter. The construction of these sensing platforms is illustrated in Fig. 10.

While DNA binding can be profiled using biosensor arrays, such as commercial DNA chips [74], there is no amino acid profiling tool. One of the reasons is the importance of the 3D structure required for amino acids to form a binding pocket. As small bioactive molecules of interest are often profiled against many biological targets, profiling the binding pattern of small biologically active molecules against an array of amino acid sequences could provide important information for the small bioactive molecules. This important information may first be built from known drugs or other commercial bioactive molecules of which the bioactive properties are proven such as commercial drugs, herbicides and insecticides.

The possibility of directly gating channels formed by alamethicin or other membrane channel-forming peptides in lipid bilayers by small bioactive molecules provides the possibility of profiling amino acid binding to a small bioactive molecule.

The technical challenge is to generate a suitable platform and provide the computational processing power to make use of the information generated.

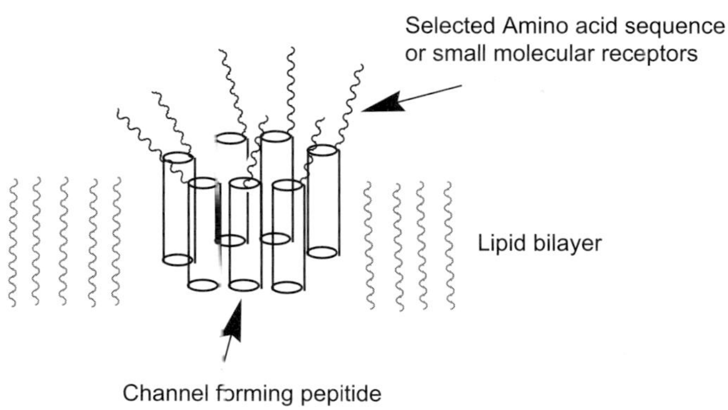

Fig. 10. Chemically modified alamethicin channels as sensing mechanism.

Recognising the importance of membrane proteins including ion channels, pumps and receptors as drug targets, Molecular Devices commercialised their Ion Works HT™ system that is capable of handling 3000 assays per day and measuring ion channel activity directly [75]. It measures ion channel activity directly by measuring channel current of a single cell, considered to be complementary to and even advantageous over the existing high-throughput cell-based drug screening assays, such as florescent dyes, radiolabelled probes, flux assays, binding assays that all measure ion channel activity indirectly. The change of current through ion channels caused by interaction with a drug is measured and IC50 values calculated (Fig. 11 [75] and Fig. 12 [76]).

Study of membrane proteins in tBLMs is also racing ahead, probably for the same reason. A recent advance made use of a mercaptoethanol spacer in a peptide-tBLM to study proton transportation of H^+-ATP synthase from chloroplasts that was shown to lose its activity in a pure thiolipopeptide [77]. The mixing ratio of membrane-forming compounds is important to form membrane architectures that can accommodate the bulk of the protein without altering the 3D conformation that is critical for maintaining protein functionality. Polymer-supported planar lipid bilayers were used to minimise the interaction of intact protein with the solid support. This 'suspension' of the protein in the planar

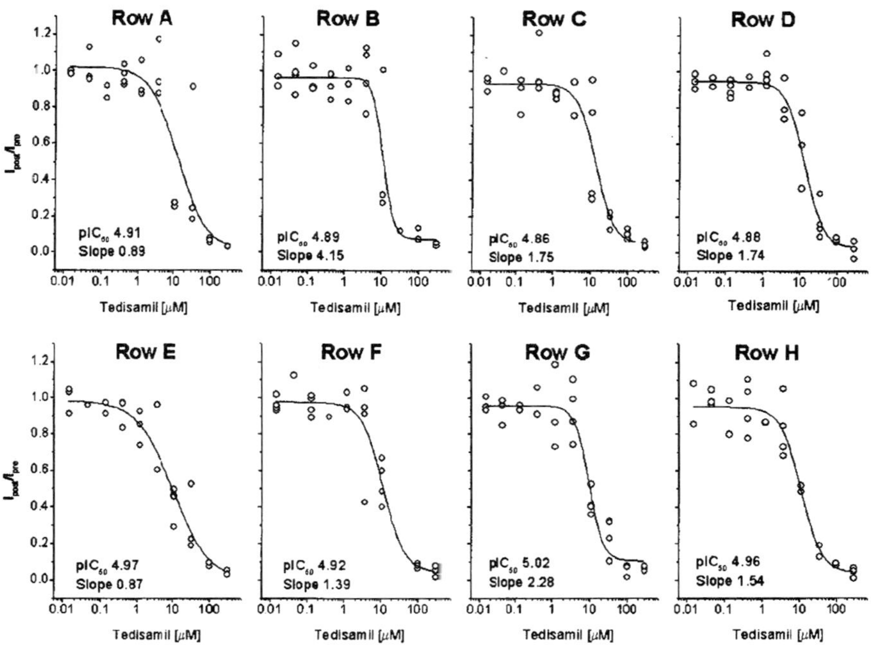

Fig. 11. Measurement of IC50 values of eight different compounds using Ion Works™ HT (with the permission of Saga Publications) [75].

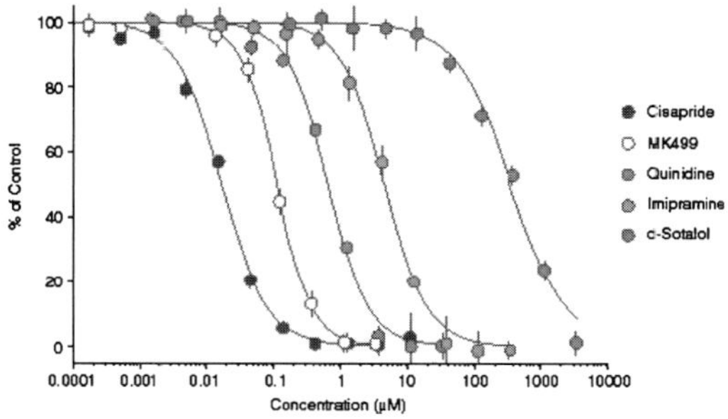

Fig. 12. Concentration dependence of hERG channel block by cisapride, MK-499, quindine, imipramine and d-sotalol (with the permission of Mary Ann Liebert, Inc.) [76].

bilayers, therefore, 'preserved' the functionality of the protein [78]. The attractiveness of the technology will largely depend on whether it will be possible to create a generic platform membrane architecture for several or many ion channel proteins. Each ion channel protein may require a unique membrane architecture to maintain protein functionality. Unlike suspended BLMs, tethered bilayers do not naturally have the flexibility to accommodate large proteins while maintaining its integrity and 3D configuration. This presents a serious limitation to use tethered bilayers as a generic platform to study large membrane proteins [8]. However, they can be used to study the core transmembrane part of these proteins that may be important for a class of proteins such as the 'signature sequence' of K channels [79].

Other applications include attaching the bioactive molecules onto the alamethicin ion channels to find the target proteins in a similar manner to the blocking of biotinylated alamethicin channels by streptavidin.

4.7. Amiloride and analogue concentration detection range

Over the concentration range of 0.01–1000 μM, 64 amiloride concentrations were tested on tBLM containing the same amount of alamethicin channels. The sensitivity of amiloride detection by tBLM containing alamethicin channels is in the μM range (Figs 13 and 14) [6], similar to the Ion Works HT™ system commercialised by Molecular Devices (Figs 11 and 12) [75,76]. A similar concentration range was used for the testing of a number of amiloride analogues illustrating the capability of the sensor to differentiate different blockers.

The relative channel-blocking potency of amiloride analogues is shown in the following sequence (experimental results see Fig. 15):

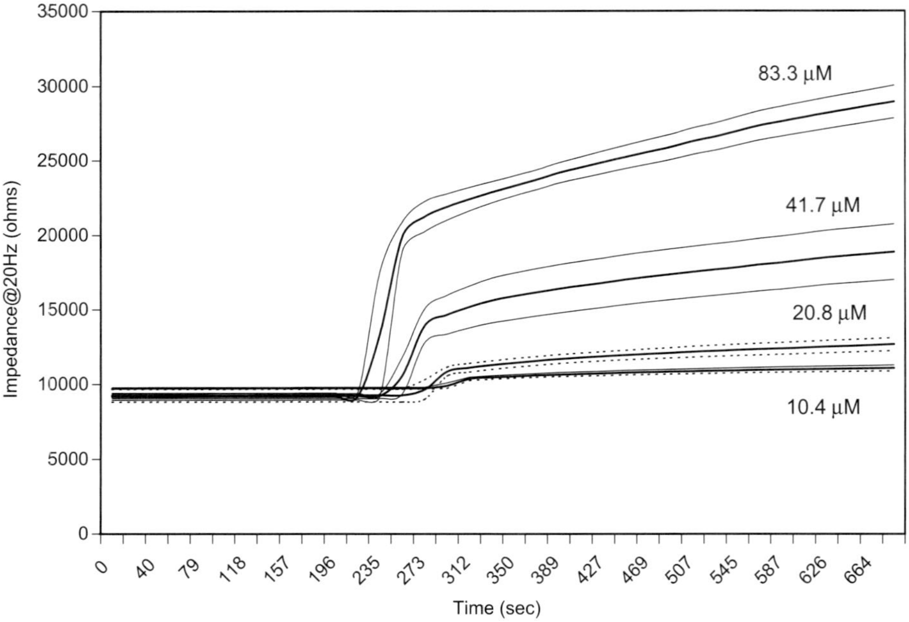

Fig. 13. Impedance change of tethered bilayer containing alamethicin channels when adding amiloride inhibitor (the thicker lines represent an average of four measurements and the thinner lines represent the standard deviations) [6].

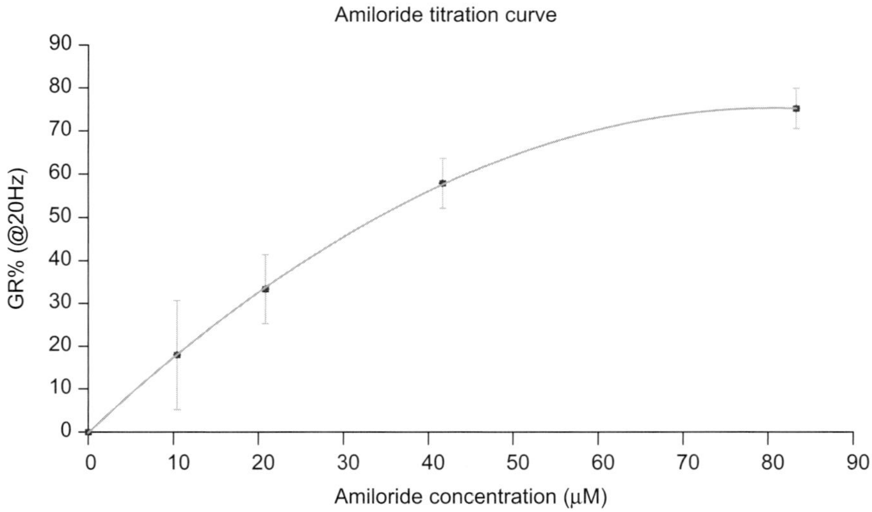

Fig. 14. Amiloride inhibition gating percentage (GR%) titration curve in tethered bilayer membrane [6].

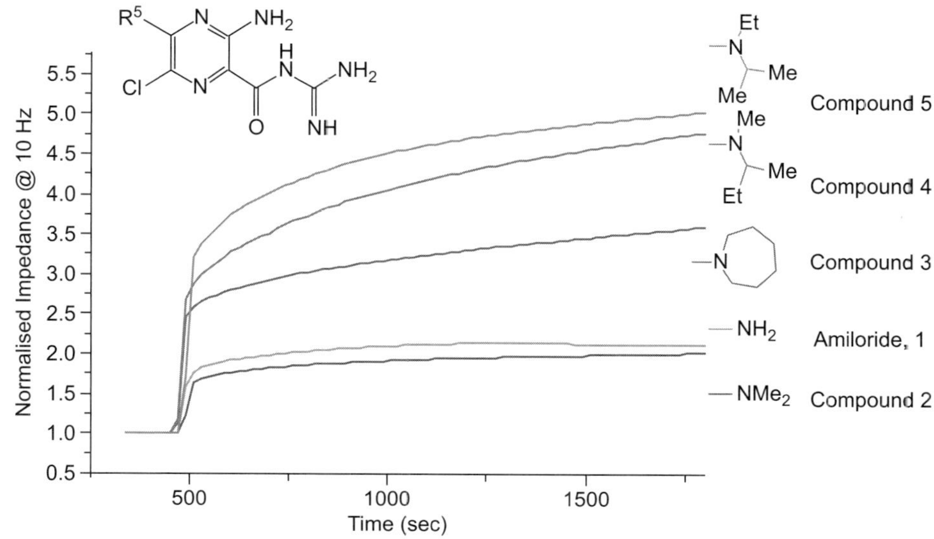

Fig. 15. Impedance change of tethered bilayer membrane containing alamethicin channels when adding different amiloride analogs (R^5=–NH_2, amiloride) at 12.5 μM final concentration [6].

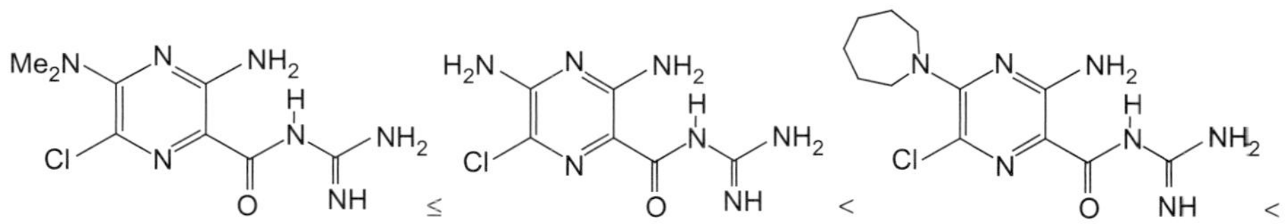

Compound 2 ≤ Amiloride, 1 < Compound 3 <

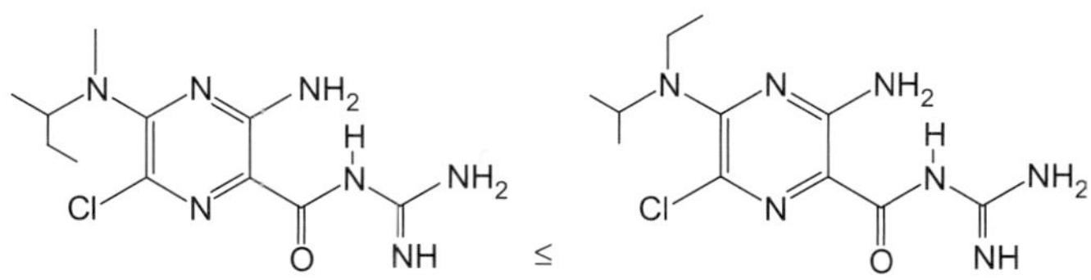

Compound 4 ≤ Compound 5

The substitution of R^5 position of amiloride (molecular structure shown in Fig. 15) plays an important role in improving the alamethicin channel-blocking ability of amiloride analogues. Compound 3 is a significantly better blocker than compounds 1 and 2, and compounds 4 and 5 are significantly better than Compound 3. Although Compound 2 has a larger R^5 group, than amiloride, the NH_2 group may become bulkier when surrounded by water molecules. Further computational studies will reveal the dimensions of the molecules and the nature of their interactions with alamethicin ion channels.

5. SUMMARY OF OPTIMISATION PROCESS FOR TETHERED BILAYER LIPID MEMBRANE SENSOR FOR AMILORIDE ANALOGUES

Figure 16 summarises the optimisation process of the amiloride and analogue sensor containing alamethicin ion channels. The optimisation took many steps starting with a chosen tBLM experimenting with a matrix of different alamethicin concentrations. Within these concentrations, two concentrations were chosen. Using these two alamethicin concentrations, a matrix of experiments of voltage applications were carried out to determine a suitable voltage application range. This was followed by a matrix of experiments at higher alamethicin concentrations and with applied voltage in order to achieve permanent insertion of the channels. These sensors containing alamethicin that cannot be removed by washing could be used to detect amiloride and its analogues. Sixty-four data points between 0.01 and 1000 μM were used to optimise the amiloride detection concentration range. This finally defined the window for the comparison of amiloride analogues channel blocking potency.

6. SUMMARY OF OUTLOOK FOR FUTURE RESEARCH AND APPLICATIONS

To control the reliability of reference material in analytical chemistry, chemical metrology plays an important role. Chemical metrology and advanced statistical process control have been applied in the semiconductor industry [80]. Adopting some aspects of this methodology to biosensor technology will no doubt enhance the process of sensor production.

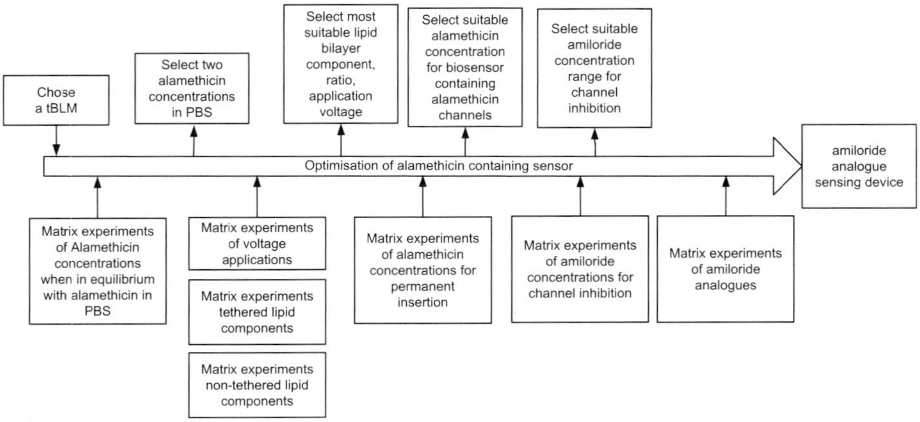

Fig. 16. Optimisation process for amiloride and analogues sensor containing alamethicin ion channels.

Table 2. Membrane sensing platform and applications

Membrane type										
Tethered or suspended	Monolayer or bilayer	Synthetic or native	Ion channel/ ion carrier/ Ion transporter	Analyte	Measurement method	Ion measures	Gating mechanism	Application	Commercial system	Reference
Tethered	Bilayer	Synthetic	Modified gramicidin A	Ferritin TSH digtroxine IgG *E. Coli* DNA IgG HCG troponin	Impedance spectroscopy	Na^+	Immunoresponse shifting the half membrane ion channels alignment	Diagnostic sensor DNA detection array	SensiDx™ system	[16–18]
Tethered	Bilayer	Synthetic	Valinomycin	K^+	Impedance spectroscopy	K^+, Na^+		K^+ sensor		[13–15]
Tethered	Bilayer	Synthetic	Alamethicin	Amiloride and analogues	Impedance spectroscopy	Na^+, Cl^-	Direct blocking of ion channels	Small molecule amiloride detection		[6]
Tethered	Bilayer	Synthetic	H^+-ATP synthase	H^+	Impedance spectroscopy	H^+				[77]
Tethered	Bilayer	Synthetic	Cytochrome *c* oxidase	Cytochrome *c*	Impedance, SWV, CA	H^+				[49]
Suspended	Bilayer	Native	Voltage-dependent sodium channel hNav1.3	Tetradotoxin Lidocaine Phenytoin Carbamzepine Veratridine Amitryptiline Fluozetine Verapramil	Impedance measurement	Na^+	Direct blocking of ion channels	Pharmacologic detection	IonWorks HT™ systems	[75]
Suspended	Bilayer	Native	Voltage-dependent potassium channel hKv1.5	Tedisamil	Impedance measurement	K^+	Direct blocking of ion channels	Pharmacologic detection	IonWorks HT™ systems	[75]
Suspended	Bilayer	Native	Voltage gated potassium ion channel – hERG K^+	Cisapride, MK-499, Quindine, Imipramine, d-Sotalol	Impedance measurement	K^+	Direct blocking of ion channels	Pharmacologic detection	Planar-Array-Based Voltage Clamp	[76]
Tethered	Bilayer	Synthetic	t-SNAREs	GFP-VAMP	TRIFM measurement Lateral diffusion measurements				Polymer suppor-ted	[78]

Some interesting areas for tBLM with small peptide transmembrane ion channels are small molecule interactions and studying the small signature sequences of large proteins.

Cornell [17] also described future opportunities in patterning and miniaturisations of tethered membrane technologies.

7. SUMMARY OF EXISTING MEMBRANE SENSING PLATFORM AND APPLICATIONS

Some of the membrane-sensing platforms and applications discussed in this chapter are summarised (Table 2).

ACKNOWLEDGEMENTS

I would like to thank Ms Gillian Woodhouse, whom I forgot to acknowledge in the previous paper, for assisting me with titration experiments, and Professor Sir J. W. Cornforth for assuring my ability as a scientist. The never fading faith in tBLM biosensor technology from Dr Bruce Cornell has always been an inspiration. I would like also to thank A/Professor Karuso for inviting me to the Chemistry Department at Macquarie University and Dr Scott Martin for proofreading.

REFERENCES

[1] A.L. Ottova, H.T. Tien, Supported planar BLMs (lipid bilayers), formation, methods of study, and applications, Interf. Catal. 2003; 421–459.
[2] A. Ottova, V. Tvarozek, H.T. Tien, Supported Planar Lipid Bilayers (s-BLMs, sb-BLMs, etc.), Membrane Science and Technology Series 7, 2003 (Planar Lipid Bilayers (BLMs) and their applications), Elsevier, Amsterdam, New York, pp. 917–961.
[3] A. Ottova, H.T. Tien, The Lipid Bilayer Principle and Molecular Electronics, NATO Science Series , II: Mathematics, Physics and Chemistry, Vol. 96, 2003 (Molecular Electronics: Bio-Sensors and Bio-Computers), IOS Press, Amsterdam, pp. 197–225.
[4] H.T. Tien, A. Ottova, The Lipid Bilayer Concept: Experimental Realization and Current Applications, Membrane Science and Technology Series 7, 2003 (Planar Lipid Bilayers (BLMs) and Their Applications), Elsevier, Amsterdam, New York, pp. 1–73.
[5] N. Matsuno, M. Murawsky, J. Ridgeway, J. Cuppoletti, Solid support membranes for ion channel arrays and sensors: application to rapid screening of pharmacological compounds, Biochim. Biophys. Acta 1665 (1–2) (2004) 184–190.
[6] P. Yin, C.J. Burns, P.D.J. Osman, B.A. Cornell, A tethered bilayer sensor containing alamethicin channels and its detection of amiloride based inhibitors, Biosens. Bioelectron. 18 (4) (2003) 389–397.
[7] http://www.neuro.wustl.edu/neuromuscular/mother/chan.html
[8] E.-K. Sinner, W. Knoll, Functional and tethered membranes, Curr. Opin. Chem. Biol. 5 (6) (2001) 705–711.
[9] H.W. Huang, Y. Wu, Lipid-alamethicin interactions influence alamethicin orientation, Biophys. J. 60 (1991) 1079–1087.
[10] S-Y. Lee, R. MacKinnon, A membrane-access mechanism of ion channel inhibition by voltage sensor toxins from spider venom, Nature 430 (6996) (2004) 232–235.

[11] J.F. Jellett, L.J. Marks, J.E. Stewart, M.L. Dorey, W. Watson-Wright, J.F. Lawrence, Paralytic shellfish poison (saxitoxin family) bioassays: automated endpoint determination and standardization of the in vitro tissue culture bioassay, and comparison with the standard mouse bioassay, Toxicon 30 (10) (1992) 1143–1156.
[12] A.L. Plant, Supported hybrid bilayer membranes as rugged cell membrane mimics, Langmuir 15 (15) (1999) 5128–5135.
[13] B. Raguse, V.L. Braach-Maksvytis, B.A. Cornell, L.G. King, P.D.J. Osman, R.J. Pace, L. Wieczorek, Tethered lipid bilayer membranes: formation and ionic reservoir characterization, Langmuir 14 (1998) 648–659.
[14] C. Steinem, A. Janshoff, K. von dem Bruch, K. Reihs, J. Goossens, H.-J. Galla, Valinomycin-mediated transport of alkali cations through solid supported membranes, Bioelectrochem. Bioenerg. 45 (1) (1998) 17–26.
[15] A.T.A. Jenkins, R.J. Bushby, N. Boden, S.D. Evans, P.F. Knowles, Q.Y. Liu, R.E. Miles, S.D. Ogier, Ion-selective lipid bilayers tethered to microcontact printed self-assembled monolayers containing cholesterol derivatives, Langmuir 14 (17) (1998) 4675–4678.
[16] B.A. Cornell, V.L.B. Braach-Maksvytis, L.G. King, P.D.J. Osman, B. Raguse, L. Wieczorek, R.J. Pace, A biosensor that uses ion-channel switches, Nature 387 (6633) (1997) 580–583.
[17] B.A. Cornell, Membrane-based biosensor, Optical Biosensors, Present and Future, Vol. 15, 2002 Chapter 15, p. 457.
[18] Ambri Ltd.,126 Greville Street, Chatswood NSW 2067, Australia, website: http://www.ambri.com/
[19] D.S. Everhart, K. Clark, Self-assembling monolayers. Alkane thiols on gold, Handbook of Applied Surface and Colloid Chemistry 2 (2002) 99–116.
[20] J.I. Siepmann, I.R. Mcdonald, Simulations of self-assembled monolayers of thiols on gold, Thin Films (San Diego) 24 (1998) 205–226.
[21] F. Schreiber, Structure and growth of self-assembling monolayers, Prog. Surf. Sci. 65 (5–6) (2000) 151–256.
[22] B. Raguse, R. J. Pace, L.G. King, B.A. Cornell V.L. Braach-Maksvytis, Self assembly of sensor membranes. International Patent Publication, 1997, WO 97/01092.
[23] W.P. Redwood, F.R. Pfeiffer, J.A. Weisbach, T.E. Thompson, Physical properties of bilayer membranes formed from a synthetic saturated phospholipid in *n*-decane Biochim. Biophys. Acta 233 (1) (1971) 1–6.
[24] J. Gutknecht, Proton conductance caused by long-chain fatty acids in phospholipid bilayer membranes, J. Membr. Biol. 106 (1) (1988) 83–93.
[25] M. Kates, B. Palameta, L.S. Yengoyan, Aliphatic diether analogs of glyceride-derived lipids. II. Synthesis of naturally occurring L-2,3,-di-O-3′,7′,11′,15′-tetramethylhexadecyl glycerol and its D isomer, Biochemistry 4 (8) (1965) 1595–1599.
[26] L.C. Stewart, M. Kates, Synthesis and characterization of deuterium-labelled dihexadecylglycerol and diphytanylglycerol phospholipids, Chem. Phys. Lipids 50 (1) (1989) 23–42.
[27] B. Pignataro, C. Steinem H.-J. Galla, H. Fuchs, A. Janshoff, Specific adhesion of vesicles monitored by scanning force microscopy and quartz crystal microbalance, Biophys. J. 78 (1) (2000) 487–498.
[28] C. Steinem, A. Janshoff, H.-J. Galla, Evidence for multilayer formation of melittin on solid-supported phospholipid membranes by shear-wave resonator measurements, Chem. Phys. Lipids 95 (1) (1998) 95–104.
[29] C. Steinem, A. Janshoff, K. von dem Bruch, K. Reihs, J. Goossens, H.-J. Galla, Valinomycin-mediated transport of alkali cations through solid supported membranes, Bioelectrochem. Bioenerg. 45 (1) (1998) 17–26.
[30] L. Cubonova, P. Smigan, Unique lipids and structures of membranes in archaebacteria, Chemicke Listy 98 (2) (2004) 75–79.

[31] G.D. Sprott, Structures of archaebacterial membrane lipids, J. Bioenerg. Biomembr. 24 (6) (1992) 555–566.
[32] T. Benvegnu, M. Brard, D. Plusquellec, Archaeabacteria bipolar lipid analogues: structure, synthesis and lyotropic properties, Curr. Opin. Colloid Interf. Sci. 8 (6) (2004) 469–479.
[33] R.O. Fox Jr., F.M. Richards, A voltage-gated ion channel model inferred from the crystal structure of alamethicin at 1.5-.ANG. resolution, Nature 300 (5890) (1982) 325–330.
[34] C. North, M. Barranger-Mathys, D.S. Cafiso, Membrane orientation of the N-terminal segment of alamethicin determined by solid-state ^{15}N NMR, Biophys. J. 69 (6) (1995) 2392–2397.
[35] A.L. Plant, Self-assembled phospholipid/alkanethiol biomimetic bilayers on gold, Langmuir 9 (11) (1993) 2764–2767.
[36] A. Kessel, D.P. Tieleman, N. Ben-Tal, S.W. George, Implicit solvent model estimates of the stability of model structures of the alamethicin channel, Eur. Biophys. J. 33 (1) (2004) 16–28.
[37] H. Duclohier, G. Alder, K. Kociolek, M.T. Leplawy, Channel properties of template assembled alamethicin tetramers, J. Pept. Sci. 9 (11–12) (2003) 776–783.
[38] B. Bechinger, Structure and function of channel-forming peptides: magainins, cecropins, melittin and alamethicin, J. Membr. Biol. 156 (1997) 197–211.
[39] P.C. Biggin, M.S.P. Sansom, Interaction of alpha-helices with lipid bilayers: a review of simulation studies, Biophys. Chem. 76 (3) (1999) 161–183.
[40] D.P. Tieleman, B. Hess, M.S.P. Sansom, Analysis and evaluation of channel models: simulations of alamethicin, Biophys. J. 83 (5) (2002) 2393–2407.
[41] V. Braach-Maksvytis, B.J. Raguse, Highly impermeable “soft” self-assembled monolayers, Am. Chem. Soc. 122 (2000) 9544.
[42] C.A. Naumann, W. Knoll, C.W. Frank, Hudered diffusion in polymer-tethered membranes: a monolayer study at the air-water interface, Biomacromolecules 2 (4) (2001) 1097–1103.
[43] C.A. Naumann, O. Prucker, T.J. Lehmann Ruhe, W. Knoll, C.W. Frank, The polymer-supported phospholipid bilayer: tethering as a new approach to substrate-membrane stabilization, Biomacromolecules 3 (1) (2002) 27–35.
[44] C.L. Erdelen, R. Haeussling,R. Naumann, H. Ringsdorf, H. wolf, J. Yang, M. Liley, J. Spinke, W. Knoll, Self-assembled disulfide-functionalized amphiphilic copolymers on gold, Langmuir 10 (4) (1994) 1246–1250.
[45] E. Ostuni, L. Yan, G.M. Whitesides, The interaction of proteins and cells with self-assembled monolayers of alkanethiolates on gold and silver, Colloids Surf. B: Biointerf. 15 (1) (1999) 3–30.
[46] D.S. Everhart, K. Clark, Self-assembling monolayers. Alkane thiols on gold, In K. Holmberg (Ed.), Handbook of Applied Surface and Colloid Chemistry, Vol. 2, John Wiley & Sons, Chichester, 2002, pp. 99–116.
[47] J.P. Folkers, P.E. Laibinis, G.M. Whitesides, J. Deutch, Phase behavior of two-component self-assembled monolayers of alkanethiolates on gold, J. Phys. Chem. 98 (2) (1994) 563–571.
[48] M.Y. Tsai, J.C. Lin, Surface characterization and platelet adhesion studies of self-assembled monolayer with phosphonate ester and phosphonic acid functionalities, J. Biomed. Mater. Res. 55 (4) (2001) 554–565.
[49] R. Naumann, E.K. Schmidt, A. Jonczyk, K. Fendler, B. Kadenbach, T. Liebermann, A. Offenhausser, W. Knoll, The peptide-tethered lipid membrane as a biomimetic system to incorporate cytochrome *c* oxidase in a functionally active form, Biosens. Bioelectron. 14 (7) (1999) 651–662.
[50] Ping Yin, Unpublished experimetal work conducted at AMBRI.

[51] I. Vodyanoy, J. Hall, T.M. Balasubramanian, G.R. Marshall, Two purified fractions of alamethicin have different conductance properties, Biochim. Biophys. Acta 684 (1982) 53–58.
[52] H. Bruckner, M. Przybylski, Isolation and structural characterisation of polypeptide antibiotics of the peptaibol class by high-performance liquid chromatography with field desorption and fast atom bombardment mass spectrometry, J. Chromatogr. 296 (1984) 263–275.
[53] D.C. Jaikaran, P.C. Biggin, H. Wenschuh, M.S. Sansom, G.A. Woolley, Structure–function relationships in helix-bundle channels probed via total chemical synthesis of alamethicin dimers: effects of a Gln7 to Asn7 mutation, Biochemistry 36 (45) (1997) 13873–13881.
[54] S. You, S. Peng, L. Lien, J. Breed, M.S.P. Sansom, G.A. Woolley, Engineering stabilized ion channels: colvalent dimers of alamethicin, Biochemistry 35 (20) (1996) 6225–6232.
[55] C. Kaduk, H. Duclohier, M. Dathe, H. Wenschuh, M. Beyermann, G. Molle, M. Bienert, Influence of proline position upon the ion channel activity of alamethicin, Biophys. J. 72 (5) (1997) 2151–2159.
[56] Australian Membrane Biotechnology Research Institute, the predecessor of AMBRI Limited, www.ambri.com
[57] P. Biggin, J. Breed, H.S. Son, M.S.P. Sansom, Simulation studies of alamethicin-bilayer interactions, J. Biophys. 72 (2, Pt. 1) (1997) 627–636.
[58] G.A. Woolley, B.A. Wallace, Model ion channels: gramicidin and alamethicin, J. Membr. Biol. 129 (1992) 109–136.
[59] Y. Jiang, A. Lee, J. Chen, M. Cadene, B.T. Chait, R. MacKinnon, Crystal structure and mechanism of a calcium-gated potassium channel, Nature, 417 (6888) (2002) 515–522 Comment in: Nature, 417 (6888), (2002), 501–502.
[60] C. Whitehouse, D. Gidalevitz, M. Cahuzac, R.E. Koeppe II, A. Nelson, Interaction of gramicidin derivatives with phospholipid monolayers, Langmuir 20 (21) (2004) 9291–9298.
[61] G.V. Miloshevsky, P.C. Jordan, Permeation in ion channels: the interplay of structure and theory, Trends Neurosci. 27 (6) (2004) 308–314.
[62] W.L. Duax, B.M. Burkhart, W.A. Pangborn, V. Pletnev, Z. Wawrzak, M. Glowka, Ion association, transport, and gating in gramicidin channels in lipid bilayers, Adv. Supramol. Chem. 8 (2002) 245–282.
[63] R.E. Koeppe II, J.M. Berg, K.O. Hodgson, L. Stryer, Gramicidin A crystals contain two cation binding sites per channel, Nature (London, UK) 279 (5715) (1979) 723–725.
[64] T.R. Kleyman, E.J. Cragoe Jr., Amiloride and its analogs as tools in the study of ion transport, J. Membr. Biol. 105 (1) (1988) 1–21.
[65] S. Futaki, Z. Youjun, Y. Sugiura, Detecting a tag on a channel opening: blockage of the biotinylated channels by streptavidin, Tetrahedron Lett. 42 (8) (2001) 1563–1565.
[66] T. Rink, H. Bartel, G. Jung, W. Bannwarth, G. Boheim, Effects of polycations on ion channels formed by neutral and negatively charged alamethicins, Eur. Biophys. J. 23 (3) (1994) 155–165.
[67] A. Victoria, Measurements in miniature, Chem. World June (2004) 42–45.
[68] L. Ababshahi, F.J. Schmitz, Brominated tyrosine metablites from an unidentified sponge, J. Org. Chem. 52 (1987) 3584.
[69] E. Quinoa, P. Crews, Phenolic constituents of *Psammaplysilla*, Tetrahedron Lett. 28 (28) (1987) 3229–3232.
[70] A.D. Rodriguez, R.K. Akee, P.J. Scheuer, Two bromotyrosine–cysteine derived metabolites from a sponge, Tetrahedron Lett. 28 (42) (1987) 4989–4992.
[71] Y. Park, Y. Liu, J. Hong, C.-O. Lee, H. Cho, D.-K. Kim, K.S. Im, J.H. Jung, New bromotyrosine derivatives from an association of two sponges, *Jaspis wondoensis* and *Poecillastra wondoensis*, J. Nat. Prod. 66 (11) (2003) 1495–1498; N.B. Pham, M.S. Butler, R.J. Quinn, Isolation of psammaplin A 11′-sulfate and bisaprasin 11′-sulfate from the marine sponge *Aplysinella rhax*, J. Nat. Prod. 63 (3) (2000) 393–395;

J. Shin, H.-S. Lee, Y. Seo, J.-R. Rho, K.W. Cho, V.J. Paul, New bromotyrosine metabolites from the sponge *Aplysinella rhax*, Tetrahedron 56 (46) (2000) 9071–9077; G.M. Nicholas, L.L. Eckman, S. Ray, R.O. Hughes, J.A. Pfefferkorn, S. Barluenga, K.C. Nicolaou, C.A. Bewley, Bromotyrosine-derived natural and synthetic products as inhibitors of mycothiol-S-conjugate amidase, Bioorg. Med. Chem. Lett. 12 (17) (2002) 2487–2490; J.N. Tabudravu, V.G.H. Eijsink, G.W. Gooday, M. Jaspars, D. Komander, M. Legg, B. Synstad, D.M.F. van Aalten, Psammaplin A, a chitinase inhibitor isolated from the Fijian marine sponge *Aplysinella rhax*, Bioorg. Med. Chem. 10 (4) (2002) 1123–1128; D. Kim, I.S. Lee, J.H. Jung, C.O. Lee, S.-U. Choi, Psammaplin A, a natural phenolic compound, has inhibitory effect on human topoisomerase II and is cytotoxic to cancer cells, Anticancer Res. 19 (5B) (1999) 4085–4090; D. Kim, I.S. Lee, J.H. Jung, S.-I. Yang, Psammaplin A, a natural bromotyrosine derivative from a sponge, possesses the antibacterial activity against methicillin-resistant *Staphylococcus aureus* and the DNA gyrase-inhibitory activity, Arch. Pharmacol. Res. 22 (1) (1999) 25–29; A. Suzuki, K. Matsunaga, H. Shin, J. Tabudrav, Y. Shizuri, Y. Ohizumi, Bisprasin, a novel Ca(2+) releaser with caffeine-like properties from a marine sponge, *Dysidea* spp. acts on Ca(2+)-induced Ca(2+) release channels of skeletal muscle sarcoplasmic reticulum, J. Pharmacol. Exp. Ther. 292 (2) (2000) 725–730; C. Jimenez, P. Crews, Novel marine sponge derived amino acids, 13. Additional psammaplin derivatives from *Psammaplysilla purpurea*, Tetrahedron 47 (12–13) (1991) 2097–2102; S.W. Remiszewski, The discovery of NVP-LAQ824: from concept to clinic, Curr. Med. Chem. 10 (22) (2003) 2393–2402; J.H. Jung, C.J. Sim, C.-O. Lee, Cytotoxic compounds from a two-sponge association, J. Nat. Prod. 58 (11) (1995) 1722–1726.

[72] I.C. Pina, J.T. Gautschi, G.-Y.-S. Wang, M.L. Sanders, F.J. Schmitz, D. France, S. Cornell-Kennon, L.C. Sambucetti, S.W. Remiszewski, L.B. Perez, K.W. Bair, P. Crews, Psammaplins from the sponge *Pseudoceratina purpurea*: inhibition of both histone deacetylase and DNA methyltransferase, J. Org. Chem. 68 (10) (2003) 3866–3873.

[73] G.M. Nicholas, L.L. Eckman, S. Ray, R.O. Hughes, J.A. Pfefferkorn, S. Barluenga, K.C. Nicolaou, C.A. Bewley, Bromotyrosine-derived natural and synthetic products as inhibitors of mycothiol-S-conjugate amidase, Bioorg. Med. Chem. Lett. 12 (17) (2002) 2487–2490.

[74] http://www.affymetrix.com

[75] K. Schroeder, B. Neagle, D.J. Trezise, J. Worley, IonWorks HT: a new high-throughput electrophysiology measurement platform, J. Biomol. Screen. 8 (2003) 50–64.

[76] L. Kiss, P.B. Bennett, V.N. Uebele, K.S. Koblan, S.A. Kane, B. Neagle, K. Schroeder, High throughput ion-channel pharmacology: planar-array-based voltage clamp, Assay Drug Dev. Technol. 1 (2003) 127–135.

[77] R. Naumann, T. Baumgart, P. Graber, A. Jonczyk, A. Offenhausser, W. Knoll, Proton transport through a peptide-tethered bilayer lipid membrane by the H^+-ATP synthase from chloroplasts measured by impedance spectroscopy, Biosens. Bioelectron. 17 (1–2) (2002) 25–34.

[78] M.L. Wagner, L.K. Tamm, Tethered polymer-supported planar lipid bilayers for reconstitution of integral membrane proteins: silane–polyethyleneglycol–lipid as a cushion and covalent linker, Biophys. J. 79 (3) (2000) 1400–1414.

[79] http://www.hhmi.org/research/investigators/mackinnon.html

[80] A.P. Shanmugasundram, M. Sarfaty, A. Schwarm, J. Paik, Integrated metrology and advanced process control in semiconductor manufacturing, Appl. Mater. Proc. Electrochem. Soc. 2002, pp. 2002–2002.

[81] B.M. Burkhart, N. Li, D.A. Langs, W.A. Pangborn, W.L. Duax, The conducting form of gramicidin A is a right-handed double-stranded double helix, Proc. Natl Acad. Sci. USA 95 (22) (1998) 12950–12955.

[82] D. Bali, L. King, S. Kim, Syntheses of new gramicidin a derivatives, Aust. J. Chem. 56 (4) (2003) 293–300.

CHAPTER 3

NMR Structure Determination of Proteins in Bilayer Lipid Membranes: The FXYD Family Proteins

Carla M. Franzin and Francesca M. Marassi*

The Burnham Institute, 10901 North Torrey Pines Road, La Jolla, CA 92037, USA

Contents

Abstract

Solid-state NMR spectroscopy enables the structures of membrane peptides and proteins to be determined in lipid bilayers. Regardless of fold, all membrane proteins adopt three-dimensional structures with a unique direction in space defined by the membrane environment. Because this directionality is an intrinsic characteristic of membrane protein structure and function, it is highly desirable to carry out structure determination within the context of bilayer lipid membranes. Solid-state NMR spectroscopy of uniaxially oriented planar bilayer lipid samples is ideally suited for this purpose. This article outlines the methodology for membrane protein structure determination using solid-state NMR of oriented planar lipid bilayer samples. Recent developments in sample preparation, recombinant bacterial expression systems for the preparation of isotopically labeled membrane proteins, pulse sequences for high-resolution spectroscopy, and structural indices that guide the structure assembly process, have greatly extended the capabilities of the technique, and are described. The methods are illustrated with examples from the FXYD proteins, a family of auxiliary regulatory subunits of the Na^+,K^+-ATPase.

---

* Corresponding author. Tel.: +1 858 713 6282; Fax: +1 858 713 6268;
E-mail: fmarassi@burnham.org

ADVANCES IN PLANAR LIPID BILAYERS AND LIPOSOMES, VOLUME 2
ISSN 1554-4516 DOI: 10.1016/S1554-4516(05)02003-X

1. INTRODUCTION

Membrane proteins regulate some of the most basic cellular functions, including transport, signaling, and adhesion, and therefore it is not surprising that they constitute approximately 30% of all expressed genes. Because of their biomedical importance, they are also major targets for drug discovery and structure determination initiatives, however, despite their importance, only 140 structures of membrane proteins have been deposited in the PDB (Protein data bank), compared to the approximately 23,000 coordinates deposited for globular proteins to date (http://blanco.biomol.uci.edu/Membrane_Proteins_xtal.html, and http://www.rcsb.org/pdb/). The hydrophobic character of membrane proteins and their requirement for lipids, make them difficult to crystallize for X-ray crystallographic analysis, and the lipids required for solubilization lead to slow overall reorientation rates and broad lines for solution NMR spectroscopy.

Overall, four main types of folds are observed for the membrane-associated domains of proteins: transmembrane α-helices, membrane interfacial α-helices, transmembrane β-barrels, and membrane interfacial β-sheets [1]. Regardless of fold, all membrane proteins adopt three-dimensional structures with a unique orientation in the lipid bilayer membrane, and since this directionality is an intrinsic element of membrane protein structure and function, it is highly desirable to determine structures within the context of lipid bilayer membranes.

Solid-state NMR spectroscopy of planar oriented lipid bilayer samples is ideally suited for this task because it enables membrane protein structures to be determined in a native-like environment [2]. Since the lipid bilayers are oriented with respect to the applied magnetic field, the samples preserve the intrinsic directional character of membrane proteins. For membrane proteins that can be expressed, isotopically labeled, and purified chromatographically, reconstitution in oriented lipid bilayers gives single-line spectra with line widths that rival those from single crystals [3], and with characteristic patterns that directly reflect protein structure and topology [4–6]. The structures of a variety of membrane peptides and proteins have been investigated using this approach and several atomic-resolution structures have been determined and deposited in the PDB (Fig. 1) [7–13].

In this chapter, we describe the methods for solid-state NMR structural studies of membrane proteins in planar oriented lipid bilayers, with special focus on the FXYD membrane proteins that have been recently identified as auxiliary regulatory subunits of the Na^+,K^+-ATPase. The spectra provide the first view of these protein structures in membranes, and provide the starting point for three-dimensional structure determination in their native environment.

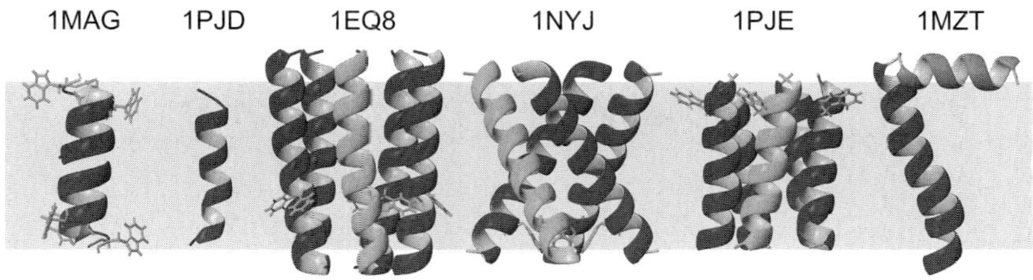

Fig. 1. Structures of membrane proteins determined by solid-state NMR in planar oriented bilayers. The PDB file numbers above each structure correspond to: (1MAG) gramicidin; (1PJD) 6th transmembrane domain of Ste2p α-mating factor receptor; (1EQ8) M2 pore-forming domain of acetylcholine receptor; (1NYJ) pore-forming domain of influenza M2 proton channel; (1PJE) pore-forming domain of HIV-1 Vpu; (1MZT) membrane-bound bacteriophage fd coat protein [7–13].

2. THE FXYD FAMILY PROTEINS

The FXYD family proteins are expressed abundantly in tissues that perform fluid and solute transport (breast/mammary gland, kidney, colon, pancreas, prostate, liver, lung, and placenta), or that are electrically excitable (muscle, nervous system), where they function to regulate the flux of transmembrane ions, osmolytes, and fluids [14]. The protein sequences are highly conserved through evolution, and are characterized by a 35-amino acid FXYD homology (FH) domain, which includes the transmembrane (TM) domain (Fig. 2). The short motif PFXYD (Pro, Phe, X, Tyr, Asp), preceding the transmembrane domain, is invariant in all known mammalian examples, and identical in other vertebrates, except for the proline. Residue X is usually Tyr, but can also be Thr, Glu, or His. In all these proteins, conserved basic residues flank the TM domain, the extracellular N-termini are acidic, and the cytoplasmic C-termini are basic.

FH — PFXYD + TM ++

Protein	Sequence	Length
PLM (FXYD1)	MESPKEHDPFTYDYQSLQIGGLVIAGILFILGILIVLSRRCRCKFNQQQRTGEPDEEEGTFRSSIRRLSTRRR	72
Gamma (FXYD2)	MTELSANHGGSAKGTENPFEYDYETVRKGGLIFAGLAFVVGLLILLSKRFRCGGSKKHRQVNEDEL	65
Mat8 (FXYD3)	MNDLEDKNSPFYYDWHSLQVGGLICAGVLCALGIIIVLSAKCKCKFGQKSGHHPGETPPLITPGSAQS	67
CHIF (FXYD4)	MNGPVDKGSPFYYDWESLQLGGLIFGGLLCIAGIALALSGKCKCRRNHTPSSLPEKVTPLITPGSAST	67
Ric (FXYD5)	—DTPQTLKPSGFHEDDPFFYDEHTLRKRGLLVAAVLFITGIIILTSGKCRQLSRLCRNHCR	157
Php (FXYD6)	AEKEKEMDPFHYDYQTLRIGGLVFAVVLFSVGILLILSRRCKCSFNQKPRAPGDEEAQVENLITANATEPQKAEN	75
FXYD7	NEPDPFYYDYNTVQTVGMTLATILFLLGILIVISKKVKCRKADSRSESPTCKSCKSELPSSAPGGGGV	67

Fig. 2. Amino acid sequences of mammalian FXYD membrane proteins. The FXYD homology (FH) domain encompasses the FXYD consensus sequence, and the transmembrane (TM) domain is flanked by conserved positively charged residues. Conserved Gly residues in the TM domain are highlighted in gray.

PLM (Phospholemman), is one of the best characterized members of the FXYD family, and the major substrate of hormone-stimulated phosphorylation by cAMP-dependent protein kinase A and C in the heart [15]. CHIF (corticosteroid-hormone-induced factor) is upregulated by aldosterone and corticosteroids in mammalian kidney and intestinal tracks, where it regulates Na^+ and K^+ homeostasis [16]. Mat8 (Mammary tumor protein 8 kDa) is expressed in breast, prostate, lung, stomach, and colon, as well as in human breast tumors, breast tumor cell lines, and prostate cancer cell lines, after malignant transformation by oncogenes [17,18], and other FXYD proteins are also induced by oncogenic transformation. All three proteins, PLM, Mat8, and CHIF, induce ionic currents in Xenopus oocytes, and PLM also forms ion channels in phospholipid bilayers [16,17,19]. The identification of several FXYD family members, including PLM and CHIF, as regulators of the Na^+,K^+-ATPase, points to a mechanism for regulation of the pump that involves the expression of an auxiliary subunit [14]. Recently, we described the recombinant expression, purification, and sample preparation in lipid micelles and bilayers, for three members of the FXYD family: PLM, Mat8, and CHIF [20]. The solid-state NMR spectra in lipid bilayers provide the first view of their structures and topologies, and are described below.

3. SAMPLE PREPARATION

3.1. Protein expression

NMR structural studies typically require samples containing milligram quantities of isotopically labeled proteins. Smaller peptides can be prepared by automated solid phase peptide synthesis, however, this is impractical for larger proteins and for the preparation of uniformly labeled samples, where efficient expression systems are essential. The ability to express membrane proteins in bacteria enables a wide variety of isotopic labeling schemes to be incorporated in the NMR experimental strategy. Selective labeling by amino acid type is accomplished by growing the bacteria harboring the protein gene on media where only one type of amino acid is labeled and all others are not. Uniform labeling, where all the nuclei of one or several types (^{15}N, ^{13}C, ^{2}H) are labeled, is accomplished by growing the bacteria on media containing ^{15}N-labeled ammonium sulfate, or ^{13}C-labeled glucose, or D_2O or any combination of these.

Several *Escherichia coli* expression systems and mutant cell strains have been developed for membrane protein expression and purification [21–25]. Many involve the use of fusion proteins, the formation of inclusion bodies, and the incorporation of engineered N-terminal His-tags for metal affinity chromatography, to greatly simplify protein isolation and purification. After inclusion body isolation, fusion protein purification and cleavage, the target membrane protein is finally purified, and then reconstituted into lipid bilayers for NMR studies.

Our approach to the production of PLM, CHIF, and Mat8, utilized the *E. coli* pMMHa fusion protein expression vector, which directs the synthesis of the fusion protein His_9-TrpΔLE-FXYD [20,24]. This vector has been used successfully for the production of other membrane or toxic proteins ranging in size from 80 to 150 amino acids [26]. The TrpΔLE fusion partner, from the Trp leader amino acid sequence, is very effective at forming inclusion bodies and is thus protected from proteolysis. The fusion protein is not toxic to the *E. coli* host cells, and is expressed at levels up to 20% of total cellular protein in *E. coli* strain C41(DE3) [21], grown on M9 minimal media for isotopic labeling. Intact FXYD proteins are liberated from the fusion partner using CNBr (cyanogen bromide), which cleaves specifically after Met residues [27]. The use of chemical cleavage eliminates the difficulties such as poor specificity and enzyme inactivation often encountered with protease treatment of membrane proteins in detergents.

The protein content of cells isolated before and after IPTG induction, is shown in Fig. 3A (lanes 1 and 2), where fusion protein over-expression is marked by the appearance of an intense band near 21 kDa. After protein expression, the inclusion bodies enriched in fusion protein were separated from the *E. coli* lysate by a series of wash and centrifugation steps (Fig. 3A, lane 3), the fusion protein was isolated by Nickel affinity chromatography (Fig. 3A, lane 4), and each FXYD protein was cleaved from the fusion partner using CNBr (Fig. 3A, lane 5). This yields a fragment near 14 kDa, corresponding to His_9-TrpΔLE, and a smaller fragment corresponding to either PLM at 8.4 kDa, CHIF at 7.1 kDa, or Mat8 at 7.1 kDa. Finally, the proteins were purified by size exclusion chromatography, followed by reverse-phase chromatography (Fig. 3A, lanes 6, 7, 8). Typically, 2.5 mg of purified protein are obtained from 1 l of cell culture in ^{15}N-labeled minimal media.

The mass spectra shown in Fig. 3B, demonstrate the high degree of purity obtained with this method. The major peaks have masses that correspond exactly to the FXYD proteins PLM, CHIF, and Mat8, the small peaks at half mass arise from doubly charged species, and those at double mass from a small fraction of FXYD dimer. In all cases, the spectra show no evidence of degradation or chemical modifications. The CD spectra obtained for the proteins in SDS (sodium-*n*-dodecyl sulfate) micelles are shown in Fig. 3C. For all three FXYD family members, the two minima at 208 and 222 nm are characteristic of α-helical structures, and the helical content, estimated from the CD spectra with the k2d program (www.embl-heidelberg.de/~andrade/k2d/) [28], is approximately 40%. These purified recombinant proteins were used for NMR structural studies in lipid micelles and lipid bilayers.

The solution NMR spectra of PLM, CHIF, and Mat8, provide valuable information about their structure and membrane architecture. These spectra are

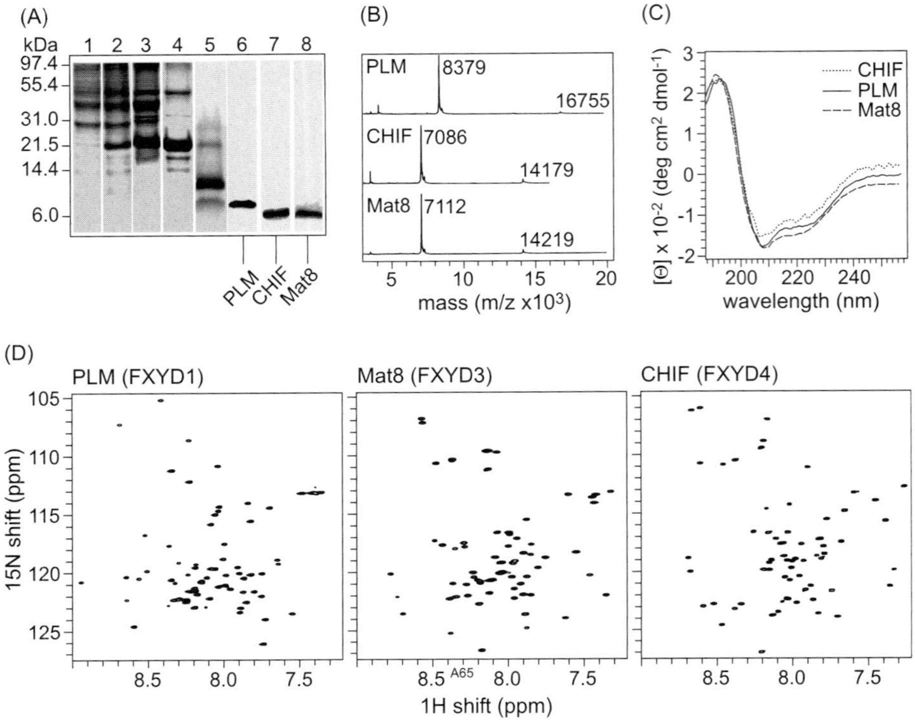

Fig. 3. Expression and characterization of the FXYD family proteins PLM, CHIF, and Mat8. (A) Protein fractions at different stages of the purification protocol were analyzed on Coomassie stained 16% SDS-PAGE. Lane 1: entire cell before IPTG induction. Lane 2: entire cell after IPTG induction with the fusion protein band at 20 kDa. Lane 3: isolated inclusion bodies enriched in fusion protein. Lane 4: fusion protein isolated by Ni affinity chromatography. Lane 5: His_9-TrpΔLE fusion partner and PLM resulting from CNBr cleavage of the fusion protein. Lane 6: purified PLM. Lane 7: purified CHIF. Lane 8 purified Mat8. (B) MALDI-TOF mass spectra of purified PLM (8379 Da), CHIF (7086 Da), and Mat8 (7112 Da). (C) CD spectra of purified PLM, CHIF, and Mat8 in SDS micelles. The samples contained 20 μM FXYD protein, 500 mM SDS, 20 mM sodium phosphate, 1 mM sodium azide, pH 5. (D) Two-dimensional $^1H/^{15}N$ HSQC spectra of uniformly ^{15}N-labeled PLM, CHIF, and Mat8, in SDS micelles at 40 °C. The samples contained 1–2 mM FXYD protein, 500 mM SDS, 20 mM sodium phosphate, 10 mM DTT, 1 mM sodium azide, pH 5. The ^{15}N and ^1H chemical shifts are referenced to 0 ppm for liquid ammonia and tetramethylsilane.

the starting points for higher dimensional solution NMR experiments and structure determination. Solution NMR spectroscopy of the FXYD proteins in lipid micelles can be used to determine their secondary structures and three-dimensional folds, because the proteins reorient fast enough in solution to give isotropic spectra with relatively narrow line widths. The two-dimensional $^1H/^{15}N$ HSQC (heteronuclear single quantum correlation) spectra of uniformly ^{15}N-labeled FXYD proteins in aqueous SDS micelles are shown in Fig. 3D. Each resonance represents a single ^{15}N-labeled site of the protein, and is characterized by ^{1}H and ^{15}N chemical shift frequencies that reflect the local environment. The samples were optimized to obtain narrow resonance line widths and high resolution, and the presence of one well-defined resonance for each amide site in the protein is indicative of a high quality micelle sample. Resonances from the Gly (^{15}N shift = 105–110 ppm), Trp indole (^{1}H shift = 10 ppm), and Gln and Asn side chain (^{1}H shift = 6.5–7.5 ppm) nitrogens are clearly resolved. The limited chemical shift dispersion reflects the helical structures of these proteins, which we have determined by means of additional ^{1}H/^{15}N double- and ^{1}H/^{15}N/^{13}C triple-resonance solution NMR experiments.

3.2. Preparation of oriented lipid bilayer samples

Glass-supported oriented phospholipid bilayers containing membrane proteins accomplish the principal requirements of immobilizing and orienting the protein for solid-state NMR structure determination. The planar lipid bilayers are supported on glass slides, and oriented in the NMR probe so that the bilayer normal is parallel to the field of the magnet, as shown in Fig. 4A. The choice of lipid can be used to control the lateral spacing between neighboring phospholipid molecules as well as

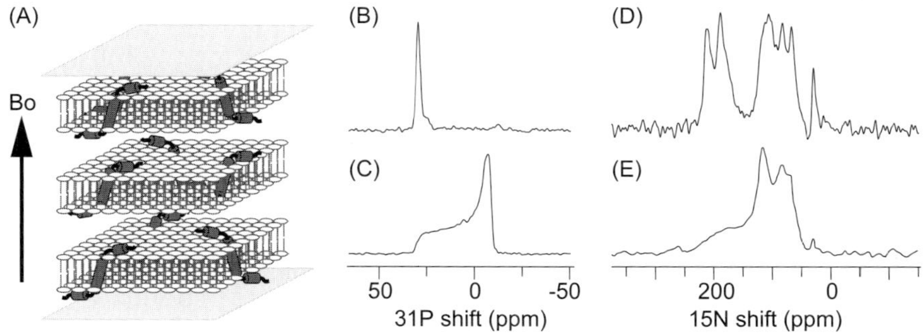

Fig. 4. Effect of sample orientation on the solid-state NMR spectra of isotopically labeled proteins. (A) The glass-supported phospholipid bilayer samples are oriented in the NMR probe so that the bilayer normal is parallel to the direction of the magnetic field (Bo). (B) Oriented phospholipid bilayers give single-line one-dimensional ^{31}P chemical shift NMR spectra, while (C) spherical lipid bilayer vesicles give powder patterns. (D) The one-dimensional ^{15}N chemical shift NMR spectrum of uniformly ^{15}N-labeled FXYD CHIF in oriented lipid bilayers displays multiple resonances, compared to (E) the powder pattern that is obtained for the same protein in unoriented lipid bilayer vesicles. The ^{15}N chemical shifts are referenced to 0 ppm for liquid ammonia.

the vertical spacing between bilayers. The use of phospholipids with unsaturated chains leads to more expanded and fluid bilayers, and the addition of negatively charged lipids increases inter-bilayer repulsions leading to larger interstitial water layers between bilayer leaflets.

There are two methods for preparing samples of membrane proteins in lipid bilayers oriented on glass slides: one is deposition from organic solvents followed by evaporation and lipid hydration [20], and the other is fusion of reconstituted unilamellar lipid vesicles with the glass surface [3]. The method of choice depends on the individual protein, and we have found that the first method generally works well for the FXYD family proteins. The choice of solvents in the first method, and of detergents in the second, is critical for obtaining highly oriented lipid bilayer preparations. In all cases the thinnest available glass slides are utilized to obtain the best filling factor in the coil of the probe. Glass slides purchased from Paul Marienfeld (Germany) are 0.06–0.08 mm thick, and may used directly after washing in detergent and rinsing in distilled water. The typical dimensions for a stacked lipid–glass slide sample are $11 \times 11 \times 5\ mm^3$. With carefully prepared samples it is possible to obtain ^{15}N resonance line widths of less than 3 ppm [3]. Notably, these line widths are less than those typically observed in single crystals of peptides, demonstrating that the proteins in the bilayers are very highly oriented, with mosaic spreads of less than about 2°.

The FXYD samples were prepared by mixing ^{15}N-labeled protein and phospholipids (DOPC, 1,2-dioleoyl-*sn*-glycero-3-phosphocholine, and DOPG, 1,2-dioleoyl-*sn*-glycero-3-[phospho-rac-(1-glycerol)]) in organic solvents, spreading the solution on the surface of the glass slides, and removing the solvents under vacuum. The glass slides were then stacked and equilibrated for 24 h at 40 °C, followed by 24 h at room temperature, in a chamber containing a saturated solution of ammonium phosphate, which provides an atmosphere of 93% relative humidity. The sample was wrapped in parafilm and then sealed in thin polyethylene film prior to insertion in the NMR probe [20].

The degree of phospholipid bilayer alignment can be assessed with solid-state ^{31}P NMR spectroscopy of the lipid phosphate headgroup. The ^{31}P NMR spectra obtained for the CHIF-containing samples are characteristic of a liquid–crystalline bilayer arrangement, in both oriented (Fig. 4B) and unoriented samples (Fig. 4C). The spectrum from the oriented sample has a single peak near 30 ppm, as expected for highly oriented planar bilayers. These spectra demonstrate that highly aligned samples of FXYD proteins in lipid bilayers can be prepared for NMR structure determination.

The one-dimensional ^{15}N chemical shift NMR spectra of ^{15}N-labeled membrane proteins in oriented bilayers display significant resolution with identifiable peaks at frequencies throughout the range of the ^{15}N amide chemical shift. For example, the spectrum in Fig. 4D was obtained from a sample of uniformly ^{15}N-labeled CHIF in oriented bilayers. The resonance intensity near 200 ppm arises from backbone amide sites in the CHIF transmembrane helix that have their NH bonds nearly perpendicular to the plane of the membrane, while the intensity near 80 ppm is from sites in the N- and C-termini of the protein, with NH bonds nearly parallel to the membrane surface. The peak near 35 ppm results from the amino groups of the lysine side chains and the N-terminus. This spectrum is strikingly different from that of an unoriented sample, which provides no resolution among resonances (Fig. 4E). Most of the backbone sites are structured and immobile on the timescale of the ^{15}N chemical shift interaction (10 kHz), contributing to the characteristic amide powder pattern between about 220 and 60 ppm. Some of the CHIF backbone sites, probably near the N- and C-termini, are mobile, and give rise to the resonance band centered near 120 ppm. Therefore, while certain resonances near 120 ppm, in the spectrum of oriented CHIF, may reflect specific orientations of their corresponding sites, some others arise from mobile backbone sites.

4. NMR SPECTRA OF PROTEINS IN PLANAR ORIENTED LIPID BILAYERS

4.1. One-dimensional spectra

The one-dimensional ^{15}N chemical shift spectra of PLM, Mat8, and CHIF (Fig. 5) provide a first glimpse of the FXYD protein architecture in membranes. A preliminary analysis of the solid-state NMR data is possible since both CD and NMR spectroscopy in micelles show that the overall secondary structure of these proteins is α-helical. Membrane proteins in lipid bilayers are largely immobile on NMR timescales, therefore, their resonances are not motionally averaged but have frequencies that reflect the orientation of their respective sites relative to the direction of the magnetic field. In our samples, the lipid bilayer plane is perpendicular to the magnetic field direction, therefore each resonance frequency reflects the orientation of its corresponding protein site in the membrane [2].

In each of the three spectra, the resonance intensity near 200 ppm is from the transmembrane helices of PLM, Mat8, and CHIF, while the intensity near 80 ppm is from sites in amphipathic helices and loops at the N- and C-termini. For all three proteins, the rather narrow dispersion of ^{15}N resonances centered at 200 ppm indicates that the transmembrane helix crosses the lipid bilayer membrane with a very small tilt angle.

Amide hydrogen exchange rates are useful for identifying residues that are involved in hydrogen bonding, and that are exposed to water. Typically, the amide hydrogens in transmembrane helices have very slow exchange rates due to their strong hydrogen bonds in the low dielectric of the lipid bilayer environment, and their ^{15}N chemical shift NMR signals persist for days after exposure to D_2O [29]. Transmembrane helices that are in contact with water because they participate

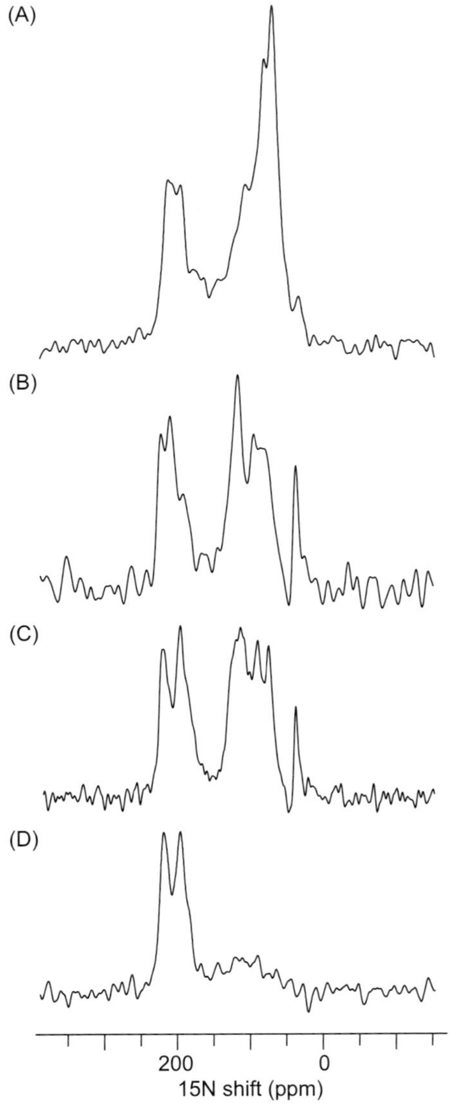

Fig. 5. Solid-state NMR ^{15}N chemical shift spectra of uniformly ^{15}N-labeled FXYD proteins, PLM, Mat8, and CHIF, in planar oriented lipid bilayers. (A) PLM; (B) Mat8; (C, D) CHIF. Resonances near 200 ppm are from amino acid residues in the transmembrane helix of each protein. Amide hydrogens in the transmembrane helix are resistant to hydrogen exchange, and are visible in both (A, B, C) the spectra obtained after hydration with H_2O, and (D) the spectrum of CHIF obtained after exposing the sample to D_2O, where the other resonances disappear. The ^{15}N chemical shifts are referenced to 0 ppm for liquid ammonia.

in channel pore formation, and other water-exposed helical regions of proteins, have faster exchange rates, and their NMR signals disappear on the order of hours [30]. When the CHIF sample was exposed to D_2O, the amide hydrogens in the transmembrane helix did not exchange while those in the rest of the protein did, and their resonances disappeared from the spectrum (Fig. 5D). This indicates that the CHIF transmembrane helix is not water-exposed, but forms a tight hydrogen bonding network that is resistant to hydrogen exchange.

4.2. Two-dimensional spectra and Pisa Wheels

When membrane proteins are incorporated in planar lipid bilayers that are oriented in the field of the NMR magnet, the frequencies measured in their multi-dimensional solid-state NMR spectra contain orientation-dependent information that can be used for structure determination [2]. The PISEMA (polarization inversion with spin exchange at the magic angle) experiment gives high resolution, two-dimensional, ^{1}H–^{15}N dipolar coupling/^{15}N chemical shift correlation spectra of oriented membrane proteins where the individual resonances contain orientation restraints for structure determination [31]. PISEMA spectra of membrane proteins in oriented lipid bilayers also provide sensitive indices of protein secondary structure and topology because they exhibit characteristic wheel-like patterns of resonances, called *Pisa Wheels*, that reflect helical wheel projections [32] of residues in both α-helices and β-sheets [4–6]. When a *Pisa Wheel* is observed, no assignments are needed to determine the tilt of a helix, and a single resonance assignment is sufficient to determine the helix rotation in the membrane. This information is extremely useful for determining the supramolecular architectures of membrane proteins and their assemblies.

The shape and position of the *Pisa Wheel* in the spectrum depends on the protein secondary structure and its orientation relative to the lipid bilayer surface, as well as the amide N–H bond length and the magnitudes and orientations of the principal elements of the amide ^{15}N chemical shift tensor. This direct relationship between spectrum and structure makes it possible to calculate solid-state NMR spectra for specific structural models of proteins, and provides the basis for a method of backbone structure determination from a limited set of uniformly and selectively ^{15}N-labeled samples [13,33].

The *Pisa Wheels* calculated for an α-helix and a β-strand oriented at varying degrees in a lipid bilayer are shown in Fig. 6. When the helix or strand cross the membrane with their long axes exactly parallel to the lipid bilayer normal and to the magnetic field direction (0°), all of the amide sites have an identical orientation relative to the direction of the applied magnetic field, and therefore all of the resonances overlap with the same dipolar coupling and chemical shift frequencies. Tilting the helix or strand away from the membrane normal introduces

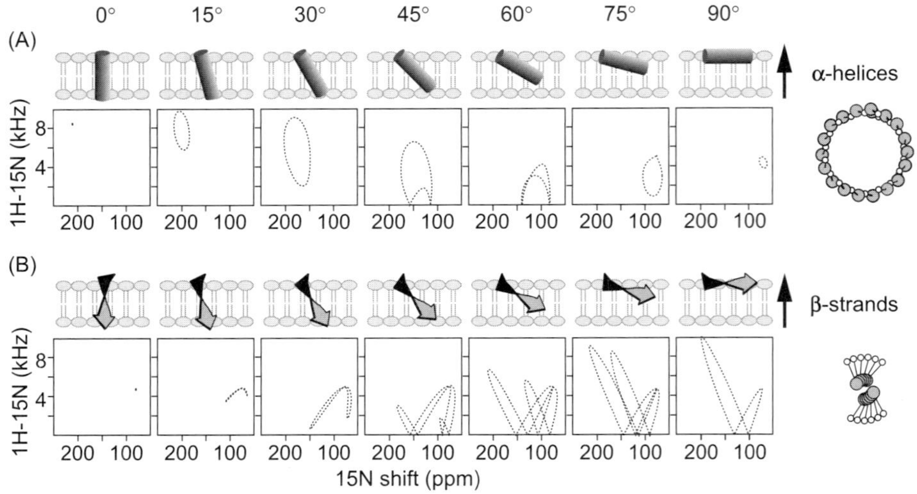

Fig. 6. Helices and strands in oriented planar lipid bilayers give characteristic solid-state NMR spectra called *Pisa Wheels*. The ^{1}H–^{15}N dipolar coupling /^{15}N chemical shift PISEMA spectra were calculated for (A) an ideal α-helix with uniform dihedral angles (phi/psi = −65°/−40°), and (B) an ideal β-strand with uniform dihedral angles (phi/psi = −135°/140°), at different tilts relative to the magnetic field direction and the membrane normal, marked by arrows. The spectra reflect the helical wheel representations of each structure shown on the right. The ^{15}N chemical shifts are referenced to 0 ppm for liquid ammonia.

variations in the orientations of the amide NH bond vectors in the magnetic field, and leads to dispersion of the ^{1}H–^{15}N dipolar coupling and ^{15}N chemical shift frequencies, manifest in the appearance of *Pisa Wheel* resonance patterns in the spectra. Since helices and strands yield clearly different resonance patterns, with circular wheels for helices and twisted wheels for strands, these spectra represent signatures of secondary structure [6]. The spectra also demonstrate that it is possible to determine the tilt of an α-helix or β-strand in lipid bilayers without resonance assignments.

Pisa wheels have been observed in the PISEMA spectra of many uniformly ^{15}N-labeled α-helical membrane proteins [8,10–13,34] including the FXYD protein CHIF (Fig. 8). Since the *Pisa Wheels* also reflect helical wheel representations of the protein, it is possible to obtain the helix or strand rotation in the membrane once a single resonance assignment is obtained. This is illustrated in Fig. 7 for the spectra of the Acetylcholine receptor M2 channel pore (AchR M2), and the integral membrane porin from *Rhodobacter capsulatus*.

The pore-lining transmembrane α-helix of AchR M2 crosses the lipid bilayer with an angle of 12° and is rotated about its long helix axis so that the hydrophilic residues face the channel pore interior [8]. This leads to the assembly

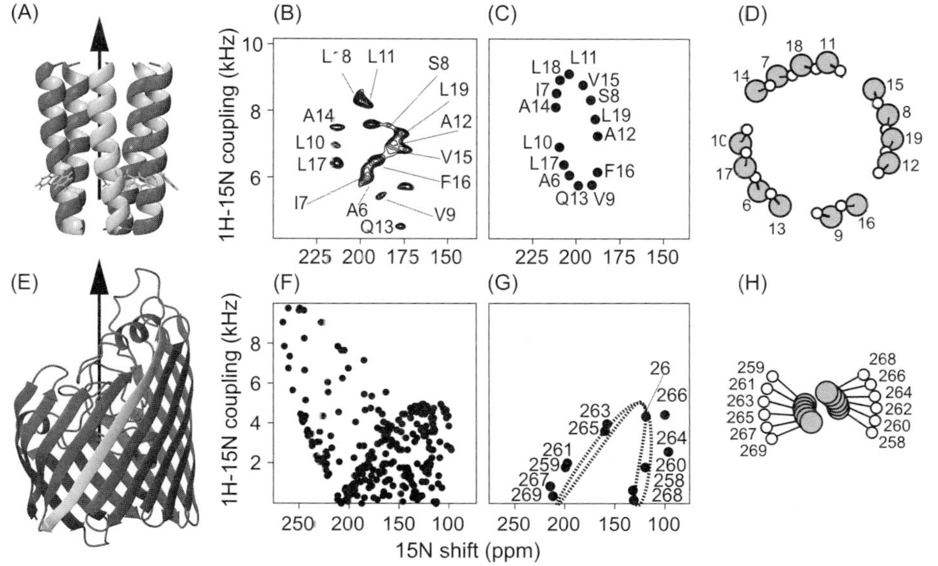

Fig. 7. The PISEMA spectra of oriented membrane proteins reflect their structures and topologies. Spectra were calculated for (A–D) the pentameric pore of AchR M2 (PDB file 1EQ8) [8], and (E–H) integral membrane porin from *R. capsulatus* (PDB file 2FOR) [35]. (A) The individual helices in the pentameric pore of the AchR M2 are tilted by 12° relative to the magnetic field direction and the membrane normal. The resonance patterns in (B) the experimental and (C) the calculated PISEMA spectra reflect (D) the helical wheel projection of the M2 helix. (C) The spectrum was calculated for an ideal α-helix with uniform dihedral angles (phi/psi = −65°/−40°) and a tilt of 12°. (E) Porin forms a β-barrel pore in the membrane. The resonance patterns in (F, G) the spectrum calculated from the PDB coordinates of porin trace out twisted *Pisa Wheels* characteristic of β-strands. (G) The spectrum extracted for β-strand number 14 overlays on the twisted *Pisa Wheel* that was calculated for an ideal β-strand with uniform dihedral angles (phi/psi = −135°/140°) and a tilt of 35° (dotted line). The spectrum reflects (D) the helical wheel projection of β-strand number 14. The arrows mark the direction of the magnetic field and the lipid bilayer normal. The ^{15}N chemical shifts are referenced to 0 ppm for liquid ammonia.

of a symmetric, pentameric, funnel-like pore, shown in Fig. 7A. All of these structural features are immediately apparent prior to complete structure determination, from the assigned PISEMA spectrum of AchR M2 (Fig. 7B) [4]. In both the experimental spectrum (Fig. 7B), and the spectrum calculated for an ideal α-helix, with uniform dihedral angles (phi, psi = −65, −40) and a tilt of 12° (Fig. 7C), the resonances trace a *Pisa Wheel* pattern that is identical to the helical wheel projection of the peptide shown in Fig. 7D. The helical wheel is arranged so that its amide nitrogen atoms coincide with their corresponding resonances

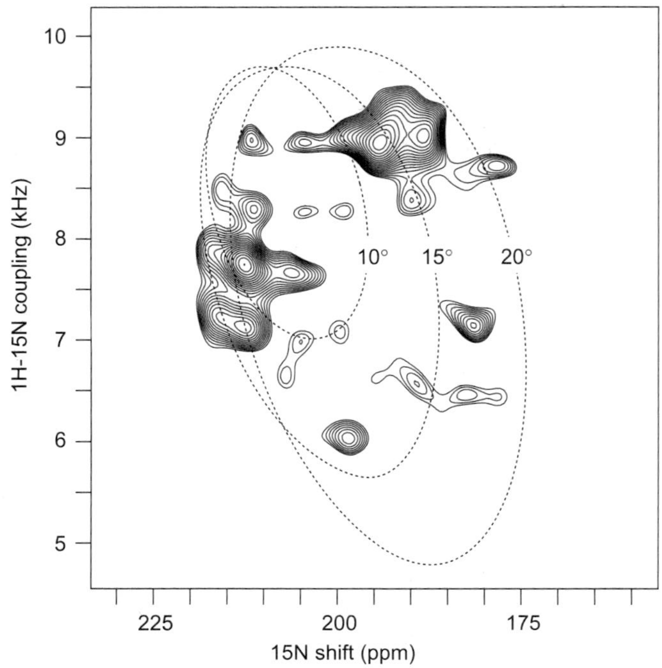

Fig. 8. The CHIF transmembrane helix has a tilt of 15°. The experimental PISEMA spectrum of CHIF is superimposed with the *Pisa Wheel* spectra (dotted lines) calculated for ideal α-helices with 3.6 residues per turn, identical backbone dihedral angles for all residues (phi, psi = −57°, −47°), and tilts of 10°, 15°, and 20°, relative to the lipid bilayer normal. The ^{15}N chemical shifts are referenced to 0 ppm for liquid ammonia.

in the experimental spectrum, and this arrangement is exactly as predicted by the experimentally assigned resonances. Thus, the polarity of the *Pisa Wheel* provides a direct measure of the helix rotation about its long axis within the membrane. In addition, since a single resonance assignment is sufficient to index the *Pisa Wheel*, a single assignment is also sufficient to determine helix rotation in the membrane.

The structure of porin from *R. capsulatus* consists of 16 β-strands arranged to form a β-barrel pore through the membrane (Fig. 7F) [35]. The PISEMA spectrum calculated from the PDB coordinates of the structure (Fig. 7F) displays only modest resolution among its 269 resonances (301 residues excluding the N-terminus and the 38 prolines), and demonstrates the need for higher dimensional spectroscopy [34]. However, the majority of resonances fall on a common twisted *Pisa Wheel*, characteristic of β-strands tilted by 30–60°. This is evident for the PISEMA spectrum of the individual β-strand number 14, which is highlighted in the structure. Strand number 14 has a tilt of about 35°, and this angle is immediately apparent from the spectrum which maps onto the twisted *Pisa Wheel* calculated for an ideal β-strand with uniform dihedral angles (phi, psi = −135, 140) and a tilt of 35° (Fig. 7G). Residues from opposite sides of the strand occupy different wings of the spectrum in a manner similar to that of its helical wheel representation (Fig. 7H). Residues 258, 260, 262, 264, 266, and 268 face the polar interior of the protein, and occupy the upfield wing of the twisted Pisa wheel, while residues 259, 262, 263, 265, 67, and 269 face the hydrophobic lipid core, and occupy the downfield wing. This segregation of resonances from pore-facing or lipid-facing residues is seen in the spectra from all strands.

5. TRANSMEMBRANE HELIX OF FXYD CHIF IN PLANAR LIPID BILAYERS

The two-dimensional PISEMA spectrum CHIF in lipid bilayers is shown in Fig. 8. Each amide site in the protein contributes one correlation peak, characterized by ^{1}H–^{15}N dipolar coupling and ^{15}N chemical shift frequencies that reflect NH bond orientation relative to the membrane. For CHIF, the *Pisa Wheel* that is observed in the region from 6 to 10 kHz and 180 to 220 ppm of the PISEMA spectrum, provides definitive evidence that the protein associates with the lipid bilayer as a transmembrane helix. To estimate the tilt of the CHIF transmembrane helix we compared the experimental spectrum with those calculated for an ideal α-helix, with 3.6 residues per turn and identical backbone dihedral angles for all residues (phi, psi = −57°, −47°), tilted at 10, 15, and 20° relative to the lipid bilayer normal. This comparative analysis demonstrates that the CHIF helix is tilted by about 15° in the membrane (or 75° from the membrane surface).

6. CONCLUSIONS

Solid-state NMR is a powerful approach for the structure determination of proteins in planar lipid bilayers. The method requires the production of milligram quantities of isotopically labeled protein, followed by protein reconstitution in oriented lipid bilayers. Substantial progress has been made in both the areas of recombinant protein expression and oriented lipid sample preparation, so that the spectra of several membrane proteins have now been recorded for structure determination, and the high-resolution structures of six membrane proteins have been determined in bilayers [7–13].

The data presented in this study demonstrate that the solid-state NMR spectra of uniformly ^{15}N-labeled proteins, associated with planar oriented lipid bilayers, yield characteristic patterns of resonances, or *Pisa Wheels*, that provide powerful and visually accessible indices of membrane protein secondary structure and topology. Pisa wheels are observed for many α-helical membrane proteins in lipid bilayers, and we predict that they are also likely to be observed for oriented β-stranded proteins, although this case remains to be examined experimentally. Although the definitive answer about the topology of a membrane protein can only come from its three-dimensional structure, obtained from the resolution and sequential assignment of resonances from all residues, *Pisa Wheels* enable valuable structural information to emerge from the spectroscopic studies prior to complete three-dimensional structure determination, and also provide the basis for structure determination methods [10,13].

The expression and purification of the recombinant FXYD family proteins PLM, CHIF, and Mat8, enable NMR structural studies to be performed in membrane environments, since the proteins can be isotopically labeled, and obtained in quantities that are suitable for NMR structure determination. The $^{1}H/^{15}N$ HSQC spectra in SDS micelles display excellent resolution, and demonstrate the feasibility of determining their structures in the alternative environment of lipid micelles. Both the solution and solid-state NMR spectra reflect helical structures for all three proteins, and the solid-state NMR spectra of CHIF in planar oriented lipid bilayer samples provide the first view of the topological features of its transmembrane helix. The FXYD membrane proteins regulate ion, osmolyte, and fluid homeostasis in a variety of tissues, and are emerging as auxiliary tissue-specific and physiological state-specific subunits of the Na^{+},K^{+}-ATPase. The ability to produce milligram quantities of pure FXYD proteins also opens the door for functional studies that, together with structure determination, can provide important structure-activity correlations. For example, the reconstitution of Na^{+},K^{+}-ATPase activity in the presence of FXYD proteins would be an important step in understanding the mechanism of pump regulation, while the incorporation of FXYD proteins in lipid bilayers would enable ion channel activities to be characterized by measuring specific ionic currents.

ACKNOWLEDGEMENTS

Our research was supported by grants from the National Institutes of Health (R01CA82864), the Department of the Army Breast Cancer Research Program (DAMD17-00-1-0506, DAMD17-02-1-0313), and the California Breast Cancer Research Program (8WB0110). The NMR studies utilized the Burnham Institute NMR Facility, supported by a grant from the National Institutes of Health (P30CA30199), and the Biomedical Technology Resources for Solid-State NMR of Proteins at the University of California San Diego, supported by a grant from the National Institutes of Health (P41EB002031).

REFERENCES

[1] S.H. White, W.C. Wimley, Annu. Rev. Biophys. Biomol. Struct. 28 (1999) 319–365.
[2] F.M. Marassi, Concepts Magn. Reson. 14 (2002) 212–224.
[3] F.M. Marassi, A. Ramamoorthy, S.J. Opella, Proc. Natl Acad. Sci. USA 94 (1997) 8551–8556.
[4] F.M. Marassi, S.J. Opella, J. Magn. Reson. 144 (2000) 150–155.
[5] J. Wang, J. Denny, C. Tian, S. Kim, Y. Mo, F. Kovacs, Z. Song, K. Nishimura, Z. Gan, R. Fu, J.R. Quine, T.A. Cross, J. Magn. Reson. 144 (2000) 162–167.
[6] F.M. Marassi, Biophys. J. 80 (2001) 994–1003.
[7] R.R. Ketchem, W. Hu, T.A. Cross, Science 261 (1993) 1457–1460.
[8] S.J. Opella, F.M. Marassi, J.J. Gesell, A.P. Valente, Y. Kim, M. Oblatt-Montal, M. Montal, Nat. Struct. Biol. 6 (1999) 374–379.
[9] K.G. Valentine, S.F. Liu, F.M. Marassi, G. Veglia, S.J. Opella, F.X. Ding, S.H. Wang, B. Arshava, J.M. Becker, F. Naider, Biopolymers 59 (2001) 243–256.
[10] J. Wang, S. Kim, F. Kovacs, T.A. Cross, Protein Sci. 10 (2001) 2241–2250.
[11] S.H. Park, A.A. Mrse, A.A. Nevzorov, M.F. Mesleh, M. Oblatt-Montal, M. Montal, S.J. Opella, J. Mol. Biol. 333 (2003) 409–424.
[12] A.C. Zeri, M.F. Mesleh, A.A. Nevzorov, S.J. Opella, Proc. Natl Acad. Sci. USA 100 (2003) 6458–6463.
[13] F.M. Marassi, S.J. Opella, Protein Sci. 12 (2003) 403–411.
[14] K.J. Sweadner, E. Rael, Genomics 68 (2000) 41–56.
[15] C.J. Palmer, B.T. Scott, L.R. Jones, J. Biol. Chem. 266 (1991) 11126–11130.
[16] B. Attali, H. Latter, N. Rachamim, H. Garty, Proc. Natl Acad. Sci. USA 92 (1995) 6092–6096.
[17] B.W. Morrison, J.R. Moorman, G.C. Kowdley, Y.M. Kobayashi, L.R. Jones, P. Leder, J. Biol. Chem. 270 (1995) 2176–2182.
[18] M.H. Vaarala, K. Porvari, A. Kyllonen, P. Vihko, Lab. Invest. 80 (2000) 1259–1268.
[19] J.R. Moorman, C.J. Palmer, J.E. John 3rd, M.E. Durieux, L.R. Jones, J. Biol. Chem. 267 (1992) 14551–14554.
[20] K.J. Crowell, C.M. Franzin, A. Koltay, S. Lee, A.M. Lucchese, B.C. Snyder, F.M. Marassi, Biochim. Biophys. Acta 1645 (2003) 15–21.
[21] B. Miroux, J.E. Walker, J. Mol. Biol. 260 (1996) 289–298.
[22] C. Di Guan, P. Li, P.D. Riggs, H. Inouye, Gene 67 (1988) 21–30.
[23] D.B. Smith, K.S. Johnson, Gene 67 (1988).
[24] J.P. Staley, P.S. Kim, Protein Sci. 3 (1994) 1822–1832.
[25] A. Pautsch, J. Vogt, K. Model, C. Siebold, G.E. Schulz, Proteins 34 (1999) 167–172.
[26] S.J. Opella, C. Ma, F.M. Marassi, Methods Enzymol. 339 (2001) 285–313.
[27] E. Gross, B. Witkop, J. Am. Chem. Soc. 83 (1961) 1510–1511.
[28] M.A. Andrade, P. Chacon, J.J. Merelo, F. Moran, Protein Eng. 6 (1993) 383–390.
[29] C.M. Franzin, J. Choi, D. Zhai, J.C. Reed, F.M. Marassi, Magn. Reson. Chem. 42 (2004) 172–179.
[30] C. Tian, P.F. Gao, L.H. Pinto, R.A. Lamb, T.A. Cross, Protein Sci. 12 (2003) 2597–2605.
[31] C.H. Wu, A. Ramamoorthy, S.J. Opella, J. Magn. Reson. A 109 (1994) 270–272.
[32] M. Schiffer, A.B. Edmunson, Biophys. J. 7 (1967) 121–135.
[33] F.M. Marassi, S.J. Opella, J. Biomol. NMR 23 (2002) 239–242.
[34] F.M. Marassi, C. Ma, J.J. Gesell, S.J. Opella, J. Magn. Reson. 144 (2000) 156–161.
[35] M.S. Weiss, A. Kreusch, E. Schiltz, U. Nestel, W. Welte, J. Weckesser, G.E. Schulz, FEBS Lett. 280 (1991) 379–382.

CHAPTER 4

Cell Surface Models on Polymer Supports – From Artificial Membranes to Native Cells

Motomu Tanaka*,[1,2]

Department of Physics, Technical University of Munich, James-Franck-Straße, D-85748 Garching, Germany

Dedicated to Prof. Erich Sackmann for his 70th birthday

Contents

Abstract

Models of cell surfaces on solid substrates can serve as a quasi two-dimensional matrix that makes it possible to connect biological and artificial materials. They can provide with not only systems for scientific studies, such as principle of wetting of solid with complex fluids, stress-free reconstitution of transmembrane cell receptors, and quantitative measurement of the strength of biological adhesion, but also versatile of technological applications. Planar supported membranes can be fabricated by deposition of lipid membranes either directly on solids, or on ultrathin (typical thickness of 10–100 nm) polymer supports, which mimic the generic role of extracellular matrix and glycocalix. Furthermore, the combination of modern genetic engineering and bioorganic chemistry also enables us to functionalize such cell membrane models with many types of biomolecules. In this chapter, I highlight recent advances in methods in fabrication/manipulation of supported lipid membranes (especially, polymer supported membranes), and present their potential for practical applications such as protein purification, fast screening tests, and local detection of protein functions.

* Corresponding author. Tel.: +49-89-28912539; Fax: +49-89-28912469; E-mail: mtanaka@ph.tum.de

[1] On leave at Center of Excellence and Department of Physics, Kyoto University.

[2] Present address: Biophysical Chemistry Laboratory, University of Heidelberg, Im Neuenheimerfeld 253, D69120 Heidelberg, Germany.

ADVANCES IN PLANAR LIPID BILAYERS AND LIPOSOMES, VOLUME 2
ISSN 1554-4516 DOI: 10.1016/S1554-4516(05)02004-1

1. INTRODUCTION, SCIENTIFIC BACKGROUND

During the last 20 years, phospholipid bilayers deposited onto solids surfaces (called supported membranes in the following) [1–3] have been intensively and widely used as models of cell surfaces to study physical principles of cell–cell and cell–tissue interactions, e.g., immune response and cell adhesion [4–8]. Quasi-2D lipid membranes are the major constituent of plasma membranes, and can incorporate membrane-associated (both peripheral and integral) proteins purified from living cells and organelles. By utilizing the modern protein engineering and bioorganic chemistry, it is possible to fabricate not only 'biomimetic' materials but also new classes of supramolecular composites with even higher functions [5,9–12].

In contrast to other model membranes, such as freestanding black lipid membranes or spherical lipid vesicles [13–16], supported membranes have strong advantages. Their excellent mechanical stability owing to the underlying solid supports enables us to prepare the membranes with various lipid compositions, which is often found to be difficult for freestanding BLMs. Moreover, their ability to coat macroscopically large (order of cm^2), planar substrates allows for the applications of a variety of surface sensitive techniques to study their structures (formation of functional micro-domains) and functions (specific ligand–receptor interaction) quantitatively.

The most straightforward strategy to introduce biological functions to artificial membrane models is the incorporation of transmembrane proteins into supported membranes. A conventional method to prepare supported membranes with transmembrane proteins is the spreading of lipid vesicles with reconstituted proteins (proteoliposomes). Since the first report by Brian and McConnell in 1980's [4], there have been an increasing number of studies conducted on the spreading of proteoliposomes on hydrophilic solid substrates like glass or quartz [5,6,17–19]. However, in spite of their remarkable progresses, there remains a fundamental drawback. The close proximity of the artificial membrane to the solid support with a typical distance of 5–20 Å [20–22] does not provide a sufficiently thick water reservoir to avoid denaturation of the protein subunits facing to the solid surface. This is a particularly serious problem that prohibits the lateral diffusion of cell adhesion receptors, if their extracellular domains (some of which can range even up to several tens of nm) are exposed to the solid surface.

To overcome this problem, it is desirable to separate the membrane from the solid substrate by ultrathin (typical thickness of less than 20 nm) hydrophilic polymer supports. Such lubricating polymer 'cushions' and 'tethers' can avoid the direct contact between proteins and solid supports, and therefore, can reduce the frictional coupling at the interface [23,24]. In this chapter, I survey recent advances in the fabrication and manipulation of artificial polymer supported lipid membranes. In addition to the perspectives in their scientific applications, I also

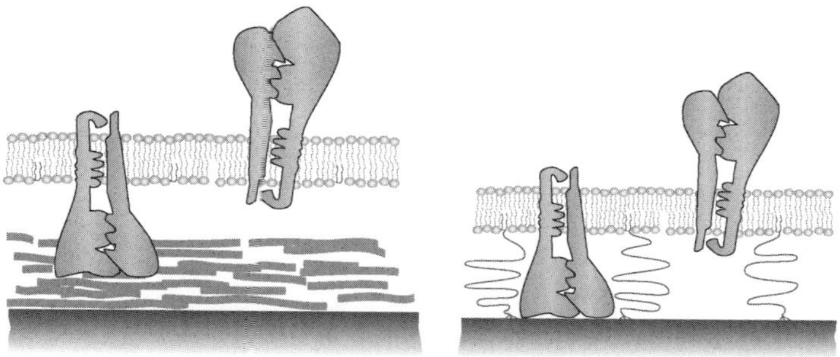

Fig. 1. Cell membrane models supported by a polymer 'cushion' (left) and polymer 'tethers' (right).

introduce the potentials of their biotechnological applications, such as fast screening tests and local detection of protein functions (Fig. 1).

2. POLYMER SUPPORTS AS EXTRACELLULAR MATRIX AND GLYCOCALIX MODELS

One of the most important tasks of polymer supports is to achieve thermodynamically and mechanically stable supported membranes by adjusting 'wetting conditions' at the interface [23,25,26]. In nature, cell–cell and cell–tissue interactions are mediated by hydrated layers of biopolymers, such as extracellular matrix (ECM) [27] and cell surface glycocalix [28]. They avoid direct, non-specific cell–cell contacts and create hydrodynamic pathways for solute transport.

In order to generally describe a contact between two plasma membranes (Fig. 2a), let us take a simple model: two apposing planes keep a stable, distinct separation distance *via* a thin interlayer. If the middle of a thin interlayer under equilibrium still retains its intrinsic bulk properties, the individual interfaces can be explained within the framework of the classical Gibbs capillary theory (Fig. 2b). A decrease in the interlayer thickness at a constant volume and constant interface areas would not cost any change in the free energy of the system except for the energy dissipation to overcome passive resistance forces like a viscous force. However, this does not hold any longer as soon as the fields of the long-range forces in thin interlayers overlap (Fig. 2c). Under this condition, any change in the interlayer thickness will cost positive or negative work. This work is originated from the net sum of attractive or repulsive forces in the overlapped interfacial region, which can range even more than several tens of nm. Long-range forces can be specified into van der Waals (dispersion) forces, electrostatic forces, steric (entropic) forces, and hydration forces. The last, hydration force, was found to stabilize colloidal dispersions [29], polar surfaces [30], and soap films [31].

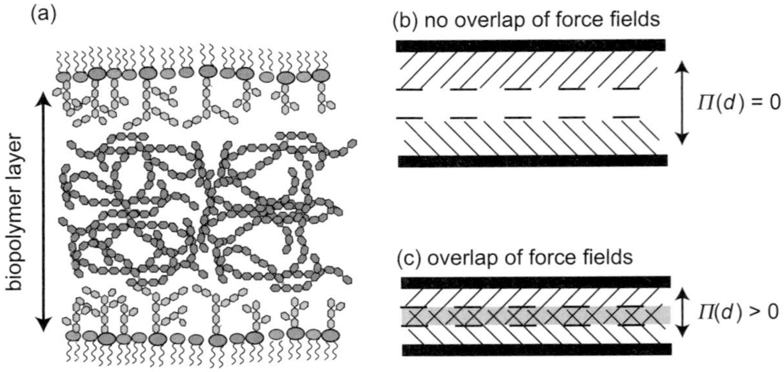

Fig. 2. (a) Schematic illustration of a cell–cell contact mediated by a hydrated biopolymer layer (glycocalix and extracellular matrix). (b) Two parallel Gibbs layers separated at a large distance, where the intrinsic properties of the bulk phase retain. (c) Two parallel planes in a close vicinity, where the long-range force fields overlap within the interlayer.

This force is a consequence of the work necessary for removing water from a hydrated layer to the bulk liquid phase.

Thermodynamics of thin wetting films can analytically be described in terms of a disjoining pressure, which is the sum of the net effect of the various interfacial forces [32]. This approach is very helpful to simplify the extremely complex calculation of each individual force contribution. The disjoining pressure Π can be measured by applying external forces to keep the interlayer at mechanical equilibrium. Namely, it can be given as the difference between the pressure in the interlayer P_1 and the pressure in the bulk layer P_0:

$$\Pi = P_1 - P_0. \tag{1}$$

The disjoining pressure can also be related to other thermodynamic parameters. For example, a change in the interlayer thickness d by δd at constant temperature T and chemical potential μ, the work done by the external force $-\Pi(d)\partial d$ equals to the increment of the Gibbs free energy δG:

$$\Pi(d) = -(\partial G/\partial d)_{T,\mu}. \tag{2}$$

The minimum of the free energy that determines the equilibrium state can be found at $\partial^2 G/\partial d^2 < 0$. Thus, to keep a distinct equilibrium distance d between two planes (in biology, plasma membrane contact mediated by a biopolymer interlayer), the disjoining pressure should fulfill the condition:

$$\partial\Pi(d)/\partial d < 0. \tag{3}$$

In case of cell adhesion, the trapping of membranes at this local minimum can be referred as 'weak adhesion', where the interaction potential may be approximated by a harmonic potential [33,34].

On the other hand, the interlayer becomes unstable when

$$\partial \Pi(d)/\partial d > 0. \tag{4}$$

Continuous thinning of the wetting layer below the free energy minimum leads to a negative disjoining pressure, which results in dewetting. Similar to the rupturing of polymer and surfactant films [35], such a phenomenon can be found as the formation of tight membrane contacts (adhesion plaque) [36,37].

To physically model cell–cell and cell–tissue contacts according to such a 'wetting principle', well defined models of ECM and glycocalix are required.

To date, one of the most widely used ECM models is the poly(ethylene glycol) (PEG) brushes [38]. For example, the covalent grafting of the PEG [39–41] or oligo-EG [42,43] chains on planar substrates and capillaries can avoid non-specific adsorption of proteins. On the other hand, incorporation of lipids with PEG chains (PEG lipids) into vesicles results in a remarkable increase in the blood circulation time [44,45].

Another class of common glycocalix model system is the dextran films, which are randomly grafted on solid surfaces. Dextran forms hydrogels can store large amount of water within the branched carbohydrate chains, and are practically utilized as a biocompatible coating material to protect the surface against non-specific adsorption of proteins [46–49]. The high resistance against protein adsorption can be attributed to high disjoining pressures exerted from steric (or entropic) interactions between the hydrated polymer chains. However, supported membranes are often found not to be stable on such hydro gel surfaces [50,51]. Although it is possible to deposit lipid membranes by the successive transfer of under and upper leaflets on the hydrated dextran films, the resulting membranes often exhibit a budding instability due to the high thermodynamic fluctuation [50]. In fact, in spite of the hydrophilicity of the surface, small vesicles do not undergo spontaneous formation of a lipid bilayer, but hover on the lubricating dextran films.

In our previous studies, we have designed two classes of polymer supports as new classes of synthetic ECM models: (a) ultrathin films (dry film thickness of 5–20 nm) of regenerated cellulose, and (b) end-grafted brushes of poly(2-alkyl-2-oxazoline) derivatives. Cellulose films are prepared by Langmuir–Blodgett (LB) deposition or spin-coating of trimethylsilylcellulose [52–54]. After the deposition, trimethylsilyl side chains are regenerated to hydroxyl group by the treatment with acid vapor. The regenerated cellulose films are stable in the presence of bulk water due to the tight hydrogen bonding network between rod-like cellulose backbones, which is in contrast to physisorbed films of other polysaccharides (dextran and hyaluronic acid) [55]. Films of poly(2-oxazoline) derivatives with trimethoxysilane coupling groups are prepared by covalent grafting onto SiO_2 surfaces, which enables the film to be stable in bulk water [56]. Poly(2-oxazoline) derivatives synthesized by living cationic polymerization

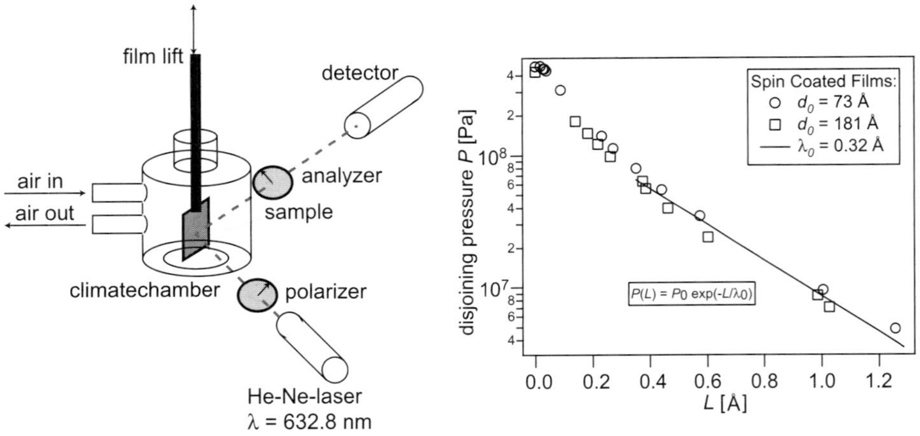

Fig. 3. (left) An ellipsometer equipped with a climate chamber. The relative humidity in the atmosphere can be precisely controlled between 4 and 98%. (right) Quantitative force–distance relationships of spin-coated cellulose films. The maximum swelling ratio and the decay length are independent from the dry film thickness.

[57,58] possess well-defined degrees of polymerization and polydispersity indices, and furthermore, different side substitutions enable us to tailor the hydrophilicity of the chains[3].

To gain quantitative force–distance relationships within the hydrated films, an ellipsometer combined with a climate chamber (Fig. 3) is used as a non-invasive technique with a high precision (± 0.1 Å) [50,55,56]. The hydration of the film can be studied by measuring film thickness at a well-defined osmotic pressure. Since the chemical potential of water in the atmosphere can be assumed to be identical to that within the polymer layer under equilibrium, the disjoining pressure can be represented as:

$$p = -\partial G/\partial d = -\left(\frac{RT}{V_m}\right)\ln\left(\frac{p_w}{p_0}\right) = -(RT/V_m)\ln X \tag{5}$$

where V_m is the molar volume of water, p_w/p_0 is the ratio between the actual vapor pressure and the saturation pressure, and X is relative humidity [59,60]. The quantitative force–distance relationship enables us to compare the hydration of different polymer films in terms of the monomer distance L and the calculated decay length λ_0. In case the monomer size is not given, one can take the normalized swelling ratio ρ and the dimensionless decay constant λ^* [54,56,61].

$$P = P_0 \exp(-L/\lambda_0), \tag{6a}$$

[3] Prior to the swelling experiments, we confirm the smooth topography of the polymer films, exhibiting root mean square roughness values of 4–6 Å within a 1×1 μm^2 AFM scan (tapping mode) [54,56].

Table 1. Dry film thickness d_0, maximum swelling ratio ρ_{max}, and decay length λ_0 of the regenerated cellulose films [54] and other polysaccharide hydro gels [55] calculated at $P<5\times10^7$ Pa

Sample	d_0 (Å)	ρ_{max}	λ_0 (Å)
LB-C-1	24	1.6	0.30
LB-C-2	56	1.4	0.32
LB-C-3	104	1.7	0.32
SP-C-1	73	1.6	0.32
SP-C-2	181	1.5	0.32
SP-DX	175	2.0	1.1
SP-HA	140	2.7	2.0

C, cellulose; DX, dextran; HA, hyaluronic acid; LB, LB film; SP, spin-coated film.

or

$$P = P_0 \exp(-\rho/\lambda^*) \tag{6b}$$

The distance between two neighboring monomers L can be represented as: $L(\rho)=a_0(\rho^{1/3}-1)$ for a cubic lattice model. a_0 is the size of a monomer, and ρ is the swelling ratio.

Table 1 summarizes the hydration of cellulose LB films and that of spin-coated films under equilibrium. Here, the size of monosaccharide unit is assumed to be $a_0\approx7.5$ Å [62]. The hydration of the cellulose films are found to be independent from the preparation methods as well as from the initial dry film thicknesses ($d_0=$ 2.5–18 nm). Actually, the maximum swelling ratio ($\rho_{max}\sim1.6$) and the exponential decay length ($\lambda_0=0.32$ Å) are almost identical for all the samples [54]. In comparison to the corresponding values obtained for other polysaccharide hydro gel films (spin-coated films of dextran and hyaluronic acid) with comparable dry film thicknesses (Table 1, bottom), the maximum swelling ratio, i.e., the water storage capability, of the cellulose films is apparently smaller ($\rho_{max}=2.0$ for dextran, 2.7 for hyaluronic acid, respectively) [55]. Moreover, the calculated decay length is 3–6 times smaller than those found for dextran ($\lambda_0=1.1$ Å) and hyaluronic acid ($\lambda_0=2.0$ Å) [55]. The clear differences in these two characteristic parameters can be attributed to the differences in the hydrogen bonding between linear, cylindrical cellulose backbones and those between more 'molten' chains of other polysaccharide films. Actually, these two parameters can be used as semi-quantitative indicators to describe the vertical force interaction in the wetting of supported membranes (Section 4)[4].

[4] Dynamic (i.e., non-equilibrium) hydration/dehydration of polymer thin films upon an abrupt change in atmospheric humidity can be studied using the self-made set-up, where one can switch the local humidity conditions in proximity of the measurement between 4 and 98%. The hydration kinetics can be attributed to the conformational degrees of freedom of polymer chains as well as to the strength of inter-molecular hydrogen bridges [54].

As a potential candidate for the hydrophilic spacer to separate a supported membrane and a solid substrate (see Section 3), the static swelling behaviors of poly(2-methyl-2-oxazoline) and poly(2-ethyl-2-oxazoline) films with different initial thicknesses were also studied following the same experimental strategy. Interestingly, the swelling behaviors of poly(2-oxazoline) brushes exhibit a clear dependence upon the chain length in terms of the maximum swelling ratio and the decay constant; however, almost no influence from the side chain substitutions could be observed [56].

3. ARTIFICIAL MODEL MEMBRANES ON POLYMER SUPPORTS

Supported membranes can be prepared by several different methods; (i) successive transfer of monolayers, (ii) fusion of lipid vesicles, and (iii) exchange of solvent. The first method includes the transfer of the proximal monolayer (under leaflet) onto a hydrophilic surface by the vertical lifting from the water subphase (LB transfer), which is followed by the transfer of the upper leaflet through horizontal dipping of the monolayer-coated substrate into the water subphase (called Langmuir Schäfer transfer) [1]. Although this is a laborious method requiring careful handling of the sample, it allows for the preparation of a supported membrane with asymmetric compositions in two leaflets. The second method is based on the rupturing of small lipid vesicles (typically, the diameter of 30–100 nm prepared by sonication or extrusion) on the surface and the fusion of membrane patches by self-assembly [63]. The third method follows the exchange of solvent from alcohol to aqueous buffer [64,65]. As mentioned in Section 1, the planar membranes that cover macroscopically large surfaces allow us for the combinations of surface sensitive techniques (optical microscopy, x-ray/neutron reflectivity, infrared spectroscopy, surface plasmon resonance, quartz crystal resonator, electrochemical techniques, etc.) to characterize the structures and functions of membranes and membrane-associated proteins[5].

Among various polymer films, ultrathin films of regenerated cellulose can serve as ideal inter-layers that facilitate the fusion of vesicles and the simultaneous spreading of supported membranes. In fact, cellulose films can assist the self-healing of local defects in the membrane, which can be monitored as a function of time using fluorescence microscopy [66] and micro-interferometry (reflection interference contrast microscopy, RICM) (Fig. 4) [67].

---

[5] Another alternative is the direct spreading of a lipid bilayer from the blotted dry lipid reservoir. However, this technique is not practically suited to coat a macroscopically large surface of a viscous polymer film with a smaller spreading coefficient β, since the spreading velocity follows the power law, $\nu = (\beta/t)^{0.5}$ [66].

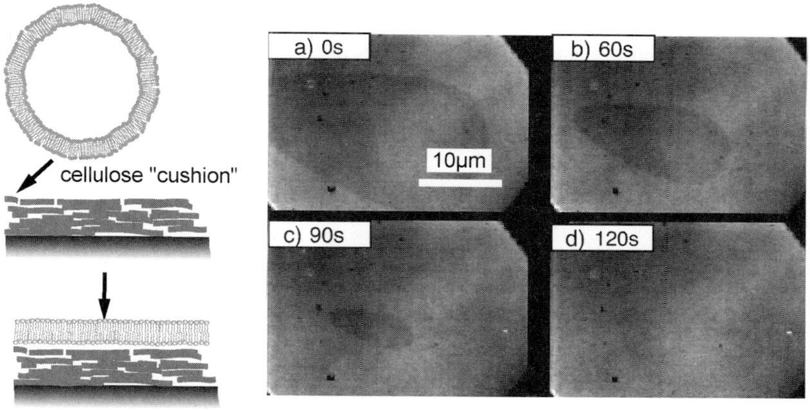

Fig. 4. (left) Schematic illustration of fusion of lipid vesicles on a cellulose support. (right) Self-healing of a local defect in a supported membrane (dark gray zone) on cellulose, monitored by micro-interferometry (RICM).

If one considers the intrinsic advantage of lipid membrane, i.e., extremely low ion permeability, the complementary hybridization of such a defect-free polymer supported membrane and electric devices would be very promising to design membrane-based biosensors. As an alternative technique to the classical patch-clamp technique, supported membranes have been drawing attentions because of their high mechanical stability. This straightforward idea, however, suffers to achieve a high membrane resistance comparable to patch-clamped membranes. The high electrical resistance of lipid membranes is often called as 'giga-seal', which is of $M\Omega cm^2$ order if normalized with the membrane area. By controlling the interactions between the membrane and electrode, we accomplished the deposition of the polymer supported membrane with a high electric resistance ($0.5\ M\Omega cm^2$) on indium tin oxide (ITO) electrodes [67], which is by a factor of $5 \sim 50$ larger than that obtained for a lipid bilayer directly deposited on ITO [68]. Electrochemical properties of the individual layers (semiconductor, organic monolayer, polymer cushion, lipid membrane, and electrolyte) can quantitatively be characterized using AC impedance spectroscopy in a wide frequency regime (10 mHz $\sim$ 100 kHz) [65,67,69–72]. Here, the leak current through the local defect can be significantly suppressed *via* self-healing on polymer lubricants, and therefore, the polymer supported membranes can serve as an electrically insulating platform to study the ion channel functions on semiconductor electrodes even at relatively low channel densities (about 10^7 channels/cm^2) [65,67]. In this system, the cellulose film is fully hydrated and possesses almost identical electrochemical properties as a bulk electrolyte. Therefore, such a configuration (electrolyte–membrane–electrolyte–semiconductor, EMES) [65,67,68] coupled to the reference and counter electrodes can be considered as an analog of a free standing BLM that partitions two electrolyte reservoirs. The kinetics of peptide uptake can be studied by monitoring the membrane conductance as a function of time upon the injection of peptide solution, while the ion selective transport across the membrane can be confirmed by exchange of ion species in the electrochemical cells. The high electric resistance of lipid membranes can also be applied to detect the changes in membrane surface potentials. To avoid the effect of the electrochemical screening in the water reservoir between the membrane and the semiconductor electrode, a lipid monolayer on a hydrophobic (and thus insulating) polymer film can be used as an analog of metal–oxide–semiconductor (MOS) configurations. Changes in surface potentials due to charging of lipids or protein coupling can be detected sensitively as changes in space charge capacitance of the semiconductor. Our recent study demonstrated a significant change in the global shape of impedance spectra according to the charging and de-charging of lipid head groups, suggesting the reachable sensitivity of up to one surface charge in 30 nm^2 (Fig. 5) [73].

More recently, we also found that the spreading of lipid vesicles with cell adhesion molecules (platelet integrin $\alpha_{IIb}\beta_3$) on cellulose thin films (thickness of about 5–10 nm) leads to a significant improvement in the homogeneity and lateral diffusivity of integrin receptors [74]. The diffusion constant of a protein in the polymer supported membrane follows the Saffman–Delbrück's law

$$D = \frac{k_B T}{8\pi\eta_m} \ln\left(\frac{\eta_c R_p^2}{\eta_m R_p} - \gamma\right) \tag{7}$$

where η_m and η_c correspond to the viscosity of the membrane and that of the polymer support. R_p and h_p stand for the radius and the height of the hydrophobic transmembrane core. For thin polymer supports (thickness h_p of less than 100 nm) with a high viscosity, this can be rewritten as [75]

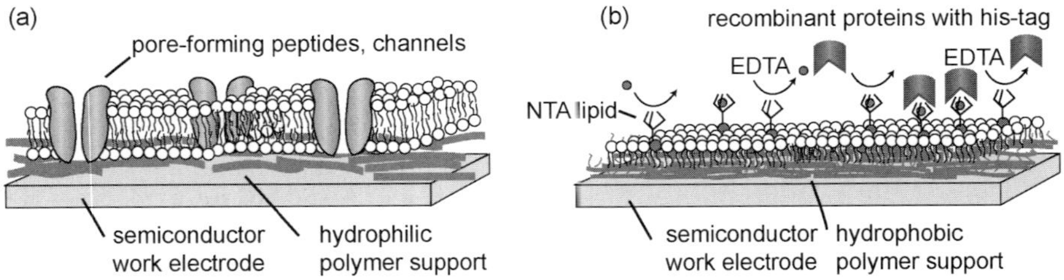

Fig. 5. Schematic views of sensor prototypes based on supported membranes. (a) Conductive sensing of pore-forming antibiotic peptides in a lipid bilayer. (b) Capacitive sensing of surface potentials using a biomimetic electrolyte–insulator–semiconductor structure using a lipid monolayer.

$$D = \frac{k_B T h_p}{\eta_c R_p^2}. \tag{8}$$

Thus, the distinct increase in the protein diffusivity observed on a cellulose surface strongly confirms that the deposition of a lubricant film with a thickness of $h_p \sim 10$ nm can increase the lateral mobility of transmembrane proteins. Furthermore, the adhesion free energy of a giant lipid vesicle bearing synthetic ligands to the polymer supported membrane with integrins seems to be 3–10 times larger than the corresponding values on the solid supported membranes [74].

An alternative strategy to separate the lipid bilayer from the solid substrate is to incorporate lipids with polymer head groups (lipopolymer 'tethers') [24] into the proximal layer facing to the substrate. Several spacers based on oligo- [76–78] and poly(ethylenoxide) [79], and oligopeptides with thiol groups [80] have been successfully applied to control the substrate–membrane spacing. As another class of polymer tethered membranes, we recently established supported membranes with variable spacer length ($n = 10$–100) by taking an advantage of living cationic polymerization of poly(2-oxazoline)s [81]. Prior to the membrane preparation, the swelling behaviors of poly(2-oxazoline) homopolymer brushes without hydrophobic lipid anchors have quantitatively been studied by ellipsometry (see Section 2) [26,56]. Here, supported membranes have been prepared by the combination of the methods (i) and (ii) described in this Section. First, a suitable lipopolymer/lipid mixture was deposited by LB transfer, and annealed to establish the covalent coupling to the surface. On the dry lipopolymer/lipid monolayer, the upper leaflet was deposited by vesicle fusion. Optimization of both preparation steps resulted in the formation of stable and defect-free membranes. Impacts of the spacer length and the lipopolymer fraction upon the lateral diffusivity of the lipids were systematically compared to see the influence of the thickness and viscosity of the lubricating layer. Moreover, first experiments on the incorporation of integrin $\alpha_{IIb}\beta_3$ into the polymer-tethered membrane suggested that the length of the polymer tether plays a crucial role in distribution and function of integrin receptors (Fig. 6) [81].

4. CELL MEMBRANES AS NATIVE SUPPORTED MEMBRANES

A completely different, and more challenging approach is to deposit 'native' cell membranes onto planar substrates. As pointed out in the previous sections, the spreading of proteoliposomes, even on polymer supports, still includes a serious technical drawback: the surface density and orientation of the transmembrane proteins can hardly be controlled. Therefore, the direct spreading of native cells onto planar substrates allows us to overcome these problems in spite of

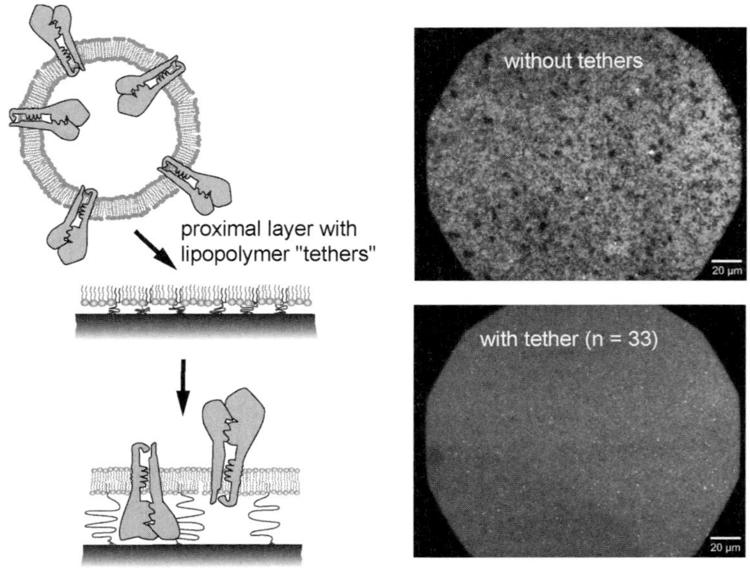

Fig. 6. (left) Schematic illustration of spreading of proteoliposomes with platelet integrin $\alpha_{IIb}\beta_3$ on a proximal layer with lipopolymer tethers. (right) Fluorescence labeling of integrin clearly demonstrates that the lubricant layer provided by lipopolymer spacers significantly improves the homogeneity of integrin receptors.

experimental difficulties, since nature stringently controls both the population and the orientation of transmembrane proteins in plasma membranes.

In 2001, we accomplished the spreading of human red blood cell (erythrocyte) membranes on cellulose thin films [82]. Erythrocyte ghosts are prepared from freshly drawn blood as reported by Schwoch [83], and incubated for 60 min with variously prepared surfaces at room temperature ($\sim$20 °C). After the incubation, the remaining ghosts and vesicles in the chamber were removed by intensive rinsing with buffer. Orientation of the erythrocyte membrane in contact with surfaces is identified with two immune-fluorescence labels: (1) the extracellular part of glycophorin can be labeled with a fluorescence labeled peanut agglutinin [82], or with a first monoclonal antibody (mouse IgG) and a second polyclonal antibody (goat anti-mouse IgG) labeled with dyes [84], while (2) the cytoplasmic domain of Band 3 can be recognized with a first monoclonal antibody (mouse IgG) and a fluorescence labeled second polyclonal goat anti-mouse IgG antibody (inside label).

When the erythrocytes are incubated with planar glass cover slides, no adsorption or rupture of erythrocytes can be observed. This can be attributed to the repulsive interaction between the negatively charged cell glycocalix (owing to the sialic acid residues) and the negatively charged glass surface. To reduce the interfacial repulsion between cells and the surface, two types of polymer films are deposited on the glass cover slides: physisorbed films of cationic poly-lysine, and LB films of cellulose derivatives. In contrast to the intactness found on glass surfaces, the inside labeling with antibodies indicates the homogeneous wetting of erythrocyte membranes on cellulose surfaces, exhibiting no boundaries between individual patches of cell membranes. Since the outside labeling results in no fluorescence signal, we conclude that all the adherent cells rupture and expose their cytoplasmic domain to bulk electrolyte. This was the first demonstration of transforming three-dimensional (3D) cells into quasi-2D films by simply adjusting the interfacial interaction. This strategy is now being extended to other cell types and microsomes extracted from the living organelle (e.g., sarcoplasmic reticulum extracted from rabbit muscle cells (Fig. 7) [84]).

The significant improvement in the membrane–surface interaction by a thin polymer film with nm order thickness can be understood partially in terms of the reduction in electrostatic repulsion due to the deposition of a neutral polysaccharide layer that is thicker than the electrostatic screening length. Another possible contribution would be the osmotic pressure balance between the hydrated cellulose layer and the cell surface glycocalix. To gain the deeper insight into the wetting conditions, quantitative measurements of interaction potentials at cell surface interface will be necessary. On the other hand, the inside labeling of the same cell membranes on poly-lysine shows a very inhomogeneous surface coverage, which cannot be improved even by a prolonged incubation time or thermal annealing. The outside labeling shows no fluorescence over the whole area, confirming that the dark region in the figure corresponds to the defects. This can be understood as the dewetting of cell membranes because of the strong attraction between weakly anionic cell glycocalix and strong polycation (poly-lysine) surfaces. Actually, such a stable, strong coupling had been often reported

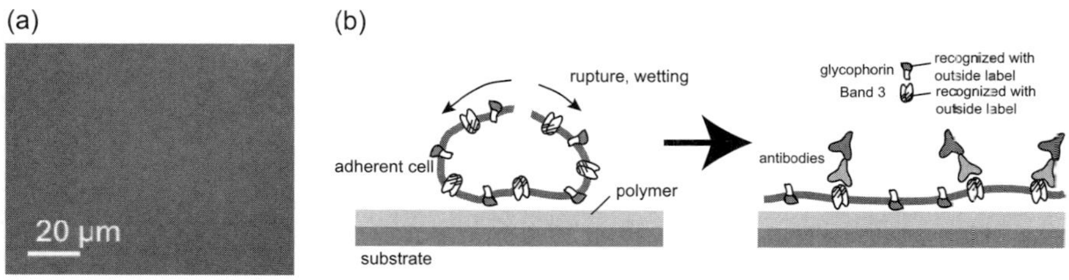

Fig. 7. (a) Immunofluorescence image of human erythrocyte membranes spread on cellulose. The pair of antibodies that identify the cytoplasmic domain of Band 3 (inside labels) are applied. The incubation with the antibodies for the extracellular domain of glycophorin (outside labeling) results in no fluorescence. (b) The postulated mechanism of the spreading of cell membranes on polymer supports.

for the adsorbed polyelectrolytes [85] or charged proteins [86] on oppositely charged surfaces.

The formation of stable supported membranes on soft polymer films, i.e., cell–cell contact stabilized *via* hydrated biopolymer layers, can be generally considered as wetting of solids with complex fluids. Irrespective of the preparation method, the stratified films are only stable if the conditions of complete wetting are fulfilled for each individual layer, that is, the free energy should be gained by the spreading process. The change in the surface free energy associated with the spreading of a thin film is determined by the so-called spreading coefficient S,

$$S = \sigma_{\mathrm{SL}} - \sigma_{\mathrm{SM}} - \sigma_{\mathrm{ML}} \geq 0. \quad (9)$$

σ_{SL}, σ_{SM}, and σ_{ML} correspond to the tensions at surface/liquid, surface/membrane, and membrane/liquid interfaces (measured in mN/m), respectively [25,26,87].

However, this is not a sufficient condition to ensure the stability of lipid membranes on polymer supports. As explained in Section 2, the interaction potential between the solid and the membrane must be weakly repulsive. If the interaction is attractive, i.e., if the disjoining pressure is negative, the composite film will become unstable resulting in the formation of regions of tight local contact. On the contrary, the membranes cannot maintain the stable planar geometry and tend to form budding of small vesicles when they are strongly repelled by a highly positive disjoining pressure [50].

Such a rather simple 'wetting concept' can even be applied to understand the fundamental principle of cell adhesion. The gain in the interfacial free energy W can be represented as an analog of that of a partially wetting liquid droplet. Here, the balance of surface tensions at the three phase contact line can be described by the Young's law [88–90]:

$$W = \sigma(1 - \cos\theta_{\mathrm{c}}), \quad (10)$$

where σ is the lateral membrane tension and θ_{c} is the contact angle. Similarly, the contact curvature $1/R_{\mathrm{c}}$ is determined from the balance of bending moments:

$$W = \kappa/R_{\mathrm{c}}^2, \quad (11)$$

where κ is the bending elasticity of the membrane [91]. Thus, the free energy of adhesion can be determined by measuring the geometric parameters θ_{c} and R_{c} using micro-interferometry [36] or fluorescence interference contrast microscopy (FLIC) [22].

Recently, Marx *et al.* [92] experimentally demonstrated the oscillatory transition of membrane between state of strong and weak adhesion. In case of a weak adhesion, where the membrane is attracted with a generic interaction, the membrane is separated from the surface at a distance of about 30 nm, showing a strong flickering. An increase in the interfacial attraction causes a tension induced

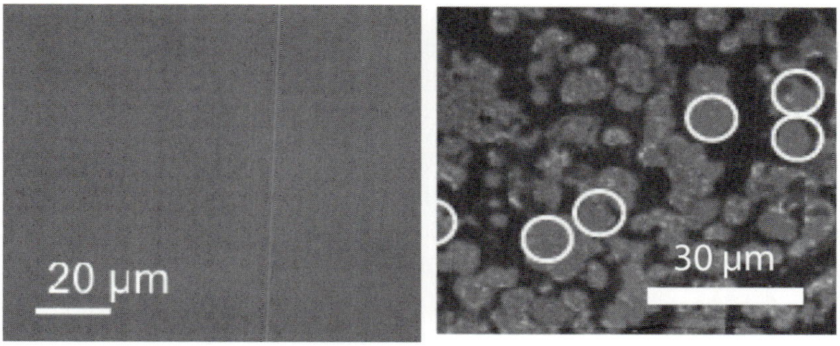

Fig. 8. Wetting of human erythrocyte membranes on a cellulose surface (left) and dewetting of the same cell membranes found on a polycation (polylysine) surface (right).

by adhesion, resulting in a suppression of the membrane undulation. In fact, such a "tension-induced" transition from the weak adhesion to the strong adhesion can be represented as the first order wetting/dewetting phase transition [37].

From this generalized viewpoint, the significant improvement in the membrane–surface interaction found on cellulose films can be attributed to the reduction in electrostatic repulsion (owing to the film thickness larger than the electrostatic screening length) and the osmotic pressure balance between the hydrated cellulose layer and the cell surface glycocalix. On the contrary, the dissipation of human erythrocyte membranes in contact with a strong polyelectrolyte film can be understood as the dewetting of cell membranes on oppositely charged polyelectrolyte surfaces, which might even cause a negative disjoining pressure (Fig. 8) [26].

5. 'LOCAL' BIOFUNCTIONALIZATION OF SUPPORTED MEMBRANES

Plasma membranes in nature are known to express intrinsically asymmetric distribution of lipids in their cytoplasmic and extracellular leaflets, as well as to form laterally organized functional micro-domains enriched with certain types of lipids (e.g., glycosphingolipids, cholesterol, and sphingomyelins). Such domains, called as 'rafts', are postulated to play key roles in complex cellular functions such as endocytic traffic, signal transduction [93]. Moreover, dynamic accumulation of ligand–receptor pairs can be observed in cell adhesion processes (such as extravasations of leukocytes into tissues at inflammation sites), which eventually results in a firm adhesion [36,94]. At the initial step, the adhesion areas (or focal adhesion sites) including clusters of ligand–receptor pairs are separated by regions of weak adhesion. Here, I highlight several experimental approaches to laterally confine biofunctional molecules in supported membranes, such as (i) control of self-assembling processes, (ii) accumulation of membrane-associated proteins by electrophoresis, and (iii) micro-patterning of supported membranes.

The first approach uses the self-assembling process to control the shape and size of the biofunctional domains. For example, 'rafts' of glycosphingolipids can be formed in artificial vesicles [95,96] and in supported membranes [97,98] by taking a similar lipid composition to the native one, suggesting that lipid compositions determine the formation of micro-domains. Indeed, such micro-domains can also be made by doping non-biological, synthetic ligands (for example, ligands with fluorinated lipid anchors) into the matrix phospholipids [99]. The synthetic ligands form clusters in supported membranes as well as in a giant vesicle, whose size and distribution can systematically be controlled by the lipid composition. Furthermore, such artificially designed (and thus non-biological) micro-domains can even enhance the cell adhesion strength with respect to the ligand/matrix mixtures with biological (alkyl) lipid anchors (Fig. 9).

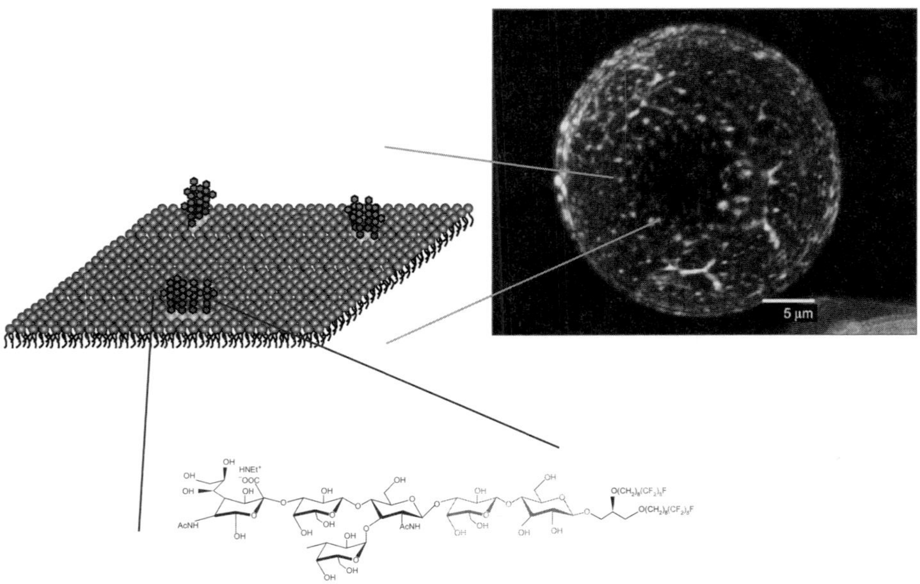

Fig. 9. 3D reconstruction of fluorescently labeled micro-domains of synthetic, i.e., non-biological, sialyl Lewis^X glycolipids with partially fluorinated anchors, reconstituted in a giant vesicle of phospholipids (DMPC). Owing to a phase separation of alkyl and fluoroalkyl lipid anchors, the size and distribution of carbohydrate clusters can flexibly be controlled. Supported membranes with sialyl Lewis^X micro-domains enable one to adjust the biospecific function of carbohydrate ligands against dynamic cell adhesion.

The second approach utilizes 'fluid' supported membranes as a quasi-2D matrix to separate membrane proteins under lateral electrical fields. To date, this technique has mainly been used to accumulate charged lipid molecules [100,101], adsorbed proteins [102,103], or DNA bound to the membrane [104]. To accumulate various transmembrane proteins, however, the technical difficulties still exist in the friction between proteins and membranes, and the electro-osmosis that competes against the electrophoretic forces. The first problem arises from the fact that the mean velocity of proteins in a membrane is determined by the lateral diffusion constant. The second problem becomes rather serious for the transmembrane proteins with large extracellular domains. The charged extracellular domains (vertical height can be up to several 10 nm) experience a significant electro-osmotic flow in the opposite direction to the electrophoresis [102,103]. The first problem can be overcome by deposition of polymer supports that increases the lateral diffusivity of the membrane components. For example, lipid monolayers on hydrophobic 'hairs' of polymers showed a diffusion constant more than 10 times larger than that on alkylsilanes [73]. Furthermore, a more recent study demonstrated that the genetically engineered proteins (e.g., green fluorescent proteins (GFPs) with a molecular weight of 27 kDa) coupled to a lipid monolayer show a similar diffusion constant to those of lipids (1–2 $\mu m^2/s$), which allows for the electrophoretic accumulation of surface-tagged GFPs under a DC electric field (Fig. 10) (J. Hermann, M. Fischer, S.G. Boxer, M. Tanaka, unpublished results). Upon the application of the electrophoretic potential (typically about 10 V/cm), the proteins undergo electrophoretic accumulation to the diffusion barrier, reaching to a new equilibrium state. The electrophoretic movement of proteins can be described

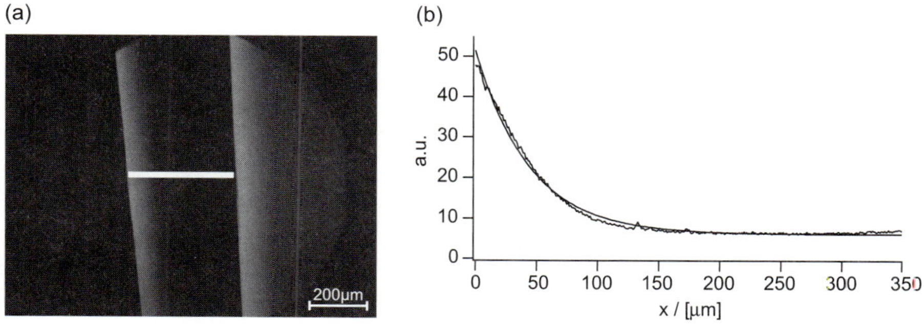

Fig. 10. Electrophoretic accumulation of histidin-tagged GFP in a polymer supported lipid monolayer. The stationary protein gradient (a) can be analyzed with a simple differential equation, yielding a mean velocity of proteins, $\langle v \rangle = 0.033\ \mu m/s$ (b) (J. Hermann, M. Fischer, S.G. Boxer, M. Tanaka, unpublished results).

using a simple differential equation:

$$\frac{\partial c(\vec{r},t)}{\partial t} = D\frac{\partial^2 c(\vec{r},t)}{\partial \vec{r}^2} + \langle v\rangle \frac{\partial c(\vec{r},t)}{\partial \vec{r}} \tag{12}$$

where $c(\vec{r},t)$ stands for the protein concentration as a function of position and time, D is the diffusion constant, and $\langle v\rangle$ is the mean velocity of the protein. Since the time-dependent term becomes zero at the equilibrium, one can solve the protein concentration gradient at $\partial c(\vec{r},t)/\partial \vec{r} = 0$. It is noteworthy that the proteins do not only experience the electrophoretic force but also the counter acting electro-osmotic flow. In fact, the opposite contributions from two forces can be adjusted by changing the zeta potential difference between proteins and the membrane surface (J. Hermann, M. Fischer, S.G. Boxer, M. Tanaka, unpublished results).

The third approach, the micro-patterning of supported membranes includes a large potential for the parallel monitoring of interactions between the partitioned membranes and the analytes (antibodies, drugs, etc.) [3]. However, the confinement of transmembrane proteins in defined membrane micro-patterns still remains as a challenging task. One of the promising ways is to process the micro-patterns of polymer cushions that locally express the membrane-compatible domains [54]. For instance, the fact that the spreading of cell membranes is possible on cellulose films but not on glass slides (see Section 4) [82] suggests the possibility to spread (wet) cell membranes selectively on cellulose micro-patterns [84]. Cellulose micro-patterns are processed by UV photolithography, where the illumination of the cellulose capped with a metal photomask by deep UV light (emission lines at $\lambda = 244$, 194, and 185 nm) leads to the ablation of cellulose backbones and alkylsilanes, resulting in a hydrophilic glass surface [54]. Immunofluorescence labeling clearly demonstrates that the area coated with cellulose can be wet with erythrocyte membranes, while no adherent or ruptured cells could be found on the intact glass surface. Another possibility is to deposit the protein barriers such as bovine serum albumin (BSA) onto homogeneous cellulose films using the micro-contact printing technique, as reported on glass slides [105,106]. The patterns of the 'stamped' proteins are confirmed to be stable under water more then several days, and serve as effective barriers to isolate the deposited cell membranes [84]. Such simple strategies are further applied for other native microsome membranes (e.g., sarcoplasmic reticulum of rabbit muscles), and allow for the localization of membrane proteins in different geometries (stripes, grids, etc.) in various length scales. Moreover, both types of membrane micro-patterns retain their structure for more than a week, verifying their thermodynamic and mechanical stability. Thus, the rather processing methods established here allow for further manipulation by the addition of other native cell membranes to design

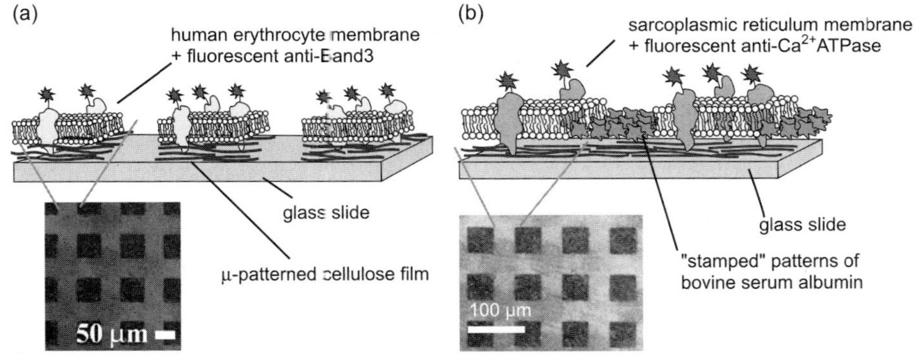

Fig. 11. Two types of micro-arrays of 'native' supported membranes on biocompatible templates [84]. (a) Human erythrocyte membranes deposited on a lithographically micro-patterned cellulose film. Cytoplasmic domain of the proteins is recognized with fluorescent antibodies. (b) Immunofluorescence image of the of sarcoplasmic reticulum membranes on a cellulose film with 'stamp-inked' micro-patterns of bovine serum albumin.

hetero-biofunctional surfaces to study the function of membrane proteins under more native environments (Fig. 11).

The supported membranes that locally express the biological functions are scientifically applicable as artificial plasma membrane models, where the local density and aggregation size of ligands and receptors can be controlled in a well defined manner. Actually, synthetically designed, non-biological carbohydrate ligands are found to realize even more enhanced functionalities than the natural ligands by controlling the phase separation with the surrounding matrices [99]. This strongly indicates that the combination of protein engineering and (bio)organic chemistry enables not only the fabrication of biomimetic materials, but also the tailoring of novel classes of more functional supramolecular composites. Furthermore, the electrophoretic accumulation of recombinant proteins coupled to a membrane surface provide with a tool to manipulate the lateral density of proteins in fluid supported membranes (Fig. 12) (J. Hermann, M. Fischer, S.G. Boxer, M. Tanaka, unpublished results).

On the other hand, the pre-patterning of biocompatible polymer supports allows us for the position selective spreading of 'native' cell membranes and microsomes as quasi-2D films in restricted geometries, retaining the orientation and lateral density of transmembrane proteins in their native state (Fig. 11) [26]. In fact, such micro-arrays of native membranes include a versatile of biotechnological applications. For example, the use of micro-pipettors will enable the parallel monitoring of interactions between the partitioned membranes and the analytes (antibodies, drugs, etc.) by spotting different analytes onto individual corrals. Moreover, the partitioning of membranes prohibits the diffusion of

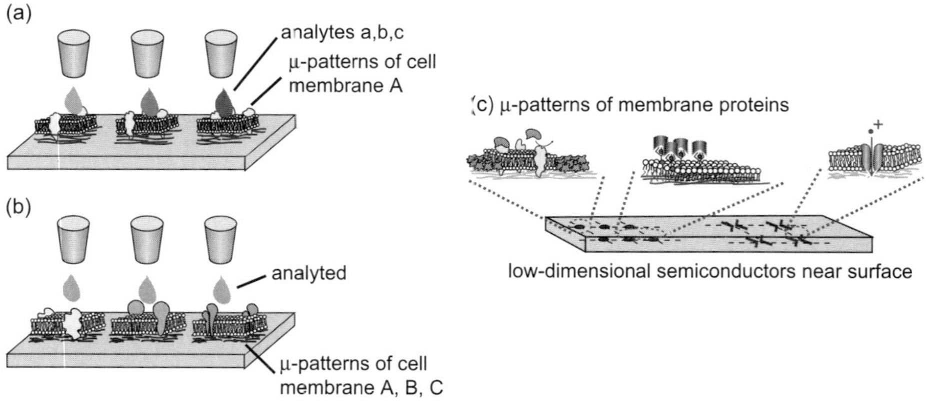

Fig. 12. Potential biotechnological applications of polymer supported membranes micro-arrays. (a,b) Application of micro-patterns of native cell membranes coupled to conventional spotters for the fast screening test. (c) Membrane-semiconductor hybrids: (left) isolated arrays of near surface quantum dots or quantum wells as optical probes, (right) arrays of high electron mobile transistors to detect the change in potential near the surface.

proteins across the separating barriers and thus includes a large potential for complimentary coupling of bio-organic molecular systems and low-dimensional semiconductors [107–109]. The minitualization of quasi-2D cell membranes to the lateral dimensions of array-integrated semiconductor devices is an interdisciplinary scientific endeavor to extend the concept of membrane-based biosensors on bulk semiconductors towards the local sensing of protein functions in the order of several molecules.

ACKNOWLEDGEMENTS

I am indebt to Erich Sackmann for many scientific inspirations and fruitful comments, and to my colleagues who contributed to this work, including O. Purrucker, F. Rehfeldt, M. Tutus, A. Wong, J. Hermann, H. Hillebrandt, K. Adlkofer, M. Hochrein, S. Gönnenwein, and S. Kaufmann. I am also thankful to my collaboration partners, R. Jordan, M. Fischer (Dept. Chem. TUM), G. Wegner (MPI Mainz), G. Abstreiter, M. Stutzmann (WSI, TUM), and S.G. Boxer (Dept. Chem. Stanford). This work has been supported by grants from the Deutsche Forschungs Gemeinschaft (SFB563), NSF-MRSEC, and the Fonds der Chemischen Industrie. Last but not least, I am grateful to the post-doctoral fellowships from Japan Society for the Promotion of Science (JSPS) and Alexander von Humboldt Foundation, and the habilitation fellowship from the DFG (Emmy Noether Program).

REFERENCES

[1] L.K. Tamm, H.M. McConnell, Supported phospholipid bilayers, Biophys. J. 47 (1985) 105–113.

[2] E. Sackmann, Supported membranes: scientific and practical applications, Science 271 (1996) 43–48.

[3] J.T. Groves, S.G. Boxer, Micropattern formation in supported lipid membranes, Acc. Chem. Res. 35 (2002) 149–157.

[4] A.A. Brian, H.M. McConnell, Allogeneic stimulation of cytotoxic T cells by supported planar membranes, Proc. Natl Acad. Sci. USA 81 (1984) 6159–6163.

[5] P. Chan, M.B. Lawrence, M.L. Dustin, L.M. Ferguson, D.E. Golan, T.A. Springer, Influence of receptor lateral mobility on adhesion strengthening between membranes containing LFA-3 and CD2, J. Cell Biol. 10 (1991) 245–255.

[6] E.-M. Erb, K. Tangemann, B. Bohrmann, B. Müller, J. Engel, Integrin αIIbβ3 reconstituted into lipid bilayers is nonclustered in its activated state but clusters after fibrinogen binding, Biochemistry 36 (1997) 7395–7402.

[7] A. Kloboucek, A. Behrisch, J. Faix, E. Sackmann, Adhesion-induced receptor segration and adhesion plaque formation: a model membrane study, Biophys. J. 77 (1999) 2311–2328.

[8] S.Y. Qi, J.T. Groves, A.K. Chakraborty, Synaptic pattern formation during cellular recognition, Proc. Natl Acad. Sci. USA 98 (2001) 6548–6553.

[9] L. Moroder, R. Romano, W. Guba, D.F. Mierke, H. Kessler, C. Delporte, J. Winand, J. Christophe, New evidence for a membrane-bound pathway in hormone–receptor binding, Biochemistry 32 (1993) 13551–13559.

[10] L. Schmitt, C. Dietrich, R. Tampé, Synthesis and characterization of chelator-lipids for reversible immobilization of engineered proteins at self-assembled lipid interfaces, J. Am. Chem. Soc. 116 (1994) 8485–8491.

[11] C. Dietrich, O. Boscheinen, K.D. Scharf, L. Schmitt, R. Tampe, Functional immobilization of a DNA-binding protein at a membrane interface via histidine tag and synthetic chelator lipids, Biochemistry 35 (1996) 1100–1105.

[12] Y.D. Vankar, R.R. Schmidt, Chemistry of glycosphingolipids-carbohydrate molecules of biological significance, Chem. Soc. Rev. 29 (2000) 201–216.

[13] P. Mueller, W.C. Wescott, D.O. Rudin, H.T. Tien, Methods for formation of single bimolecular lipid membranes in aqueous solution, J. Phys. Chem. 67 (1963) 534–535.

[14] A.D. Bangham, M.M. Standish, N. Miller, Cation permeability of phospholipid model membranes – effect of narcotics, Nature 208 (1965) 1295.

[15] M. Montal, P. Mueller, Formation of bimolecular membranes from lipid monolayers and a study of their electrical properties, Proc. Natl Acad. Sci. USA 69 (1972) 3561–3566.

[16] A.D. Bangham, Lipid bilayers and biomembranes, Annu. Rev. Biochem. 41 (1972) 753.

[17] P.B. Contino, C.A. Hasselbacher, J.B.A. Ross, Y. Nemerson, Use of an oriented transmembrane protein to probe the assembly of a supported phospholipid bilayer, Biophys. J. 67 (1994) 1113–1116.

[18] R. Naumann, A. Jonczyk, R. Kopp, J. van Esch, H. Ringsdorf, W. Knoll, P. Gräber, Incorporation of membrane proteins in solid-supported lipid layers, Angew. Chem. Int. Ed. Engl. 34 (1995) 2056–2058.

[19] J. Salafsky, J.T. Groves, S.G. Boxer, Architecture and function of membrane proteins in planar supported bilayers: a study with photosynthetic reaction centers, Biochemistry 35 (1996) 14773–14781.

[20] S.J. Johnson, T.M. Bayerl, D.C. McDermott, G.W. Adam, A.R. Rennie, R.K. Thomas, E. Sackmann, Structure of an adsorbed dimyristoylphosphatidylcholine bilayer measured with specular reflection of neutrons, Biophys. J. 59 (1991) 289–294.
[21] B.W. Koenig, S. Krueger, W.J. Orts, C.F. Majkrzak, N.F. Berk, J.V. Silverton, K. Gawrisch, Neutron reflectivity and atomic force microscopy studies of a lipid bilayer in water adsorbed to the surface of a silicon single crystal, Langmuir 12 (1996) 1343–1350.
[22] A. Lambacher, P. Fromherz, Fluorescence interference-contrast microscopy on oxidized silicon using a monomolecular dye layer, Appl. Phys. A 63 (1996) 207–216.
[23] E. Sackmann, M. Tanaka, Supported membranes on soft polymer cushions: fabrication, characterization and applications, Trends Biotechnol. 18 (2000) 58–64.
[24] W. Knoll, C.W. Frank, C. Heibel, R. Naumann, A. Offenhäuser, J. Rühe, E.K. Schmidt, W.W. Shen, A. Sinner, Functional tethered lipid bilayers, Rev. Mol. Biotechnol. 74 (2000) 137–158.
[25] G. Elender, E. Sackmann, Wetting and dewetting of Si/SiO_2-wafers by free and lipid-monolayer covered aqueous solutions under controlled humidity, J. Phys. II 4 (1994) 455–479.
[26] M. Tanaka, F. Rehfeldt, M.F. Schneider, G. Mathe, A. Albersdörfer, K.R. Neumaier, O. Purrucker, E. Sackmann, Wetting and dewetting of extracellular matrix and glycocalix models, J. Phys.: Condens. Matter 17 (2005) S649–S663.
[27] W.D. Comper, Extracellular Matrix, Harwood Academic Publishers, Amsterdam, 1996.
[28] H.J. Gabius, S. Gabius, Glycoscience, Chapmann & Hall, Weinheim, 1997.
[29] B.V. Derjaguin, B.N. Kabanov, T.N. Voropaye, A.S. Titiyevs, Surface forces and stability of colloids and disperse systems J. Colloid Sci. 19 (1964) 113.
[30] N.V. Churaev, B.V. Derjaguin, Inclusion of structural forces in the theory of stability of colloids and films, J. Colloid Interface Sci. 103 (1985) 542–553.
[31] J.S. Clunie, J.F. Goodman, P.C. Symons, Solvation forces in soap films, Nature 216 (1967) 1203.
[32] B.V. Derjaguin, N.V. Churaev, Surface forces, Consultants Bureau, New York, 1987.
[33] R. Lipowsky, U. Seifert, Adhesion of membranes: a theoretical perspective, Langmuir 7 (1991) 1867–1873.
[34] P.S. Swain, D. Andelman, The influence of substrate structure on membrane adhesion, Langmuir 15 (1999) 8902–8914.
[35] D. Langevin, Dynamics of surfactant layers, Curr. Opin. Colloid Interface Sci. 3 (1998) 600–607.
[36] A. Albersdörfer, T. Feder, E. Sa'ckmann Adhesion-induced domain formation by interplay of long-range repulsion and short-range attraction force: a model membrane study, Biophys. J. 73 (1997) 245–257.
[37] E. Sackmann, R.F. Bruinsma, Cell adhesion as wetting transition?, Chem. Phys. Phys. Chem. 3 (2002) 262–269.
[38] J.M. Harris, Poly(ethyleneglycol) Chemistry, Plenum Press, New York, 1992.
[39] B.J. Herren, S.G. Shafer, J. Alstine van, J.M. Harris, R.T. Snyder, Control of electroosmosis in coated quartz capillaries, J. Colloid Interface Sci. 115 (1987) 46–55.
[40] W. Nashabeh, Z. El Rassi, Capillary zone electrophoresis of proteins with hydrophilic fused-silica capillaries, J. Chromatogr. 559 (1991) 367–383.
[41] Z. Yang, J.A. Galloway, H. Yu, Protein interactions with poly(ethylene glycol) self-assembled monolayers on glass substrates: diffusion and adsorption, Langmuir 15 (1999) 8405–8411.
[42] K.L. Prime, G.M. Whitesides, Self-assembled organic monolayers – model systems for studying adsorption of proteins at surfaces, Science 252 (1991) 1164–1167.

[43] P. Harder, M. Grunze, R. Dahint, G.M. Whitesides, P.E. Laibinis, Molecular conformation in oligo(ethylene glycol)-terminated self-assembled monolayers on gold and silver surfaces determines their ability to resist protein adsorption, J. Phys. Chem. B 102 (1998) 426–436.
[44] M.C. Woodle, D.D. Lasic, Sterically stabilized liposomes, Biochim. Biophys. Acta 1113 (1992) 171–199.
[45] D. Needham, T.J. McIntosh, D.D. Lasic, Repulsive interactions and mechanical stability of polymer-grafted lipid-membranes, Biochim. Biophys. Acta 1108 (1992) 40–48.
[46] S. Löfas, B. Johnson, A novel hydrogel matrix on gold surfaces in surface plasmon resonance sensors for fast and efficient covalent immobilization of ligands, J. Chem. Soc. Chem. Commun. 21 (1990) 1526–1528.
[47] S. Löfas, M. Malmqvist, I. Ronnberg, E. Stenberg, B. Liedberg, I. Lundstrom, Bioanalysis with surface-plasmon resonance, Sens. Actuators B 5 (1991) 79–84.
[48] E. Osterberg, K. Bergstrom, K. Holmberg, T.P. Schuman, J.A. Riggs, N.L. Burns, J.M. Vanalstine, J.M. Harris, Protein-rejecting ability of surface-bound dextran in end-on and side-on configurations – comparison to PEG, J. Biomed. Mater. Res. 29 (1995) 741–747.
[49] L. Josephson, C.H. Tung, A. Moore, R. Weissleder, High-efficiency intracellular magnetic labeling with novel superparamagnetic-Tat peptide conjugates, Bioconjugate Chem. 10 (1999) 186–191.
[50] G. Elender, M. Kühner, E. Sackmann, Functionalisation of Si/SiO_2 and glass surfaces with ultrathin dextran films and deposition of lipid bilayers, Biosens. Bioelectron. 11 (1996) 565–577.
[51] T. Baumgart, A. Offenhausser, Polysaccharide-supported planar bilayer lipid model membranes, Langmuir 19 (2003) 1730–1737.
[52] M. Schaub, G. Wenz, G. Wegner, A. Stein, D. Klemm, Ultrathin films of cellulose on silicon wafers, Adv. Mater. 5 (1993) 919–922.
[53] V. Buchholz, G. Wegner, S. Stemme, L. Ödberg, Regeneration, derivatization and utilization of cellulose in ultrathin films, Adv. Mater. 8 (1996) 399–402.
[54] F. Rehfeldt, M. Tanaka, Hydration forces in ultrathin films of cellulose, Langmuir 19 (2003) 1467–1473.
[55] G. Mathe, A. Albersdörfer, R. Neumaier, E. Sackmann, Disjoining pressure and swelling dynamics of thin adsorbed polymer films under controlled hydration conditions, Langmuir 15 (1999) 8726–8735.
[56] F. Rehfeldt, L. Pagnoni, R. Jordan, M. Tanaka, Static and dynamic swelling behavior of grafted poly(2-alkyl-2-oxazoline)s, Langmuir 18 (2002) 4908–4914.
[57] S. Kobayashi, S. Iijima, T. Igarashi, T. Saegusa, Synthesis of a nonionic polymersurfactant from cyclic imino ethers by the initiator method, Macromolecules 20 (1987) 1729–1734.
[58] R. Jordan, K. Graf, H. Riegler, K.K. Unger, Polymer-supported alkyl monolayers on silica: synthesis and self-assembly of terminal functionalized poly(*N*-propionylethylenimine)s, Chem. Commun. 1996; 1025–1026.
[59] R.P. Rand, Interacting phospholipid bilayers: measured forces and induced structural changes, Annu. Rev. Biophys. Bioeng. 10 (1981) 277–314.
[60] D.C. Rau, V.A. Parsegian, Direct measurement of forces between linear polysaccharides xanthan and schizophyllan, Science 249 (1990) 1278–1281.
[61] V.A. Parsegian, N. Fuller, P. Rand, Measured work of deformation and repulsion of lecithin bilayers, Proc. Natl Acad. Sci. USA 76 (1979) 2750–2754.
[62] G. Aspinall, The Polysaccharides, Academic Press, New York, 1982.
[63] E. Kalb, S. Frey, L.K. Tamm, Formation of supported planar bilayers by fusion of vesicles to supported phospholipid monolayers, Biochim. Biophys. Acta 1103 (1992) 307–316.

[64] C. Miller, P. Cuendet, M. Grätzel, K+ sensitive bilayer supporting electrodes, J. Electroanal. Chem. 278 (1990) 175–192.
[65] O. Purrucker, H. Hillebrandt, K. Adlkofer, M. Tanaka, Deposition of highly resistive lipid bilayer on silicon–silicon dioxide electrode and incorporation of gramicidin studied by ac impedance spectroscopy, Electrochim. Acta 47 (2001) 791–798.
[66] J. Nissen, S. Gritsch, G. Wiegand, J.O. Rädler, Wetting of phospholipid membranes on hydrophilic surfaces – concepts towards self-healing membranes, Eur. Phys. J. B 10 (1999) 335–344.
[67] H. Hillebrandt, G. Wiegand, M. Tanaka, E. Sackmann, High electric resistance polymer/lipid composite films on indium–tin-oxide electrodes, Langmuir 15 (1999) 8451–8459.
[68] S. Gritsch, P. Nollert, F. Jähning, E. Sackmann, Impedance spectroscopy of porin and gramicidin pores reconstituted into supported lipid bilayers on indium–tin-oxide electrodes, Langmuir 14 (1998) 3118–3125.
[69] K. Adlkofer, M. Tanaka, H. Hillebrandt, G. Wiegand, E. Sackmann, T. Bolom, R. Deutschmann, G. Abstreiter, Electrochemical passivation of gallium arsenide surface with organic self-assembled monolayers in aqueous electrolytes, Appl. Phys. Lett. 76 (2000) 3313–3315.
[70] H. Hillebrandt, M. Tanaka, Electrochemical characterization of self-assembled alkylsiloxane monolayers on indium–tin-oxide (ITO) semiconductor electrodes, J. Phys. Chem. B 105 (2001) 4270–4276.
[71] K. Adlkofer, M. Tanaka, Stable surface coating of gallium arsenide with octadecylthiol monolayers, Langmuir 17 (2001) 4267–4273.
[72] K. Adlkofer, W. Eck, M. Grunze, M. Tanaka, Surface engineering of gallium arsenide with 4-mercaptobiphenyl monolayers, J. Phys. Chem. B 107 (2003) 587–591.
[73] H. Hillebrandt, M. Tanaka, E. Sackmann, A novel membrane charge sensor: sensitive detection of surface charge at polymer/lipid composite films on indium–tin-oxide electrodes, J. Phys. Chem. B 106 (2002) 477–488.
[74] S. Gönnenwein, M. Tanaka, B. Hu, L. Moroder, E. Sackmann, Functional incorporation of integrins into solid supported membranes on ultrathin films of cellulose – impact on adhesion, Biophys. J. 48 (2003) 646–655.
[75] E. Evans, E. Sackmann, Translational and rotational drag coefficients for a disk moving in a liquid membrane associated with a rigid substrate, J. Fluid Mech. 194 (1988) 553–561.
[76] H. Lang, C. Duschl, H. Vogel, A new class of thiolipid for the attachment of lipid bilayers on gold surfaces, Langmuir 10 (1994) 197–210.
[77] B. Raguse, V. Braach-Maksvytis, B.A. Cornell, L.G. King, P.D.J. Osman, R.J. Pace, L. Wieczorek, Tethered lipid bilayer membranes: formation and ionic reservoir characterization, Langmuir 14 (1998) 648–659.
[78] S.M. Schiller, R. Naumann, K. Lovejoy, H. Kunz, W. Knoll, Archaea analogue thiolipids for tethered bilayer lipid membranes on ultrasmooth gold surfaces, Angew. Chem. Int. Ed. Engl. 42 (2003) 208–211.
[79] M.L. Wagner, L.K. Tamm, Tethered polymer-supported planar lipid bilayers for reconstitution of integral membrane proteins: silane-polyethyleneglycol-lipid as a cushion and covalent linker, Biophys. J. 79 (2000) 1400–1414.
[80] N. Bunjes, E.K. Schmidt, A. Jonczyk, F. Rippmann, D. Beyer, H.P.G. Ringsdorf, W. Knoll, R. Naumann, Thiopeptide-supported lipid layers on solid substrates, Langmuir 13 (1997) 6188–6194.
[81] O. Purrucker, A. Förtig, R. Jordan, M. Tanaka, Supported membranes with well-defined polymer tethers – incorporation of cell receptors, Chem. Phys. Chem. 5 (2004) 327–335.
[82] M. Tanaka, S. Kaufmann, J. Nissen, M. Hochrein, Orientation selective immobilization of human erythrocyte membranes on ultrathin cellulose films, Phys. Chem. Chem. Phys. 3 (2001) 4091–4095.

[83] G. Schwoch, H. Passow, Preparation and properties of human erythrocyte ghosts, Mol. Cell. Biochem. 2 (1973) 197–217.
[84] M. Tanaka, A.P. Wong, F. Rehfeldt, M. Tutus, S. Kaufmann, Selective deposition of native cell membranes on biocompatible micro-patterns, J. Am. Chem. Soc. 126 (2004) 3257–3260.
[85] G. Decher, H.D. Hong, J. Schmitt, Buildup of ultrathin multilayer films by a self-assembly process: III. Consecutively alternating adsorption of anionic and cationic polyelectrolytes on charged surfaces, Thin Solid Films 210–211 (1992) 831–835.
[86] T. Afshar-Rad, A.I. Baily, P.F. Luckham, W. MacNaughtan, D. Chapman, Forces between proteins and model polypeptides adsorbed on mica surfaces, Biochim. Biophys. Acta 915 (1987) 101–111.
[87] F. Brochard-Wyart, P.G. de Gennes, Dynamics of partial wetting, Adv. Colloid Interface Sci. 39 (1992) 1–11.
[88] G.I. Bell, Models for the specific adhesion of cells to cells, Science 200 (1978) 618–627.
[89] R.F. Bruinsma, Physical aspects of adhesion of leukocytes, in: T. Riste, D. Sherrington (Eds.), NATO Science Series: E: Applied Sciences, Kluwer Academic Publishers, Dordrecht, 1995.
[90] E. Evans, K. Ritchie, Dynamic strength of molecular adhesion bonds, Biophys. J. 72 (1997) 1541–1555.
[91] U. Seifert, R. Lipowsky, Shape transformation of vesicle – phase-diagram for spontaneous-curvature and bilayer-coupling models, Phys. Rev. A 42 (1990) 4768–4771.
[92] S. Marx, J. Schilling, E. Sackmann, R.F. Bruinsma, Helfrich repulsion and dynamical phase separation of multicomponent lipid bilayers, Phys. Rev. Lett. 88 (2002) 138102.
[93] K. Simons, E. Ikonen, Functional rafts in cell membranes, Nature 387 (1997) 569–572.
[94] T.A. Springer, Traffic signals on endothelium for lymphocyte recirculation and leukocyte emigration, Annu. Rev. Physiol. 57 (1995) 827–872.
[95] R. Schroeder, E. London, D. Brown, Interaction between saturated acyl chains confer detergent resistance on lipids and glycosylphosphatidylinositol (GPI)-anchored proteins: GPI-anchored proteins in liposomes and cells show similar behavior, Proc. Natl Acad. Sci. USA 91 (1994) 12130–12134.
[96] S.N. Ahmed, D.A. Brown, E. London, On the origin of sphingolipid/cholesterol-rich detergent-insoluble cell membranes: physiological concentrations of cholesterol and sphingolipid induce formation of a detergent-insoluble, liquid-ordered lipid phase in model membranes, Biochemistry 36 (1997) 10944–10953.
[97] C. Dietrich, L.A. Bagatolli, Z.N. Volovyk, N.L. Thompson, M. Levi, K. Jacobson, E. Gratton, Lipid rafts reconstituted in model membranes, Biophys. J. 80 (2001) 1417–1428.
[98] C. Dietrich, Z.N. Volovyk, M. Levi, N.L. Thompson, K. Jacobson, Partitioning of Thy-1, GM1, and cross-linked phospholipid analogs into lipid rafts reconstituted in supported model membrane monolayers, Proc. Natl Acad. Sci. USA 98 (2001) 10642–10647.
[99] C. Gege, M.F. Schneider, G. Schumacher, U. Rothe, G. Bendas, M. Tanaka, R.R. Schmidt, Functional micro-domains of glycolipids with partially fluorinated membrane anchors – impact on cell adhesion, Chem. Phys. Chem. 5 (2004) 216–224.
[100] M. Stelzle, R. Mielich, E. Sackmann, 2-Dimensional microelectrophoresis in supported lipid bilayers, Biophys. J. 63 (1992) 1346–1354.
[101] J.T. Groves, S.G. Boxer, H.M. McConnell, Electric field-induced reorganization of two-component supportedbilayer membranes, Proc. Natl Acad. Sci. USA 25 (1997) 13390–13395.

[102] S. McLaughlin, The electrostatic properties of membranes, Annu. Rev. Biophys. Biophys. Chem. 18 (1989) 113–116.
[103] J.T. Groves, C. Wulfing, S.G. Boxer, Electrical manipulation of glycan phosphatidyl inositol tetheredprotein in planar supported bilayers, Biophys. J. 71 (1998) 2716–2723.
[104] D.J. Olson, J.M. Johnson, P.D. Partel, E.S.G. Shaqfeh, S.G. Boxer, G.G. Fuller, Elecrophoresis of DNA adsorbed to a cationic supported bilayer, Langmuir 17 (2001) 7396–7401.
[105] L.A. Kung, L. Kam, J.S. Hovis, S.G. Boxer, Patterning hybrid surfaces of proteins and supported lipid bilayers, Langmuir 16 (2000) 6773–6776.
[106] A.R. Sapuri, M.M. Baksh, J.T. Groves, Electrostatically targeted intermembrane lipid exchange with micropatterned supported membranes, Langmuir 19 (2003) 1606–1610.
[107] K. Adlkofer, E.F. Duijs, F. Findeis, M. Bichler, A. Zrenner, E. Sackmann, G. Abstreiter, M. Tanaka, Enhancement of photoluminescence from near-surface quantum dots by suppression of surface state density, Phys. Chem. Chem. Phys. 4 (2002) 785–790.
[108] G. Steinhoff, O. Purrucker, M. Tanaka, M. Stutzmann, M. Eickhoff, $Al_xGa_{1-x}N$ – a new material system for biosensors, Adv. Funct. Mater. 13 (2003) 841–846.
[109] S. Luber, A. Adlkofer, U. Rant, A. Ulman, A. Gölzhäuser, M. Grunze, D. Schuh, M. Tanaka, M. Tornow, G. Abstreiter, Liquid phase sensors based on chemically functionalized GaAs/AlGaAs heterostructures, Physica E 21 (2004) 1111–1115.

CHAPTER 5

The Control of Membrane Properties by Synthetic Polymers

Nickolay Melik-Nubarov[1,*] and Oxana Krylova[2]

[1]*Chemistry Department, Lomonosov Moscow State University, Vorobiovy Gory, 119992 Moscow, Russia*
[2]*Membrane Transport Group, Institute of Biophysics, Johannes Kepler University, Linz, Altenberger str. 69, A-4040 Linz, Austria*

Contents

Abstract

This review describes the effects caused by synthetic polymers on model lipid membranes. Among a large variety of synthetic polymers, we focused on two quite different groups. One of them consists of polyelectrolytes, whose adsorption on lipid membranes is driven by Coulomb forces, and the other group consisting of Pluronics

---

* Corresponding author. E-mail: melik.nubarov@genebee.msu.ru

ADVANCES IN PLANAR LIPID BILAYERS AND LIPOSOMES, VOLUME 2
ISSN 1554-4516 DOI: 10.1016/S1554-4516(05)02005-3

and polyethoxylated surfactants that have amphiphilic structure and interact with lipid bilayers by hydrophobic forces. This choice was determined by the continuously increasing interest to the application of these polymers in biomedical research. It was shown that interactions of cationic polymers with oppositely charged lipid bilayers result in the formation of lipid domains in the membranes, influence the lateral diffusion of lipid molecules and proteins incorporated into the bilayer, modulate functional activity of membrane proteins and in some cases even favor asymmetrical distribution of lipids. Incorptions of amphiphilic copolymers into lipid bilayers may result in changes in the rate of flip-flop and membranes microviscosity. Polymer-membrane interactions in both cases cause changes in the membrane premeability, obviously resulting from formation of disturbances in the membrane packing. In case of Pluronics, non-covalent interactions between the membrane-bound polymer and the permeant may also contribute to the transportation of the latter through lipid bilayer. Special attention in this review is paid to the relatioship between the polymer structure and its ability to modulate membrane properties. A qualitative thermodynamic treatment of the polymer-membrane interactions allowed to explain the polymer effects on the bilayer structure and led to a list of properties that the polymer structure must possess in order to affect membrane permeability.

1. INTRODUCTION

The ability of chain molecules to form multiple bonds with surfaces and low molecular weight substances determines unrelenting interest to synthetic polymers in various fields of science and engineering. Model bilayer membranes and the membranes of living cells also may be substrates for polymer adsorption and targets for their action [1–2]. Therefore, the ability of synthetic polymers to interact with lipids and proteins in the membranes and to modulate their properties offers a challenge for their applications in medicine and fundamental biomembranes research [3–6].

One of the first and most important medical applications of synthetic polymers to date is their use as carriers for drug delivery and excipients for drug formulations. In this case the polymer should stabilize the drug in the bloodstream and prevent its non-specific uptake by reticulo-endothelial system. For this, hydrophilic and inert polymers such as polyethylene oxide (PEO), polyvinylpyrrolidone, and polyacrylamides are widely used and the corresponding data have been discussed in a number of recent reviews [7–10].

The present chapter is focused on other types of water-soluble polymers that are capable of interacting with biological membranes and modulating their properties, i.e., permeability, lateral organization of lipids and proteins, or functional activity of membrane enzymes. Such 'membrane active' polymers may be roughly divided into two groups depending on the nature of intermolecular forces that cause their interactions with lipid bilayers: polyelectrolytes that bind to the membranes *via* electrostatic forces and amphiphilic polymers whose adsorption onto lipid membranes is hydrophobically driven.

In the first group, polycations are of the main interest because cell membranes usually have a slight negative charge due to admixture of anionic lipids: phosphatidylserine (PS^-), phosphatidic acid (PA^-), phospatidylinositol biphosphate (PIP_2^{5-}) and cardiolipin (CL^{2-}), and therefore the polycations can effectively bind to them. Well-known applications of cationic polymers in medicine include stimulation of antigen-specific immune response [6] and facilitation of gene and oligonucleotides permeation into living cells [4–5].

In the second group, the major attention is focused on amphiphilic polyalkylene oxides, and, among them, on the three-block PEO and PPO copolymers of ABA type (Pluronics), which exhibit extremely low toxicity both *in vitro* and *in vivo*, and therefore are widely employed in cosmetics, pharmacy, and medicine as biologically tolerant emulsifiers [11]. Recently, it has been found that Pluronics can favor drug accumulation in tumor cells exhibiting a multiple resistance to chemotherapeutics [3,12]. This property of tumors frequently arises as a result of chemotherapy. In many cases, it is caused by integral membrane proteins that are responsible for the ATP-dependent efflux of drugs from cell cytoplasm. Pluronics can inhibit at least two of these proteins: P-glycoprotein and a multi-drug resistance associated protein, MRP1 [3]. It was suggested that this effect is not due to the direct interaction of Pluronics with these proteins, but is rather caused either by the polymer interaction with lipids surrounding the proteins or ATP depletion of the cells [3,12].

The molecular reasons for the effects caused by these polymers in biological membranes remain unknown. To shed light on polymer-induced effects, artificial lipid bilayers are frequently used as model biological membranes. Apart from elucidation of the molecular mechanisms of polymer–membranes interaction, this approach may help to formulate structural requirements for the polymer that determine its ability to interact with membranes and cause the desired effects. The latter may be applied for the design of novel polymers exhibiting high membrane activity.

2. POLYELECTROLYTES

2.1. Effect of polyelectrolytes on the membrane structure

Adsorption of polymers on artificial membranes composed of charged and zwitterionic lipids was studied by various techniques including Raman spectroscopy [13], differential scanning calorimetry [14], fluorescence spectroscopy [15,16], ^{31}P-NMR and ^{2}H-NMR technique [13,17–19], and even direct visualization using fluorescent microscopy [20,21]. Cationic polymers interact electrostatically with oppositely charged groups of the lipid bilayer, this interaction being strongly favored by the release of a large quantity of small counterions that are condensed in the vicinity of the macromolecule and lipid bilayer [22]. Ejecting of these counterions into the bulk is accompanied by a huge increase in the entropy of the system that favors the formation of multiple electrostatic interactions of the polymer with the lipid bilayer [23]. Importantly, the ability of lipid molecules to move within the outer monolayer of the membrane results in the formation of the maximal amount of electrostatic contacts of the polymer and the charged lipid component [24]. Zwitter-ionic lipids withdraw from such complexes and give place to the membrane components capable of electrostatic interactions with a polyelectrolyte. These phenomena result in the formation of domains in lipid membranes that are enriched with the charged lipid component [13–22,24]. Addition of polyelectrolytes to the oppositely charged lipid membranes results in nearly complete and very rapid adsorption of the polymers accompanied by strong aggregation of lipid vesicles [14,15]. Increase in the ionic strength and reduction of the net charge of the membrane or the polymer results in the weakening of the polyelectrolyte adsorption and destabilization of the domains [15].

Although the main features of the polyelectrolyte adsorption on lipid membranes are similar for all known systems, the composition of the formed complexes and their structure strongly depend on the nature of both components. Therefore, the following part will present an overview of the interrelation between the physico-chemical properties of the interacting species and the structure of domains formed in lipid membranes.

2.1.1. Structural rearrangements caused by polyelectrolytes added to liposomes during their preparation

Structural changes in the polyelectrolyte and the membrane after their interaction were studied in detail using multilamellar vesicles (MLVs) and polylysine (PLL). In these experiments, the lipid film was hydrated in PLL solution, so the MLVs contained the same polymer solution inside and outside of the lumens. This allowed the polymer to interact with both leaflets of the bilayer. Changes in PLL state were monitored by vibrational (IR and Raman) spectroscopy [13]. It was shown that the adsorption changed PLL conformation from random coil to a highly ordered secondary structure. The type of structure depended on the nature of the anionic lipid present in the bilayer. If the membrane negative charge was caused by PS^-, PLL formed an α-helix, while adsorption on the bilayers containing PG^{2-} was accompanied by preferred formation of β-sheet structures in PLL molecules [17].

The formation of domains may be formally considered as a three-stage process. Firstly, the polymer adsorbs due to electrostatic interactions and captures the accessible charged lipids on the membrane surface. Secondly, rapid lateral movements of the membrane components lead to the formation of clusters enriched with charged lipids and proteins near the polymer, the neutral component being partially ejected. On the third stage, the clusters fuse to form domains stabilized by the interaction between several polyelectrolyte chains [19,20].

Similar changes in the bilayer structure are induced by different polyelectrolytes: polystyrene sulfonate (PSS) [19], DNA and oligonucleotides, 25-mer oligopeptide MARCKS (151–175) [24], melittin [25] and, myelin [26]. Detailed overviews of these effects have been presented in the recent articles of Macdonald [19] and Glaser [20].

The compactness of domains highly depends on the structure of the polymer and its chain length (degree of polymerization – DP). PLL with a high DP (100) had a tendency to form more loose domains with the negatively charged PS^- than PLL with DP=30. At the same time, an increase in the chain length of sodium polystyrenesulfonate (PSS) increased the compactness of domains formed by the cationic amphiphile 1,2-dioleoyl-3-(dimethylamino)propane (DODAP) in lipid bilayers [19]. The authors explained the difference between the behavior of PLL and PSS by the fact that the lipid domains contain multiple polymer chains and their compactness depends on interpolymer interactions. In case of PSS, the domains are also stabilized by the π–π interaction of aromatic rings. Therefore, the domains formed by PSS of higher DP are more compact. On the contrary, in the case of PLL, interchain attraction is obviously driven by the dipole interactions between the secondary structures of the polymer. The interactions of α-helix macrodipoles favor either head-to-tail or side-by-side antiparallel arrangements [17]. Experimental studies of the statistics of cylinder stacking in two dimensions show that the increasing cylinder length loosens the domain structure [17]. Thus, longer PLL chains would tend to form less compact domains that could entrap more of the zwitter-ionic component POPC. In this case, a lesser enrichment of the PLL-bound domains with $POPS^-$ would occur [17,27]. This finding is important for understanding of the role of polymer chain length in the structure formation of polymer domains. This example shows that the properties of domains formed by various polyelectrolytes in lipid membranes strongly depend on the polymer structure.

Interaction of DNA with cationic liposomes gives another important example of domain formation by biopolyelectrolytes. Cationic lipids do not exist in nature, but the addition of DNA to vesicles containing synthetic lipid-like surfactants resulted in the formation of large domains detected by ^{2}H-NMR spectroscopy [27]. It was shown that the domain compactness, measured as an amphiphile/DNA ratio in domains, was reduced in the following series: one-tailed cetyltrimethylammonium bromide (CTAB) > two-tailed dioleoyldimethylaminopropane (DODAP) > 3-[*N*-(N′,N′-dimethylaminoethane)carbamoyl]-cholesterol. So, the ability of amphiphiles to form compact domains was reduced with the increase of the size of the hydrophobic part. The cholesterol-based cationic amphiphile was able to form domains containing only 70% of this substance, while CTAB formed domains consisting of more than 90% of the cationic component [27].

2.1.2. *Effects caused by polyelectrolytes in the preformed LUVs and SUVs*

The addition of PLL to the preformed small unilamellar lipid vesicles (SUVs) or large vesicles (LUVs) led to the formation of domains (demonstrated by DSC) and immediate aggregation of vesicles into micron-sized particles [14,28]. The addition of an excessive amount of PLL did not disaggregate them [28]. It may be suggested that the inability of PLL to undergo redistribution is due to its low flexibility. This assumption arises from the experiments performed with a more flexible polycation poly-(*N*-ethyl-4-vinyl pyridinium bromide) (PEVP). When added in the stoichiometric amount to negatively charged SUVs containing CL^{2-}, PEVP forms domains and aggregates SUVs just like PLL. However, the addition of excess of PEVP resulted in its complete adsorption, loosening of the domains, acquisition of a positive charge by the vesicles, and their disaggregation [29]. It was hypothesized that the ability of PEVP to change the sign of the surface charge of the liposomes was relevant to high flexibility of this polymer, its capability to form loops and tails on the surface of the vesicles [28,29].

High flexibility of PEVP has another important consequence. As it was shown by the laser microelectrophoresis technique, PEVP adsorption on the preformed EL/CL^{2-} SUVs in the amount equivalent to the charges of CL^{2-} located on the outer membrane leaflet did not neutralize the vesicles, so they remained negatively charged [29]. Only the addition of PEVP in the amount equivalent to all negatively charged species located on both outer and inner leaflets of the membrane completely neutralized the vesicles, indicating that PEVP causes transmembrane migration of anionic lipids to the outer membrane leaflet [29]. It has been hypothesized that the flexibility of PEVP is responsible for formation of membrane defects that facilitate transbilayer redistribution of lipids, but not the leakage of low molecular dye or NaCl from the liposomes [29].

In contrast, the interaction of relatively rigid PLL with the preformed EL/CL^{2-} SUVs and LUVs did not cause the transbilayer migration of negatively charged CL^{2-} [28].

When the linear charge density of PEVP was reduced threefold by partial substitution of *N*-ethyl-pyridinium groups for uncharged pyridines, the resulting polycation exhibited a pronounced hydrophobicity and was unable to induce transmembrane migration of CL^{2-} [30].

These facts allow to conclude that the flexibility of the polycation and its linear charge density are the main factors that determine the compactness of the domains and influence the structure of the bilayer.

DNA can adsorb effectively on preformed cationic vesicles [31]. Labeling the fatty acid of DODAP by DPH or BODIPY and DNA by ethydium bromide or BOBO intercalators allowed the evaluation of the average distance between the groove region of the DNA and the middle of the lipid bilayer using fluorescence energy transfer technique. This distance was found to be about 27 Å, indicating that the DNA double helix merely adsorbs on the bilayer surface and does not considerably penetrate into the membrane [31].

Reversibility of polyelectrolyte adsorption depends upon the size of the liposomes. Thus, the addition of PLL to SUVs or LUVs resulted in a rapid adsorption of the polymer observed by quenching of membrane probe fluorescence. The subsequent addition of 0.25 M NaCl or polyacrylic acid resulted in complete recovery of the SUVs fluorescence, while the fluorescence of LUVs remained practically unchanged [28]. It may be supposed that flexible membranes of LUVs could bend and envelop the polycation chain, thus making the complex insensitive to the composition of the surrounding solution. So, PLL desorption from the negatively charged liposomes was sensitive to their elasticity.

2.1.3. Structural rearrangements caused by polyelectrolytes adsorption on GUVs

The next step in modeling polymer interactions with biological membranes is the study of giant unilamellar vesicles (5–30 μm in size) (GUVs). It is possible to follow such interaction by means of optical and fluorescent microscopy [32]. It was found that PLL, cationic peptide poly-L-arginine (M_W 10,800), but not neutral poly(vinylpirrolidone), induced formation of dark spots on vesicle walls, which after several minutes grew into an internal linear structure visible by optical microscopy, henceforth referred to as 'ropes' (Fig. 1). The anionic lipid and the polymer are the major components of the ropes, neutral lipid being also included. No ropes formed with vesicles lacking the anionic lipid [33].

Importantly, the formation of the ropes was observed for PLLs with DP from 5 up to 140, while a longer PLL with DP = 1100 was unable to form ropes. This observation can be explained by the above-mentioned results of Macdonald *et al.* who showed that PLL with DP = 100 formed more loose domains in MLVs than

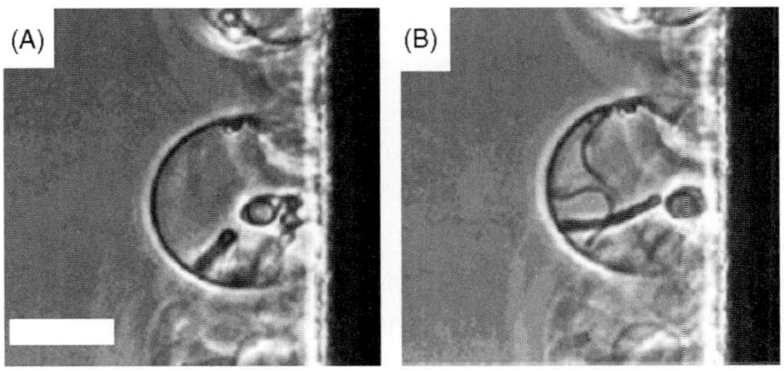

Fig. 1. Rope formation in an anionic vesicle after 30 min (A) and 60 min (B) following the addition of poly-L-lysine, M_W 29,300, to a giant vesicle composed of POPC and POPG (90:10 mol%). Bar 20 μ. According to Ref. [42].

the shorter polymer with DP = 30. One could expect that longer PLL with DP = 1100 would form even more loose domains. The membranes of GUVs are known for their extremely high elasticity. So, it may be supposed that in order to retain this state, compact and rigid domains are withdrawn from the membrane and form separate particles inside the vesicle. It may be supposed that loose domains formed by PLL with long chains do not alter the membrane elasticity significantly, and so their removal is not necessary.

Similarly, the adsorption of DNA on GUVs resulted in the formation of patches that were transformed into 'endosome-like' structures inside large GUVs [34]. This observation may be relevant to mechanisms of transfection mediated by cationic lipids or polycations.

The formation of domains in GUVs containing PIP_2 was also observed by confocal fluorescence microscopy after addition of 25-mer peptide of protein MARCKS enriched with Lys and Arg [24]. It was supposed that MARCKS induces a similar effect in cells. If so, the domain formation may be responsible for a decrease of the phospholipase C-catalyzed hydrolysis of PIP_2, which is observed in cells in the presence of MARCKS peptide. Phosphorylation of MARCKS by protein kinase C or its binding to Ca^{2+}-calmodulin results in the desorption of the protein from the membrane and restoration of the activity of phospholipase C [24]. This example demonstrates a possible role that the electrostatically driven domains play in the regulation of cell metabolism.

2.1.4. Interaction of polyelectrolytes with protein-containing lipid bilayers

The net charge of living cells is determined not only by lipids, but also by proteins present in the membranes. What are the consequences of the interaction of polycations with proteo-lipid membranes? To investigate this issue, the effect of polyelectrolytes on the functional properties of gramicidin A modified with ionizable groups on the C-terminus was studied. It is known that gramicidin forms tail-to-tail dimers in biological membranes that can conduct small inorganic ions (Na^+, K^+, Cl^-) [35]. So, the incorporation of gramicidin into membranes increases their conductivity. Photosensibilized oxidation of tryptophan residues in gramicidin leads to dissociation of gramicidin channels manifested in monoexponential decay of the transmembrane current [36]. The exponential factor of this process (τ, the characteristic time of photoinactivation) was shown to correlate well with the single-channel lifetime.

It was found that the addition of PLL to the planar bilayer lipid membranes containing unmodified neutral gramicidin A had no effect on its channel properties [37]. However, the addition of PLL to the membranes containing negatively charged O-pyromellityl-gramicidin (OPg) resulted in an increase in the membrane conductivity and a drastic deceleration of photoinactivation of gramicidin channels [37]. This effect gradually increased with the polymer concentration, reached maximum and then dropped to the initial level upon addition of a large (10–100-fold) excess of PLL. The authors supposed that the stabilization of the dimeric state of gramicidin channels was due to their clustering by oriented PLL α-helices and their dipole–dipole interactions [38]. The addition of an excess of PLL resulted in decomposition of domains and destabilization of OPg dimers.

The effect of polyelectrolyte structure on its ability to induce clustering of gramicidin channels in BLM was studied in another system, where gramicidin D was modified with a cationic group, tris(2-aminoethyl)amine (gram-TAEA). Similar to OPg, addition of gram-TAEA to BLM increased membrane conductance, and the channels underwent photoinactivation in the presence of the sensitizer [39]. Unexpectedly, addition of polyanions – polyacylate, polyglutamate, single-stranded DNA, double-stranded DNA or RNA of tobacco mosaic virus to this system had no effect on these processes. At the same time, a hydrophobically modified polyacrylate or a triple copolymer of styrene, maleic acid, and methacrylic acid (3:2:1) (König polyanion [40]) increased the membrane conductance and stabilized gram-TAEA channels. The effect of König polyanion was abolished if the gram-TAEA membrane was pretreated with a polyacrylate or DNA. It means that the latter did bind to gram-TAEA containing BLM but were unable to form domains [39]. At the same time, DNA has a strong ability to form compact domains containing low molecular weight cationic surfactants [31]. The reasons for this discrepancy remain unclear.

The formation of domains from protein molecules incorporated into lipid membrane was directly demonstrated in another model system based on proteoliposomes. Stearoylated α-chymotrypsin (sCT) was added to the preformed lipid vesicles composed from zwitter-ionic EL, and sonicated to enable protein incorporation into the liposome membrane [41,42]. These liposomes were thoroughly purified from the unbound enzyme, then mixed with polycation PEVP and treated with glutaraldehyde. SDS electrophoretic analysis of the proteoliposomes revealed the presence of large protein aggregates with molecular weight about 300 kDa (Fig. 2, lane 2), i.e., containing tens of protein molecules [42]. In the absence of PEVP, no changes of the protein molecular weight in the proteoliposomes were detected (Fig. 2, lane 1). So, the polycation adsorption on lipid membranes results in the clustering of not only lipids but also of protein molecules possessing a negative charge.

Physiological role of the formation of polycation-induced domains and transmembrane movement of lipid molecules was studied in another system resembling a biological membrane: inside-out oriented vesicles (IOVs) prepared from erythrocyte plasma membranes. The outer side of the vesicles was labeled with zwitter-ionic lipids NBD-PC or NBD-SM and their migration to the internal side of the membrane was studied. Various polycations (10-mer PLL,

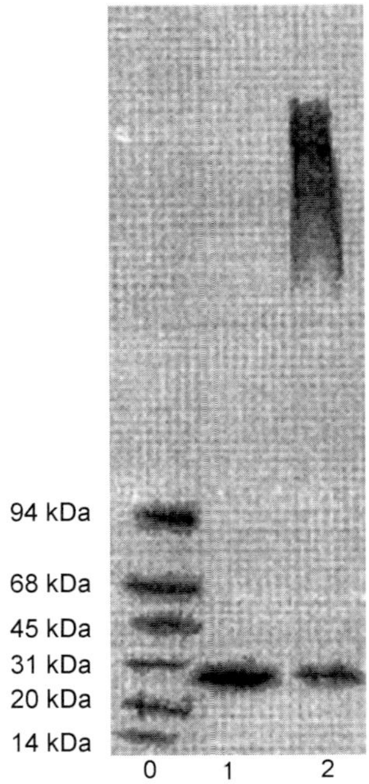

Fig. 2. Electrophoretic analysis of EL/sCT proteoliposomes (lipid/protein molar ratio 300:1) before (lane 1) and after incubation with PEVP (lane 2). The protein aggregates were fixed with glutaraldehyde prior to electrophoretic separation Lane 0 – protein standards. According to Ref. [42].

spermine, MARCKS (151–175) peptide, and neomycine) stimulated the inside transfer of zwitter-ionic lipids and the reverse transport of anionic PS^- and PIP_2^{3-} to the external leaflet [43]. The ability of the polycations to activate scrambling processes in cellular membranes should be taken into consideration in biological studies.

The data described above indicate that a polycation can bind to anionic lipids and proteins as well. Therefore, it should choose between protein anionic groups and anionic lipids while interacting with any protein-containing membrane. If the polycation can choose the target, what kind of species should it prefer? Investigation of this question in living cells faces a number of difficulties due to high complexity and undefined composition of this natural system. So, this question was studied using proteoliposomes containing either CL^{2-} and sCT, or sCT alone. Both samples were incubated with PEVP and the portion of the polymer bound to the enzyme was measured by quenching of sCT intrinsic fluorescence (Fig. 3). It turned out that the apparent dissociation constant of PEVP-CL^{2-} complex was 2–2.5 times lower than that of PEVP-sCT [42]. These data permitted to conclude that PEVP had a somewhat higher affinity

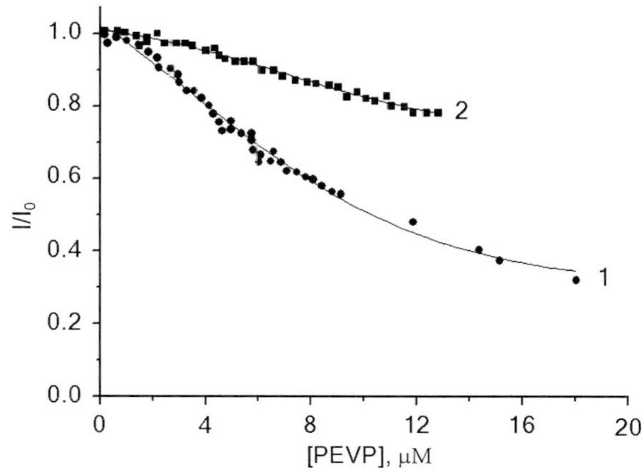

Fig. 3. Effect of PEVP on relative fluorescence intensity of proteoliposomes EL/sCT (1) and the mixture of EL/sCT and EL/CL^{2-} liposomes (2). [sCT negative groups]=2.1 μM, [CL^{2-} negative groups]=2.8 μM, [EL] =1 mg/ml. 10 mM borate, pH 8.5 According to Ref. [42].

to lipid molecules than to anionic proteins. It is apparently due to a greater ability of lipids to move within the bilayer and form more tight ionic pairs with a polycation, while the charged groups of globular proteins are strictly fixed and therefore form weaker complexes with polycations.

2.2. Effect of polycations on the permeability of model lipid membranes

2.2.1. *Permeability towards small ions*

There is a large body of evidence in the literature indicating that interactions of polycations with cells are accompanied by a leakage of K^+ from cells to the surrounding medium and a simultaneous influx of Ca^{2+} ions [44–46]. The question arises, is it the influence of polycations on the packing of lipids and proteins, leading to a formation of transient pores in cell membrane, and/or their effect on the functioning of ion channels and transport systems?

To study the permeability of lipid and proteo-lipid membranes towards small ions, the liposomes are loaded with concentrated NaCl or KCl solutions and salt leakage was followed by conductivity measurements [15]. An application of this approach to negatively charged EL/CL^{2-} liposomes showed that neither PEVP [15,29] nor PLL [28] were able to cause salt leakage from SUVs or LUVs if the osmotic pressure of the external solution was compensated by addition of appropriate amount of sucrose. The incorporation of sCT into the vesicle membrane did not alter its sensitivity towards polycation adsorption [42]. This result indicates that such proteoliposomes do not adequately model the increase in ionic permeability observed upon interaction of polycations with cell membranes [46].

At the same time, when LUVs (but not SUVs) were loaded with 0.5 M NaCl and the osmotic pressure of the surrounding solution was not compensated, addition of PLL-induced leakage of NaCl. This fact shows that structural rearrangements in the membranes of large vesicles caused by PLL adsorption, further aggravated by the osmotic pressure gradient, may destroy liposome membrane [47].

2.2.2. Permeation of weak electrolytes through lipid membranes

Weak acids or bases in neutral form can permeate biological membranes by partition–diffusion mechanism, which goes back to the Overtone's rules formulated nearly a century ago [48]. In most cases, the rate-limiting step of this process is the diffusion of the permeant across the hydrophobic region of the bilayer [49]. Therefore, this process should be sensitive to the disturbances in the membrane packing.

To check this point, a model system based on the pH-induced accumulation of anti-tumor drug doxorubicin (DOX) was used [50]. Briefly, liposome suspensions were prepared with neutral aqueous buffer solution outside (pH 7) and acidic one inside (pH 4–5). The addition of a strong fluorophore DOX to these suspensions resulted in pH-gradient-induced uptake of non-charged DOX molecules into the liposomes and their protonation inside. When DOX concentration inside the liposomes exceeded the self-quenching threshold, the fluorescence intensity in the system decreased. This process followed the first order law and reflected the kinetics of transmembrane DOX permeation (Fig. 4). So, this system provides an easy way of testing the membrane barrier properties.

The effect of the polymer on membrane permeability may be expressed as a ratio of the rate constants of DOX transport in the presence of polycation (k_P) and in its absence (k_0). The data presented on Fig. 5 show that although both PEVP (Fig. 5A) [42] and PLL (Fig. 5B) [28] increased membrane permeability, they did it in a different manner.

The effect of PEVP gradually increased until the polycation/anionic lipid charge ratio ($Z_{\pm}$) achieved unity. Further increase in $Z_{\pm}$ ratio resulted in a decrease of the polycation effect (Fig. 5A). Similar dependencies were observed for SUVs whose negative charge was caused either by CL^{2-} (curve 1) or by ganglioside GM1 (curve 2). The maximal effect of PEVP on CL^{2-}-liposomes was observed when all cardiolipin molecules migrated to the outer leaflet of the membrane. It seems unlikely that PEVP adsorption would stimulate the transmembrane migration of such bulky molecules as gangliosides. On the other hand, these glycolipids exhibit strong preference to the outward membrane leaflet in living

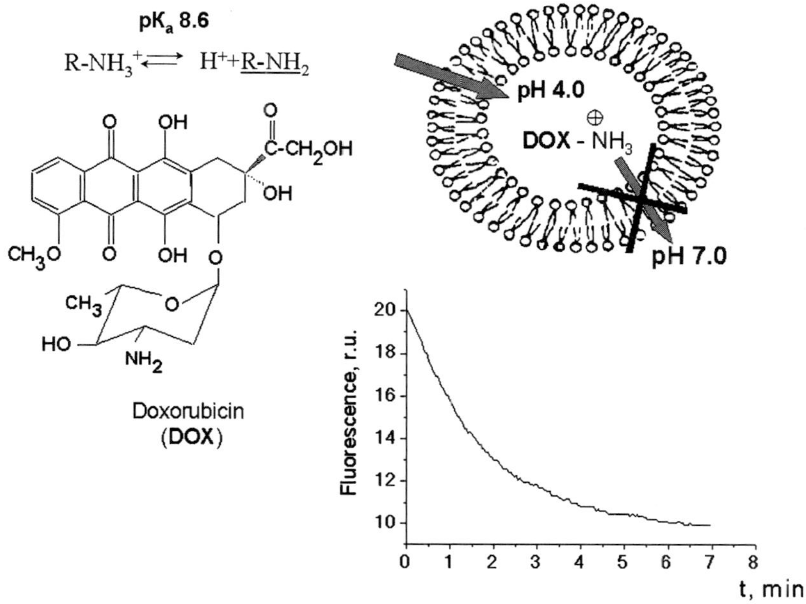

Fig. 4. pH-induced accumulation of DOX in liposomes followed by fluorescence decay.

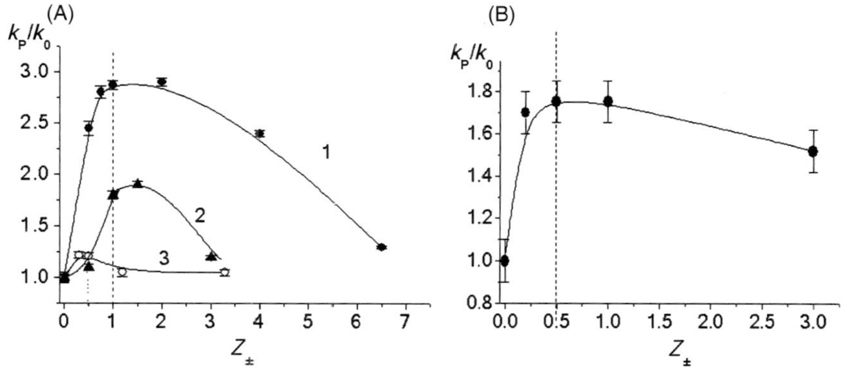

Fig. 5. Effect of PEVP (A) or PLL (B) on the rate of pH-induced accumulation of DOX in EL liposomes as a function of polymer/membrane charge ratio ($Z_{\pm}$). The liposomes contained either CL^{2-} (1 in (A) and (B)), or ganglioside (2), or hydrophobized α-chymotrypsin (3). Mole fraction of anionic groups (ν) was 0.2 for both CL^{2-} and ganglioside, and 0.08 for sCT. In all cases, the external solution contained 20 mM Tris-citrate, pH 7.5, and 0.3 M sucrose, the internal buffer solution contained 0.3 M Tris-citrate, pH 5.0. Temperature 25 °C. According to data published in Ref. [28] and [42].

cells. If the same situation is realized in SUVs, the result shown in Fig. 5A, curve 2 seems quite natural. The polycation effect on the membrane permeability achieves maximal value after addition of PEVP in the amount stoichiometric to all GM1 anionic groups present in the system, since these molecules with bulky carbohydrate chains are mainly located in the outward membrane leaflet. Nevertheless, this explanation is not more than a hypothesis and requires further investigation.

In case of proteoliposomes containing sCT, the maximal effect was induced at $Z_{\pm}=0.5$ (Fig. 5A, curve 3). It is reasonable to suppose that molecules of sCT located on the inward leaflet of the membrane remain inaccessible for PEVP added to the surrounding solution, thus the maximal adsorption of the polycation and its effect are observed at $Z_{\pm}\sim0.5$. So, the maximal effect of PEVP occurred when it was added in the amounts stoichiometric to all negatively charged groups present on the outer surface of the vesicles. The observed regularities of the influence of PEVP on membrane permeability are consistent with its previously discussed ability to form domains [15,29] (see Section 2.1).

The adsorption of PLL on EL/CL^{2-} SUVs achieved saturation at $Z_{\pm}=0.5$, and further increase in the concentration of PLL did not result in the appearance of positive charge on lipid vesicles and loosening of CL^{2-} domains in SUVs and LUVs [28,47]. At the same time, when PLL was added at $Z_{\pm}>0.5$, the membrane permeability was close to that observed at $Z_{\pm}=0.5$, so no reliable maximum for PLL effect on EL/CL^{2-} liposome permeability was observed. (Fig. 5B).

All these data indicate a distinct correlation between polycation-induced formation of domains and DOX permeation. This holds true for quite diverse anionic constituents of a membrane.

The effect of PLL on the rate of DOX permeation was strongly dependent on the polycation molecular weight (Fig. 6). Increasing of PLL length from 5 to 100 resulted in the gradual increase of its ability to accelerate DOX permeation. The PLLs with higher polymerization degrees (about 250, 500, and 1000) were less effective than PLL_{100} [17].

Such non-monotonic dependence of PLL effect on its polymerization degree agrees with the previously reported dependence of the domains compactness on PLL chain length [17]. It is reasonable to suppose that long-chained PLLs exhibit lower ability to form compact domains and therefore exert only small effects on the rate of DOX permeation.

Possible structural and biological consequences of the effect of clustering are broadly discussed in the literature [19–21]. In particular, these clusters themselves and their boundary composed of free lipids and the surrounding neighboring space composed of disturbed uncharged lipids are usually considered membrane 'defects'. The formation of clusters composed of anionic proteins and negatively charged lipids could be a reason for the increase in the membrane microviscosity

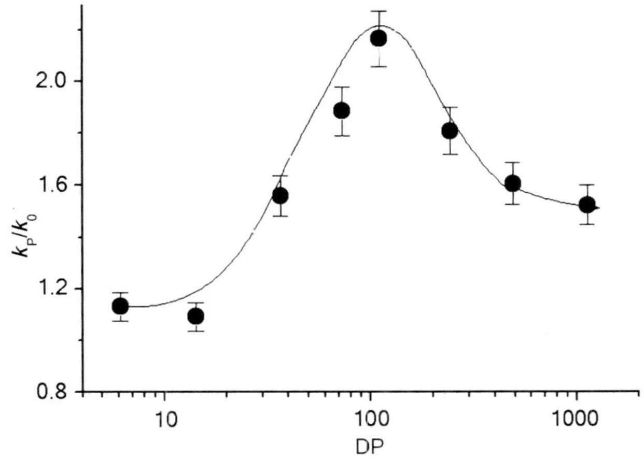

Fig. 6. Effect of PLL chain length on its ability to accelerate DOX permeation through EL/CL^{2-} liposome membrane (ν=0.25). Liposome concentration 1 mg/ml, $Z_{\pm}$=0.5, [DOX]=50 μM, 18 °C.

induced by polycation binding [51]. The appearance of defects in the cell membrane is regarded as an important event in cell and lipid vesicles fusion [45, 51]. Along this line, we also suggest that the boundaries of polycation-induced protein and lipid clusters could be responsible for the increase in the membrane permeability.

The increase in the liposome net charge achieved by a variation of the CL^{2-} content (ν) resulted in a decrease in the rate of DOX permeation in the absence of the polycation (k_0) and acceleration of DOX permeation induced by PLL (k_P) (Fig. 7A). The reduction in k_0 was consistent with the previously reported data [52]. It was likely due to a strong binding of positive DOX molecules to the outer leaflet of the negatively charged vesicular membrane. Such binding obviously shifted the equilibrium (1)

$$\mathrm{DOX-NH_2 + H^+ \rightleftarrows DOX-NH_3^+} \tag{1}$$

to the protonated form of DOX, thus decreasing the concentration of the uncharged form and decelerating its transmembrane permeation. The addition of cationic PLL to the suspension of pH-gradient EL/CL^{2-} SUVs containing protonated DOX was accompanied by formation of a strong PLL–CL^{2-} interfacial complex and release of positive DOX molecules from the vesicle surface to the solution. This, in turn, led to an increase in concentration of uncharged DOX according to equilibrium (1) and a consecutive rise of k_p value. Another possible reason for k_p increase was probably related to the mentioned PLL-induced lateral lipid segregation in the membrane and formation of domains composed of CL^{2-} species neutralized by bound PLL chains. Such patches and/or boundaries between them and

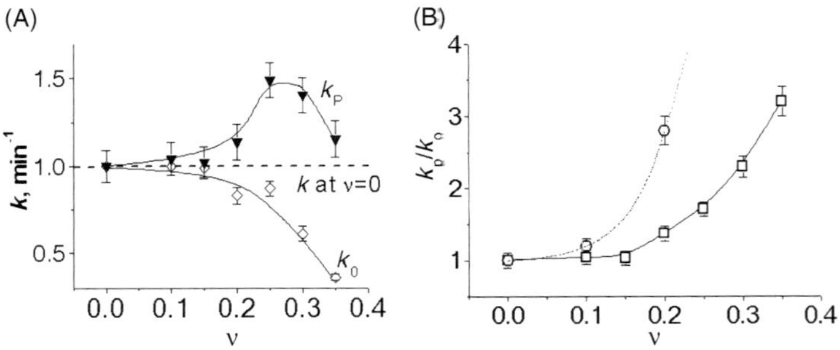

Fig. 7. (A) Effect of liposome net charge ν on the first order rate constants of DOX permeation into pH-gradient vesicles in the absence of PLL (k_0, curve 1) and in the presence of PLL ($Z_\pm = 0.5$, M_w 30,000) (k_P, curve 2); (B) dependence of k_P/k_0 on ν for PLL at $Z_\pm = 0.5$ (solid line) and PEVP (dashed line) as a function of ν. Other experimental conditions are the same as indicated in the legend to Fig. 5. According to data published in Ref. [28] and [42].

the surrounding lipid bilayer could be responsible for acceleration of transmembrane DOX permeation.

The overall effect of ν value on the PLL-mediated DOX permeation, expressed as k_p/k_0 ratio, is shown in the Fig. 7B. It is seen that the increase of ν from 0 to 0.3 resulted in nearly a 2.5-fold rise in the PLL effect on transmembrane DOX permeation. The effect of another polycation, PEVP, also increased with the growth of liposomes net charge [42]. A reason for the steeper growth of PEVP-induced acceleration of DOX permeation may be due to the fact that this polycation-induced transmembrane migration of CL^{2-} molecules in the liposomes [15,29], the effect not observed for PLL [28]. This indicates that PEVP causes more disturbances in the membrane in comparison with PLL.

Direct measurements of the vesicles zeta-potential (data not shown) demonstrated that at $Z_\pm = 1$, the inner leaflet becomes electrically neutral and does not obstruct desorption of protonated DOX molecules inside the vesicles [42]. In contrast to this, PLL interacts only with the outer CL^{2-} molecules of EL/CL^{2-} SUVs, neutralizing their charges [28]. So, the inner membrane leaflet retains negative charge and can bind protonated DOX molecules electrostatically after their translocation through the vesicular membrane. Such binding apparently decelerates desorption of DOX from the inner membrane leaflet and, lastly, results in a decrease in k_p and k_p/k_0 values.

By increasing concentration of a simple salt in the surrounding solution, the PLL effect on DOX permeation through the membrane of EL/CL^{2-} SUVs could be completely (but reversibly) suppressed (Fig. 8). However, low NaCl concentrations that strongly reduced PLL effect on DOX permeation were not sufficient for complete desorption of the polycation (Fig. 8, dashed curve) [28].

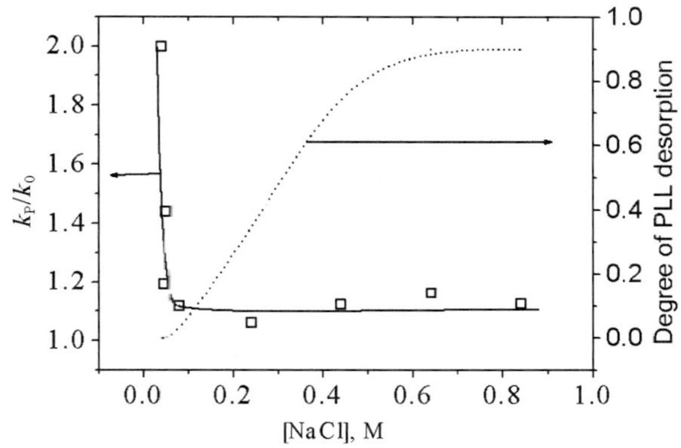

Fig. 8. Effect of NaCl on the rate of DOX permeation into EL/CL^{2-} liposomes (ν=0.2) in the presence of PLL (M_w 30,000) ($Z_{\pm}$=0.5). Dashed line represents desorption of PLL from EL/CL^{2-} liposomes. According to data published in Ref. [28].

It may be supposed that low concentrations of NaCl loosen the domains, thus decreasing the membrane permeability. Since the influence of ionic strength on an effect is generally considered an unequivocal evidence of its electrostatic nature [23], the observed reduction of PLL-induced acceleration of DOX permeation (Fig. 8) confirms the above hypothesis about the significance of the formation of domains.

So, the formation of domains in lipid membranes induced by adsorption of polyelectrolytes changes membrane permeability. This effect was observed for different polyelectrolytes and various types of charged membrane components. These effects are obviously relevant to formation of lipid-composed defective areas in lipid membranes either inside the domains, or at the interfaces between the domains and the surrounding lipid. We suggest that such defects are responsible for the increase in the membrane permeability induced by polyelectrolytes. Similar phenomena may occur in case of other types of polymers, which adsorb on lipid bilayers mainly due to hydrophobic forces.

3. AMPHIPHILIC UNCHARGED POLYALKYLENE OXIDES (PLURONICS)

Hydrophobically driven adsorption of amphiphilic polymers on the membrane of living cells leads to enhancement of flip-flop [53], formation of transient pores [54], and solubilization of lipid membranes [55]. Although the amphiphiles cause cytotoxic effect in many cases, some of them induce important physiological events when added in non-toxic concentrations. In this respect, a special attention is given to three-block copolymers of ethylene oxide and propylene oxide of

Table 1. Structure and properties of some Pluronics and Brij copolymers

Copolymer composition (trade name)	Acceleration of flip-flop ($\beta_{f\text{-}f}$), μM^{-1}	Acceleration of DOX transport (β_{DOX}), μM^{-1}	Log $P_{water\text{-}hexane}$[a]	CMC[b] (mM)
Series I				
$EO_2PO_{30}EO_2$ (L61)	0.16 ± 0.01	0.06 ± 0.01	-0.24 ± 0.037	0.11
$EO_{13}PO_{30}EO_{13}$ (L64)	0.07 ± 0.01	0.022	-1.83 ± 0.27	0.48
$EO_{76}PO_{30}EO_{76}$ (F68)	0.01 ± 0.005	N/A	-3.5 ± 0.53	0.48
Series II				
$EO_3PO_{40}EO_3$ (L81)	0.56 ± 0.045	0.23 ± 0.02	-0.11 ± 0.02	0.023
$EO_{26}PO_{40}EO_{26}$ (P85)	0.22 ± 0.06	0.1 ± 0.01	-2.65 ± 0.4	0.065
$EO_{61}PO_{40}EO_{61}$ (F87)	0.06 ± 0.01	0.02 ± 0.2	-3.19 ± 0.48	0.091
Series III				
$EO_5PO_{59}EO_5$ (L101)	1.35 ± 0.13	0.55 ± 0.06	0.11 ± 0.017	0.002
$EO_{26}PO_{56}EO_{26}$ (P105)	0.21 ± 0.03	N/A	-2.6 ± 0.4	0.008
$EO_{132}PO_{50}EO_{132}$ (F108)	0.11 ± 0.02	N/A	-3.65 ± 0.54	0.02
Series IV				
$C_{12}H_{25}EO_4$ (Brij30)	0.0035 ± 0.0012	N/A	$-1.79 \pm 0.06.4$	0.115
$C_{12}H_{25}EO_{24}$ (Brij35)	0.01 ± 0.003	N/A	3.20 ± 0.61	0.008
$C_{16}H_{33}EO_{10}$ (Brij56)	0.028 ± 0.003	N/A	2.32 ± 0.52	0.002

[a] T.V. Demina, O.O. Krylova and N.S. Melik-Nubarov, 2003, unpublished results.
[b] CMC (critical micellization concentration) values were determined in Refs. [68,69].

ABA type (also referred to as Pluronics):

$$HO{-}(CH_2\text{-}CH_2\text{-}O)_{n/2}{-}(\underset{\displaystyle CH_3}{\underset{|}{C}}H{-}CH_2\text{-}O)_m{-}(CH_2\text{-}CH_2\text{-}O)_{n/2}{-}H$$

Ethylene oxide (EO) blocks are hydrophilic and propylene oxide (PO) blocks are hydrophobic, thus ensuring the amphiphilic properties of these copolymers. The values of *n* and *m* in Pluronics may be wide-ranging, resulting in a substantial variation of the copolymer properties. The composition of some Pluronics discussed in the present article is presented in Table 1 (Section 3.2.1).

Pluronics are used now in pharmacy and medicine as immunoadjuvants, components of artificial blood and cryoprotectors, thrombosis inhibitors and modulators of neutrophile activity [11]. They were shown to increase permeability of blood–brain barrier [56], promote healing of disrupted cell membranes [57], stimulate accumulation of anthracycline antibiotics, including daunomycin and DOX, in tumor cells and demonstrate promising results in therapy of multi-drug resistant tumors [3]. Obviously, all these effects caused by Pluronics in biological systems are relevant to their 'membrane activity', i.e., the ability to interact with biological membranes. Therefore, it is important to clarify the molecular reasons for Pluronics effect on the properties of biological membranes.

3.1. Binding of Pluronics to lipid membranes

Only membrane-bound polymer may affect the properties of biological membranes. So, we measured Pluronics affinity to cells and model lipid membranes using ^{3}H-labeled Pluronics prepared for this purpose. It is likely that this way of labeling produces a negligible effect on the copolymer physico-chemical properties, and thus may provide correct information about the properties of intact copolymer.

^{3}H-Pluronics were used for the evaluation of the copolymer affinity towards membranes of different cells. Interaction of any ligands with living cells results in their adsorption on the membrane and uptake by endocytosis. Therefore, at physiological temperature (37 °C), the kinetics of Pluronics interaction with cells obeyed a biphasic law indicating a rapid binding of the copolymer to the cell surface and its slower accumulation inside the cells. At low temperature (4 °C), the endocytosis is generally suppressed. Under these conditions, saturation kinetics was observed and an ultimate level was achieved within 2 h [58]. The amount of copolymer bound to all studied cell types was proportional to its bulk concentration (Fig. 9). The surface density of the adsorbed copolymer did not depend on cells concentration. No displacement of ^{3}H-Pluronic by a large excess of unlabeled one occurred. This fact points to non-specific character of polymer interaction with cells.

Importantly, binding of Pluronics L61 and P85 to mouse myeloma SP2/0 cells (Fig. 10, lines 1, 2) was considerably higher than to normal lymphocytes (lines 3, 4) and erythrocytes (lines 5, 6). It means that the affinity of Pluronics to cell membranes highly depends on the cell type. A correlation with the membranes microviscosity was observed. The erythrocytes, whose membranes contain high amounts of cholesterol [59] and thus exhibit high microviscosity, bound Pluronics poorly. In contrast, myeloma tumor cells whose membranes are characterized by very low microviscosity are able to bind Pluronics more

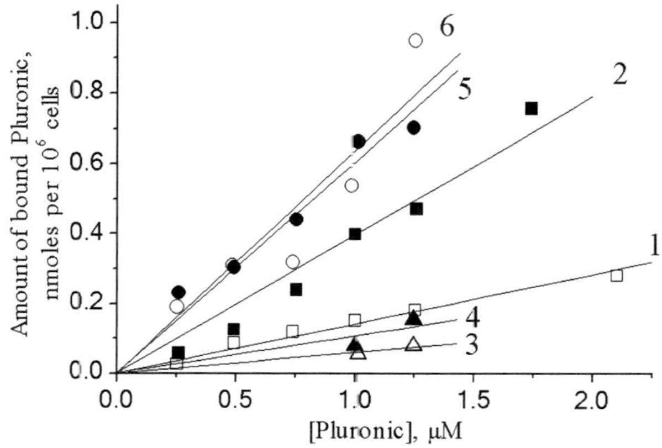

Fig. 9. Binding of ^{3}H-labeled Pluronics with lymphocytes (curves 1,2), erythrocytes (curves 3, 4) and myeloma cells SP2/0 (curves 5, 6) at 4 °C in DMEM medium. Curves 1,3,5 – Pluronic P85; curves 2,4,6 – pluronic L61. According to data published in Ref. [58].

effectively [58]. So, it may be concluded that the affinity of Pluronics to biological membranes is determined, among other factors, by chemical composition and physical state of the membranes.

This conclusion was confirmed in direct experiments on Pluronic binding with EL/CHOL SUVs differing by the cholesterol content (Fig. 10) [60,61]. In all cases, Pluronic adsorption was proportional to the amount of added copolymer.

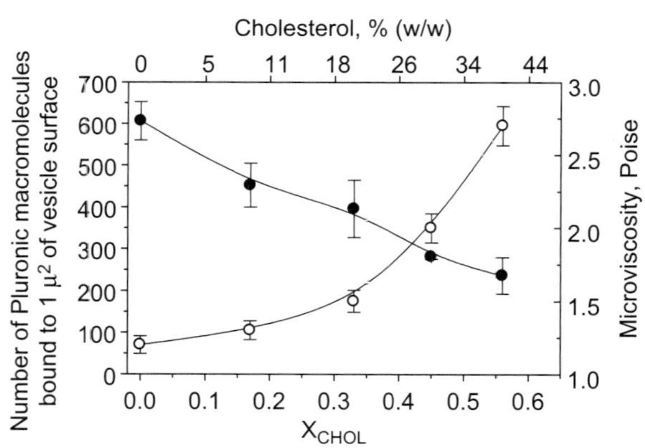

Fig. 10. Influence of SUVs composition on Pluronic binding to lipid bilayer (1) and bilayer microviscosity (2). The liposomes, 35 mg/ml, were incubated with 20 μM of Pluronic L61 at 37 °C for 48 h. Binding was measured using the equilibrium dialysis technique [61]. Microviscosity was determined by fluorescence polarization, using DPH as a probe [59]. Pluronic binding is expressed as a mean number of Pluronic macromolecules bound to 1 μ^2 of liposome surface.

The increase in CHOL mole fraction up to 0.56 (corresponding to the weight percent 40%) resulted in a threefold decrease of Pluronic binding to liposome membranes (Fig. 10, curve 1, left axis). It means that Pluronic binding to liposome membranes reduced with the increase in the membrane microviscosity (Fig. 10, curve 2, right axis). This fact gives strong evidence that the ability of Pluronic to interact with lipid bilayer is sensitive to membrane composition.

The pronounced preference of Pluronics to cholesterol-depleted membranes offers a challenge for 'physical' targeting of the polymers to the cells with reduced content of cholesterol. At the same time, these results demonstrate an opportunity to modulate the cell response to Pluronic action *via* manipulation of the cholesterol metabolism, e.g. by means of a diet.

How do Pluronics bind to lipid bilayers? Pluronics are composed of two types of alkylene oxide blocks, which seem to be very similar in structure. However, they exhibit quite different thermodynamics in solution: PEO has a lower critical temperature of mixing (LCTM) with water at about 100 °C [62], while the LCTM of PPO is about 10 °C [63]. PEO readily dissolves in water, while PPO is more thermodynamically compatible with aliphatic hydrocarbons (such as hexane) [64]. So, it is reasonable to expect that the interaction of Pluronics with lipid bilayer is a result of incorporation of PPO block in the bilayer hydrophobic region, its hydrophilic PEO blocks being exposed to the water phase. Such model of Pluronic binding to lipid bilayer was firstly proposed in the original work of I.N. Topchieva *et al.* [65] based on X-ray diffraction and calorimetry of MLVs prepared in the presence of Pluronics. Recently, these conclusions were confirmed by M. Firestone *et al.* [66] who used synchrotron SAXS for precise measurements of lamella periodicity in multi-lamellar structures formed in hydrated lipid. In these experiments, samples with low water content were prepared, so that the lipid (DMPC) formed multilamellar structures with the size of hydrophobic region about 20 Å and interlamellar spacing 57 Å. The addition of a low molecular weight Pluronic F38 containing 15 PO repeating units produced a considerable perturbation of multilamellar structure. Pluronics F68, F88, and F98 with longer PPO blocks (30, 40, and 45 repeating units) retained the multilamellar structure, but a considerable increase of inter-lamellar distance was observed. The influence of Pluronics on the multilamellar structure of DMPC was similar to that observed for PEGylated lipids indicating that PEO chains were located in the inter-lamellar lumens, while PPO blocks were inserted into the hydrophobic region of the bilayer. The results described above give important information about localization of hydrophilic PEO blocks in lipid membranes. Meanwhile, the information concerning the localization of hydrophobic PPO block remained indirect because it did not contribute to X-ray scattering. A direct information about PPO block conformation inside the membrane can be obtained by another small-angle scattering techniques (e.g. SANS) and thus clarify PPO contribution in membrane disturbances.

What are the consequences of Pluronic binding to a lipid bilayer? The biological effects of Pluronics are described in recent reviews [3,11,12], yet no detailed information concerning the Pluronic interaction with lipid membranes is available. To fill up this gap, we focused our efforts on the effects caused by Pluronics in model lipid membranes.

3.2. The effect of amphiphiles on the dynamic processes in lipid membranes

3.2.1. The influence of Pluronics on the rate of flip-flop in liposomes

To study the effects of Pluronics on the rate of flip-flop in lipid vesicles, we used a technique originally proposed by McIntyre and Sleight [67]. It is based on the ability of dithionite ions to reduce nitro groups in *N*-[(7-nitrobenz-2-oxy-1,3-diazol-4yl) dipalmitoyl] phosphatidylethanolamine (NBD-PE) to form a non-fluorescent product (Fig. 11A). The dithionite cannot permeate the bilayer, so the treatment of NBD-PE containing liposomes with excessive amounts of dithionite results in formation of asymmetrically labeled vesicles (Fig. 11A). Incubation of these asymmetrically labeled vesicles resulted in spontaneous transfer of the labeled lipid from the inner leaflet to the external one. The addition of a fresh portion of dithionite to these preincubated vesicles led to the quenching of the probe located in the outer monolayer (Fig. 11B), thus providing information about the amount of the lipid that underwent flip-flop during the incubation.

In consistence with the previously published data [67], spontaneous NBD-PE flip-flop at 25 °C was found to be a rather slow process that obeyed the first-order law satisfactorily with the effective rate constant about 0.2 h^{-1} (Fig. 12, curve 1) [61]. The amount of lipid migrated to the external leaflet asymptotically approached 50%. A minor deviation from this level was obviously due to the presence of multilayered liposomes. Their portion varied from sample to sample yet never exceeding 2–4% of the total vesicle amount.

The addition of Pluronic L61 to asymmetrically labeled liposomes resulted in a drastic acceleration of NBD-PE flip-flop (Fig. 12, curve 2, 3), the apparent rate constant of the process increased up to 0.51 h^{-1} in the presence of 5 μM L61 copolymer and to 1.28 h^{-1} in the presence of 20 μM L61 [61]. The ultimate level of the amount of the lipid migrated to the outer leaflet approached the same value as for the samples without Pluronic, indicating that Pluronic did not alter the transmembrane distribution of the lipid and only accelerated achieving the equilibrium state.

The acceleration of flip-flop induced by Pluronics was found to be proportional to its bulk concentration in the range specific to the nature of the copolymer. A representative curve for one of the copolymers studied is shown in Fig. 13.

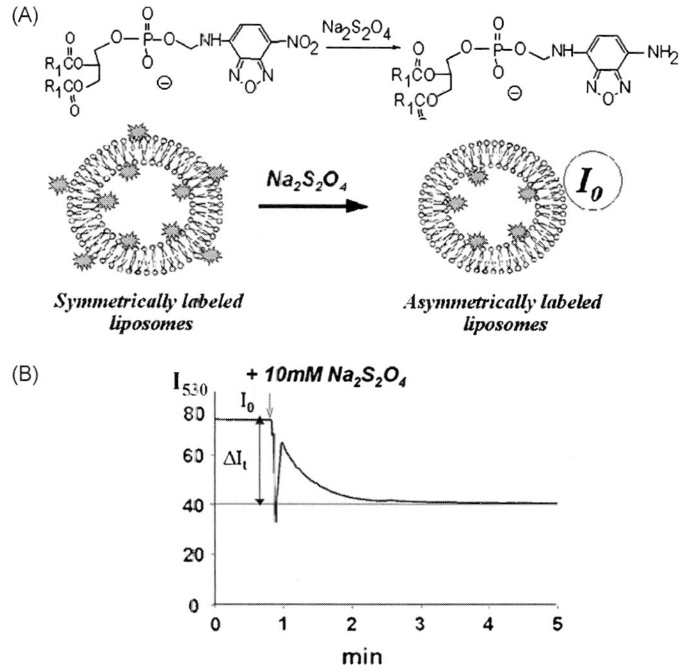

Fig. 11. Schematic of reduction of nitro groups in *N*-[(7-nitrobenz-2-oxy-1,3-diazol-4yl) dipalmitoyl] phosphatidylethanolamine (NBD-PE) in liposomes (a) and a typical kinetics of NBD-PE fluorescence decay after addition of sodium dithionite to asymmetrically labeled liposomes. The vesicles contained 0.05% NBD-PE and 99.95% EL; 10 mM Tris, 150 mM choline chloride; 1 mM EDTA; pH 7.0; 25 °C.

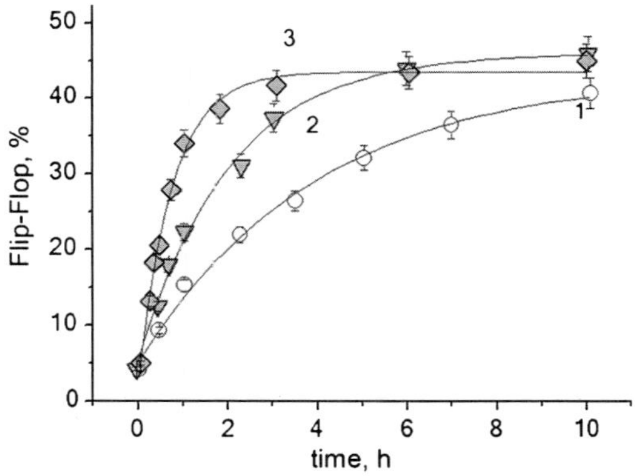

Fig. 12. Kinetics of flip-flop of NBD-PE in asymmetrically labeled EL liposomes in the absence (1) and in the presence of 5 μM (2) and 20 μM (3) of Pluronic L61. Liposome concentration 0.15 mg/ml. Other conditions are indicated in the legend to Fig. 11. According to data originally published in Ref. [61].

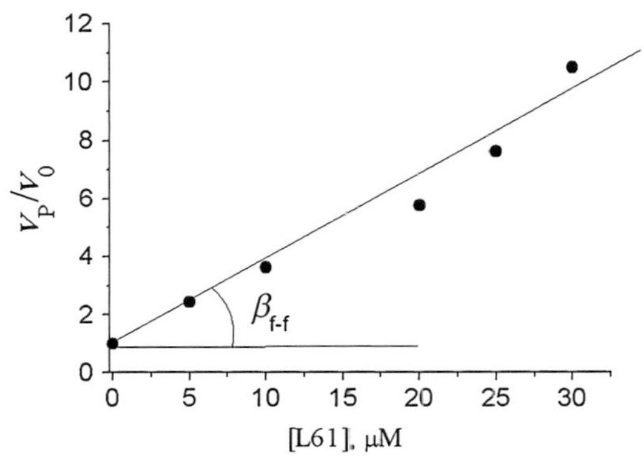

Fig. 13. Effect of Pluronic L61 on NBD-PE flip-flop in the asymmetrically labeled vesicles as a function of Pluronic concentration. Conditions of the experiment are the same as indicated in Fig. 11. According to data originally published in Ref. [61].

This dependence can be fitted by a linear equation (2):

$$\frac{v_P}{v_0} = 1 + \beta_{f\text{-}f} C_0, \tag{2}$$

where v_P/v_0 is the ratio of initial rates of flip-flop in the presence and in the absence of a polymer; C_0 is the polymer concentration, μM; $\beta_{f\text{-}f}$, μM^{-1}, is a concentration-independent ability of the polymer to accelerate flip-flop. The values of $\beta_{f\text{-}f}$ for all copolymers studied are summarized in Table 1.

To establish the main structural features that determine the ability of Pluronics to accelerate flip-flop, we studied three series of Pluronics differing in the length of their hydrophobic blocks. Each series consisted of three copolymers containing close amounts of PO repeat units and differing in the length of their hydrophilic blocks (Table 1). Variation of the length of Pluronic hydrophobic block from 30 repeat units to 50 resulted in the increase of the copolymer ability to accelerate flip-flop, whereas an increase in the length of the copolymer hydrophilic block, on the other hand, decreased their 'flippase' activity (Fig. 14). It is reasonable to suppose that the copolymer ability to accelerate flip-flop is determined, among other factors, by their affinity to lipid membrane: more hydrophobic Pluronics exhibit higher ability to bind to liposomes.

Recently, we evaluated the hydrophobicity of copolymers by measuring the partition coefficients of Pluronics in biphasic hexane–water mixtures and correlated them with $\beta_{f\text{-}f}$ parameters of the copolymers. It turned out that the increase in the length of hydrophilic PEO block at a constant length of hydrophobic PPO block resulted in the parallel decrease of the polymer $\beta_{f\text{-}f}$ [70]. At the same time, an increase in the length of PPO block shifted the line in the Log $\beta_{f\text{-}f}$ — Log$P$ coordinates in a parallel fashion (Fig. 14, cf. curves 1–3), indicating that

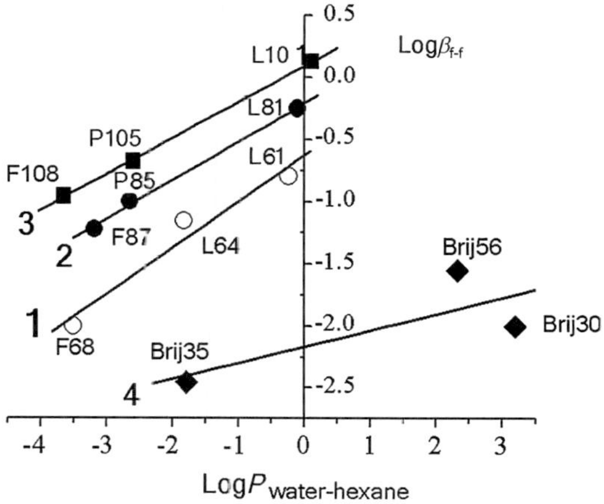

Fig. 14. Correlation between flippase activities $\beta_{f\text{-}f}$ of the copolymers and their partition coefficients in biphasic water–hexane mixtures (Log*P*). Numbers of lines correspond to the series numbers in Table 1.

the hydrophobicity of the copolymer and its affinity to lipid bilayer is not the only factor that determines the copolymer 'flippase' activity.

A similar conclusion can be drawn from the comparison of Pluronic 'flippase' activity to that of surfactants of Brij series. The latter contain PEO as hydrophilic blocks, but their hydrophobic parts are aliphatic radicals. Although the hydrophobicities of these surfactants were close to those of Pluronics, their 'flippase' activity was nearly two orders of magnitude lower (Fig. 14, curve 4). These observations indicate that the copolymer ability to accelerate flip-flop is determined to a large extent by the chemical structure and the size of its hydrophobic block.

3.2.2. The effect of copolymers on the permeability of bilayer lipid membranes

The effect of a polymer on the permeation of low molecular weight substances through lipid membranes may proceed *via* different mechanisms. It may form channels or transient pores permeable for small molecules. Another possible way is the formation of non-covalent complexes by low molecular ions and hydrophilic molecules with the polymer, which in this case may be regarded as a carrier. Lastly, permeation of substances that can penetrate into the membrane and permeate *via* simple diffusion may also be facilitated due to the copolymer influence on the diffusion coefficient of the permeant in the lipid bilayer. Of course, all three mechanisms are closely related: a lot of channel-forming substances are known to facilitate transmembrane diffusion [71]. On the other hand, many well-known ionophores cause membrane fluidization [72] and thus facilitate permeation *via* partition–diffusion mechanism. The particular mechanisms responsible for Pluronic influence on the membrane permeability towards different classes of compounds will be discussed below.

3.2.2.1. Ion permeability of lipid bilayers

The ability of Pluronic to induce ion permeability of lipid membranes was first demonstrated by measurements of ion fluxes through lipid bilayers in the presence of low molecular electrolyte [73]. For this, solvent-free membranes were formed by monolayer apposition technique [74,75], and transbilayer current was measured under voltage clamp conditions. The addition of Pluronic was accompanied by a continuous growth of the membrane permeability, which achieved a steady-state level within several minutes after polymer addition, and the membranes retained their stability within hours [73]. Such behavior points to the carrier mechanism of Pluronic action, i.e., to formation of hydrophobic complexes between the copolymer and the ions in the membrane interior (Fig. 15). From time to time, the segments with stepwise discrete jumps to higher conductance levels were, also observed (inset in Fig. 15), indicating the formation of channels or transient pores in the membrane. A quantitative analysis showed that the contribution of carrier mode to the integral membrane permeability was several orders of magnitude greater than that of channel mode [73].

Ion selectivity of Pluronic carrier activity was studied by measuring the voltage–current dependencies in the system containing KCl transmembrane gradient. Potentials were given with respect to the hyperosmotic solution; the hypo-osmotic compartment was kept electrically at ground. Application of KCl gradient in

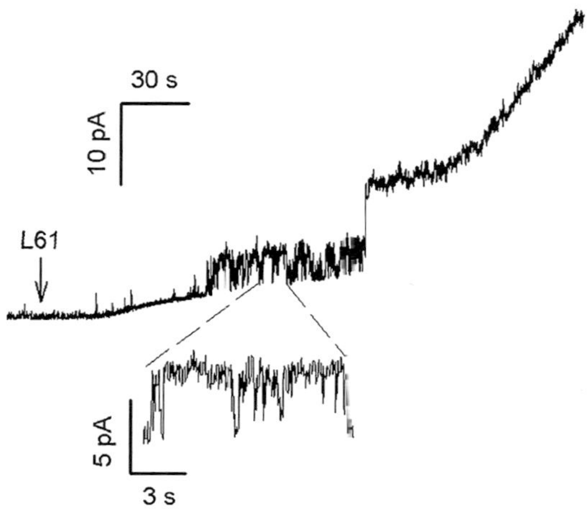

Fig. 15. Time drive of membrane permeability induced by addition of 30 μM of Pluronic L61. The buffer contained 10 mM HEPES, 10 mM TRIS and 1 M KCl, pH 7.5. To reduce noise a Gaussian filter of 7 Hz was applied [73].

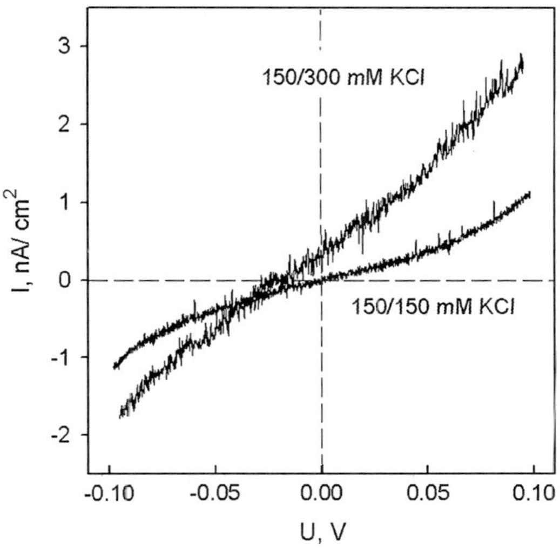

Fig. 16. Cation selectivity of L61. The current–voltage curves were measured in 50 μM aqueous solution of L61. The transmembrane KCl concentration gradient of 150/300 mM yielded a reversal potential of −17 mV indicating the ideal cation selectivity. Potentials were given with respect to the hyperosmotic solution; the hypo-osmotic compartment was kept electrically at ground. To reduce noise, a Gaussian filter of 1 Hz was applied. According to data originally published in Ref. [73].

the presence of Pluronic resulted in the shift of the voltage–current curve (Fig. 16). Reversal potential (V_0) of −17 mV was measured at twofold KCl gradient, that according to Goldman-Hodgkin-Katz equation [76] indicates an ideal cation selectivity of L61 Pluronic (Fig. 16) [72].

$$V_0 = \frac{R \times T}{F} \ln \frac{r_{ca} C_v + A_0}{r_{ca} C_0 + A_v}, \tag{3}$$

where r is the ratio of permeation coefficients of cation and anion correspondingly ($r = P_c/P_a$), C and A are the concentrations of cation and anion, indices 0 and v correspond to the opposite sides of the membrane, the potential being applied from the v-side, while 0-side was electrically ground. If $r \gg 1$, $C_v = A_v = 0.3$ M and $C_0 = A_0 = 0.1$ M, reversal potential from equation (1) should be equal −17 mV, that was really observed in the experiment. Therefore, the permeability coefficient of cations strongly exceeds that of anions in the presence of Pluronic.

Not only K^+ but also Ca^{2+} and even *N*-methyl-D-glucamine could be transported by Pluronic indicating its low substrate specificity [73]. This fact is consistent with the previously reported complexation of polyalkylene oxides with alkaline or alkaline-earth cations in organic solvents [77–78]. In particular, a strong interaction between potassium cations and PEO was observed in methanol, although such complexes did not form in water solution [79]. Such complexes of metal cations with polyalkylene oxides obviously have a donor–acceptor nature, so their formation is strongly inhibited by hydration of ether oxygen and metal cations. Therefore, Pluronics interaction with low molecular ions in water solutions is very weak and is commonly ascribed to simple cosmotropic and chaotropic effects, i.e., the effects of salts on water structure and thus on the strength of hydrophobic interactions. However, the immersion of Pluronic into the hydrophobic region of the bilayer results in the copolymer dehydration, so in this case, it is allowed to form complexes with small cations. It may be supposed that the transport of *N*-methyl-glucamine is facilitated by Pluronic due to the formation of hydrogen bonds between the sugar hydroxyls and ether groups of the copolymer.

At lower salt concentration (300 mM), Pluronic-induced membrane permeability was proportional to the copolymer concentration suggesting that single Pluronic chains were responsible for the ion current (Fig. 17A, top). At high salt concentration (1 M), the current grew with the copolymer concentration sharply and this dependence could be fitted by polynomial function of the type: $I = I_0 + ac + bc^2$, where c is the polymer concentration, I_0 the current in the absence of polymers, and a, b are constants. The quadratic dependence of ion current on Pluronic concentration strongly indicates the involvement of

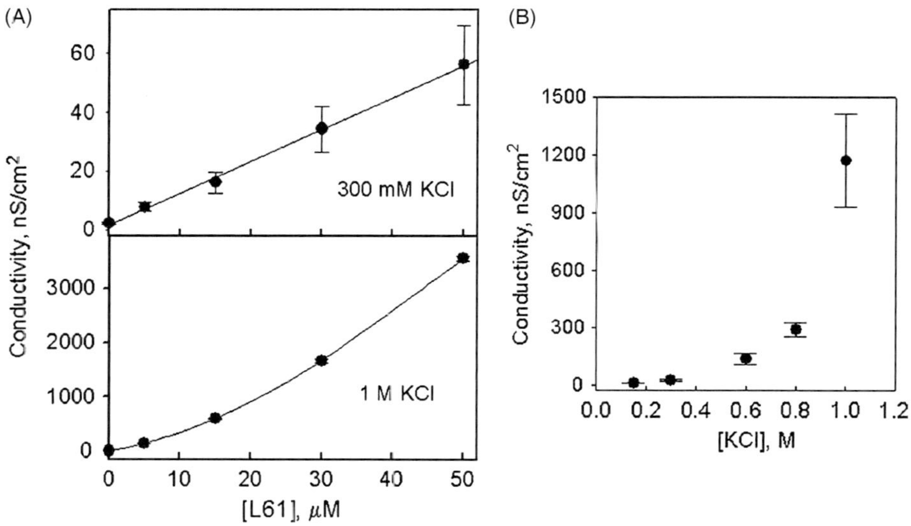

Fig. 17. Effect of Pluronic L61 on the conductivity of planar bilayer membrane. Dependence of the bilayer conductance on Pluronic concentration at 300 mM KCl (*top panel*) and 1 M KCl (*bottom panel*); (B) Dependence of the bilayer conductance on KCl concentration. Pluronic L61, 30 μM, was added to both sides of the membrane. According to data originally published in Ref. [73].

the copolymer dimers in the ion transfer. It is reasonable to suppose that an increase in the polymer content in the membrane favors its aggregation within the bilayer, so the contribution of dimers to the membrane conductivity grows with the polymer concentration.

An increase in the salt concentration resulted in the exponential growth of Pluronic-induced conductivity of lipid membrane (Fig. 17B). This result may be due to strengthening of hydrophobic interactions at higher KCl concentrations that favors polymer adsorption. In addition, this effect may be explained by changes in the stoichiometry of the ion–polymer complex as salt concentration increased. At high concentrations of KCl multiple occupancy of a polymer chain is possible.

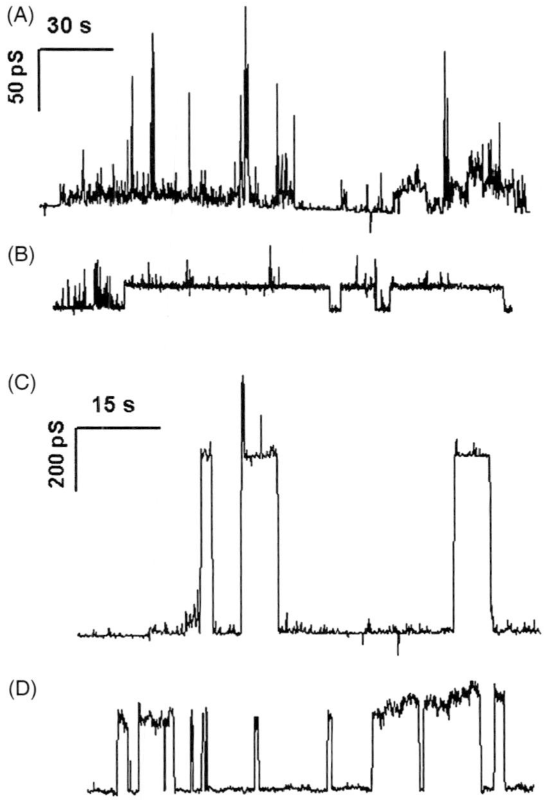

Fig. 18. Representative recordings of L61 channel activity in 1 M KCl. Both irregular current fluctuations and regular ion channels of various sizes have been observed. Corresponding copolymer concentrations and applied voltages were (A) 15 μM and +60 mV, (B) 15 μM and −90 mV, (C) 15 μM and +50 mV, and (D) 3.5 μM and +50 mV. To reduce noise, a Gaussian filter of 7 Hz was applied. According to data originally published in Ref. [73].

Generally, Pluronic effect on the ion permeability of lipid membranes (Fig. 15) may be due, apart from the carrier-like mechanism, to the formation of channels. Fig. 18 demonstrates several representative recordings of Pluronic L61 channel activity. Along with transitions between discrete conductivity levels, fast flickering spikes, irregular current fluctuations were detected (Fig. 18A,B) indicating a detergent-like mode of Pluronic action. In rare cases (about 10% of all records), recurring time-dependent activation and deactivation of Pluronic-induced ion currents have been observed (Fig. 18C,D). Such recordings may be regarded as a demonstration of gating and a proof for an ion channel mechanism. Regular ('normal') channels are characterized by open times from several seconds up to 2–3 min and by a broad spectrum of amplitudes that in 1 M KCl ranged from 20 pS to 1 nS. If present at all, the detergent-like activity was observed at the beginning of the experiments (Fig. 19, top traces). Based on Ohm's law and assuming that the channel is a cylindrical pore, the estimated pore diameters varied between 2 and 9 Å (for pore length of 50 Å and a conductivity of 110 mS/cm). Transmembrane channels are likely formed by polymer aggregates. Their size is determined by the number of Pluronics (and lipid molecules) involved. In most experiments, channel contribution to the total membrane conductance was rather low. Rare openings indicate that aggregate equilibrium is shifted toward polymer monomers and dimers (carrier mode).

So, it may be supposed that Pluronic effect on transport of ions through lipid membranes is due to formation of complexes with the permeant in the membrane interior. Although the formation of these complexes has not been yet confirmed exactly, their involvement in the transport of small ions gives indirect evidence for their formation.

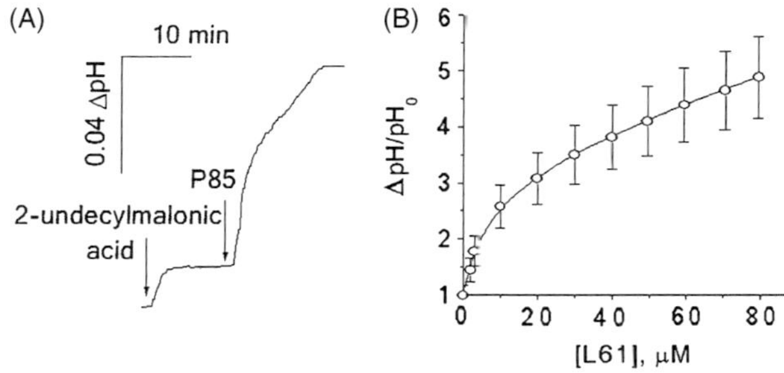

Fig. 19. (A) Effect of P85 pluronic copolymer on the flux of 2-*n*-undecylmalonic acid across the planar bilayer measured by recording transmembrane ΔpH. The concentration of 2-*n*-undecylmalonic acid on the *cis* side was 100 μM, [P85] = 43 μM on both sides of the BLM. Bathing solutions contained 1 mM TRIS, 1 mM MES, 100 mM choline chloride, 10 μM TTFB, pH 6.2. (B) Effect of the concentration of L61 Pluronic copolymer on the ratio of the 2-*n*-undecylmalonic acid fluxes in the presence of Pluronic (ΔpH) and in its absence (ΔpH_0). According to Ref. [81].

3.2.2.2. Permeability *via* partition–diffusion mechanism

Weak electrolytes commonly permeate through lipid membranes in uncharged forms *via* partition–diffusion mechanism [48,49]. This process includes partition of the solute to the membrane, its reversible conversion to the uncharged form, diffusion of the solute to the opposite side of the membrane and desorption of the ionized form of the solute from the membrane. Therefore, the transport of weak electrolytes through bilayers is accompanied by transfer of protons. Thus, it can be followed by measuring (i) pH in the vicinity of the membrane, (ii) the interleaflet boundary potential difference, or (iii) the transmembrane pH-gradients and direct detection of the solute in the internal compartment. All these approaches were used for the investigation of Pluronics effects on the permeation of weak electrolytes.

To follow the proton transfer mediated by permeation of weak acids through the lipid bilayer, pH-gradients (ΔpH) on the planar membrane were measured according to Antonenko and Yaguzhinsky [80]. Addition of a protonophore 4,5,6,7-tetrafluoro-2-(trichloromethyl)-1H-benzimidazole (TTFB) on both sides of the BLM resulted in the equalization of the pH-gradient between opposite sides of the membrane and a simultaneous appearance of the gradient of electrical potential ($\Delta\Psi$). The potential had a positive sign on the side of the membrane where the acid was added. The value of ΔpH was calculated from $\Delta\Psi$ using Nernst equation [80], and the proton flux was calculated from ΔpH according to formula:

$$J_{\mathrm{H}^+} = \frac{D_{\mathrm{buff}} B \Delta \mathrm{pH}}{2\delta},$$

where D_{buff} is the diffusion coefficient of the buffer molecules, B is the buffer capacity of the solution, and δ is the thickness of the unstirred layers near BLM. Importantly, the measurement of TTFB-mediated generation of transmembrane potential is possible only if the membrane does not contain pores permeable for H_3O^+ ions. Otherwise, the pH-gradient does not appear, and $\Delta\Psi = 0$.

Addition of 100 μM of 2-*n*-undecylmalonic acid to the *cis* side of BLM resulted in the formation of pH-gradient of about 0.016 units, indicating effective adsorption of the fatty acid on the lipid bilayer and subsequent transport of the protonated form of the acid to the opposite side of the membrane (Fig. 19) [81]. This enabled the measurement of the transmembrane pH-gradient in the presence of Pluronics that was found to increase by 0.07 units, indicating the acceleration of the transport of 2-*n*-undecylmalonic acid in the presence of Pluronic.

The effect of Pluronics on the permeation rate of 2-*n*-undecylmalonic acid was independent of the fact whether the copolymer and the acid were added to the same side of the membrane or to the opposite sides. This result clearly indicates that the Pluronic effect on the membrane permeability is determined by the incorporation of the copolymer into the hydrophobic region of the membrane, the hydrophobic block being spanned through the bilayer (Fig. 20).

L61 Pluronic had only a minor effect on the acetic acid and ammonia fluxes across the BLM [81]. This finding shows that the effect of Pluronic copolymers on the permeation of solutes is very sensitive to the solute nature. It may be proposed that the effect of copolymers is more pronounced for substances that have low membrane permeability. Thus, the transport of 2-*n*-undecylmalonic acid possessing low permeability coefficient ($P_m = 1.5 \times 10^{-8}$ cm/s [82]) accelerated considerably in the presence of Pluronics. In contrast, the translocation of highly mobile molecules of acetic acid ($P_m = 10^{-6}$ cm/s [82]) and ammonia ($P_m = 48 \times 10^{-3}$ cm/s [83]) was practically insensitive to Pluronic.

The Pluronic effect on the permeation of weak electrolytes may result from either facilitation of solutes diffusion through the hydrophobic region of the bilayer (diffusion mechanism), or from formation of non-covalent complexes with the solute in the membrane interior (carrier mechanism). The effect of Pluronics on the permeation of substances that are capable of non-covalent interactions with ether groups may be at least partially due to the carrier mechanism. The above-discussed permeation of 2-*n*-undecylmalonic acid may be facilitated by both mechanisms because the protonated form of the dicarboxylic acid

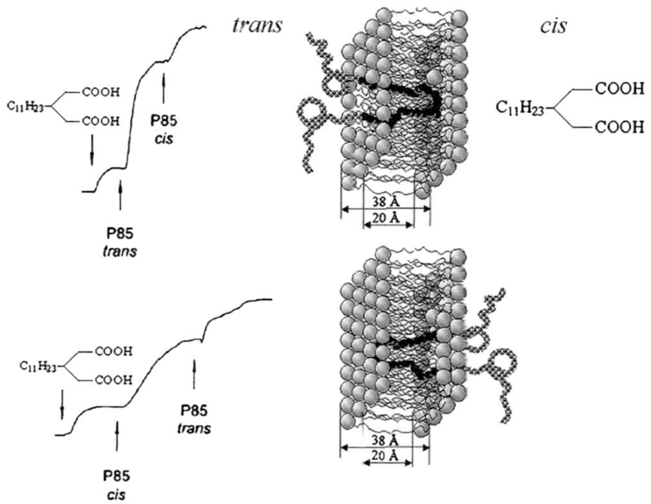

Fig. 20. Effect of P85 Pluronic copolymer added either to *trans* (A) or *cis* (B) side of BLM on the flux of 2-*n*-undecylmalonic acid across the planar bilayer measured by the recording of transmembrane ΔpH. The acid was added to the *cis*-side of the membrane. The experimental conditions are the same as in Fig. 19. According to data published in Ref. [81].

may form strong hydrogen bonds with ether oxygens of Pluronics. At the same time, if Pluronic accelerates permeation of bulky substances that cannot interact with ether groups, the existence of diffusion mechanism should be recognized.

To check the latter opportunity, the influence of Pluronic on the rate of transmembrane migration of tetraphenylboron (tetraphenylborate) (TPB^-) anions was investigated [84]. This substance was never reported to interact with polyethylene oxide (PEO). It is known that this hydrophobic anion effectively partitions into lipid bilayers. Its interaction with BLM is not only hydrophobically driven, but is also favored by the membrane dipole potential [84]. As shown earlier, the application of a voltage jump to the BLM in the presence of permeating anions

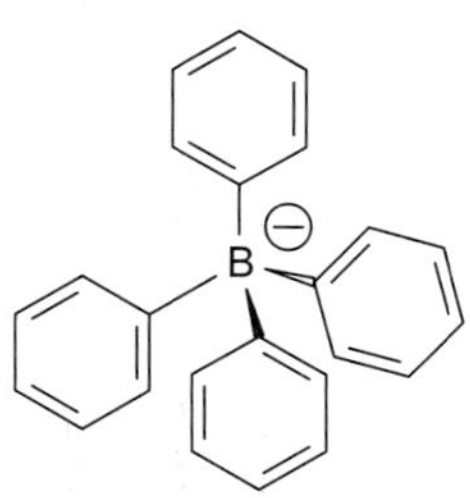

tetraphenylboron

(TPB^-) results in current relaxation kinetics with a steady-state current substantially less than the initial current [84]. This kinetics can be described by a model assuming that the rate constant of the translocation of the permeating anions across the BLM substantially exceeds that of their association with or dissociation from the BLM. Therefore, the initial transmembrane current vanishes after the depletion of TPB^- anions on one side of the membrane due to translocation of the major part of the anions to the opposite side [81]. The characteristic time (τ) of TPB^- translocation was 83 ms in our experiments (Fig. 21, curve 1). The addition of 1 μM of P85 copolymer resulted in an 11-fold decrease in τ and a fourfold increase in the initial current (Fig. 21, curve 2). The corresponding first order constant ($k = 1/\tau$) increased from 12 to 133 s^{-1} [81]. The relaxation of the current resulted in a low but definite steady-state level determined presumably by the rates of TPB^- adsorption/desorption kinetics (inset in Fig. 21). Addition of Pluronic L61 did not alter the steady-state current showing that Pluronic selectively affects the stage of TPB^- translocation across the membrane. It seems unlikely that TPB^-anion is transported through the membrane due to formation of complexes with Pluronic. At the same time, the diffusion through the hydrophobic region of lipid bilayer is the rate-limiting step in the process of a bulky TPB^--anion translocation [81]. Therefore, it is reasonable to assume that Pluronic alters the rate of TPB^- translocation mainly by facilitation of its diffusion through the lipid bilayer.

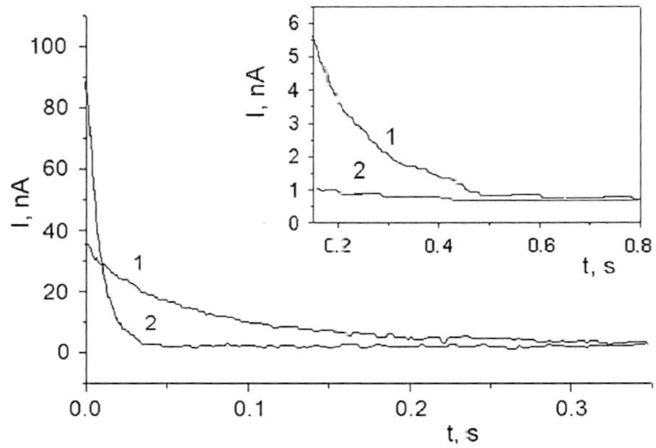

Fig. 21. Effect of P85 Pluronic copolymer (0.0005%) on the TPB-induced current transient after the application of a voltage jump (100 mV) to the BLM (curve 2). Curve 1 is a control without P85, it is well fitted by a single exponential $I = I_\infty + I_0 \exp(-t/\tau)$ with $I_0 = 30$ nA and $\tau = 83$ ms. Curve 2 is fitted by a single exponential with $I_0 = 120$ nA and $\tau = 7.5$ ms. The solution contained 10 mM TRIS, 10 mM MES, 100 mM choline chloride, pH 6.4. TPB^- concentration was 1 μM. (Inset) Filtered traces of curves 1 and 2 at the end of the recordings. According to data published in Ref. [81].

This mechanism was further confirmed in the studies of Pluronic effect on the steady-state fluorescence anisotropy ⟨r⟩ of a membrane probe diphenylhexatriene (DPH). Fluorescence anisotropy gives information about the mobility of DPH in the membrane environment. Being highly hydrophobic, DPH is located in the fatty acid region of the bilayer. The measured fluorescence anisotropy is related to the microviscosity of the lipid bilayer *via* Perren equation [59], and the microviscosity determines the diffusion coefficients of solutes in the membrane environment.

The effect of Pluronic L61 on the mobility of DPH was studied on liposomes prepared from DPPC, which undergoes phase transition at 41.5 °C [85]. The addition of L61 Pluronic to liposomes at 20 °C did not change the fluorescence anisotropy of the probe (Fig. 22, curve 1) [86]. This fact indicates that Pluronic does not interact with the lipid bilayers in the gel phase. At the temperature corresponding to the complete melting of $L\alpha$ phase (50 °C), the fluorescence anisotropy of DPH decreased indicating a considerable fluidization of liposome membranes. Addition of Pluronic under these conditions resulted in a significant decrease of DPH fluorescence anisotropy (Fig. 22, curve 2). It means that Pluronic can adsorb only on liposome membranes in the liquid crystalline state and enhance their fluidity.

Thus, the experiments performed on model membranes (Figs 10 and 22) as well as on living cells (Fig. 10) showed that Pluronic could incorporate only into the membranes with lowered microviscosity. In this case, it induces disturbances

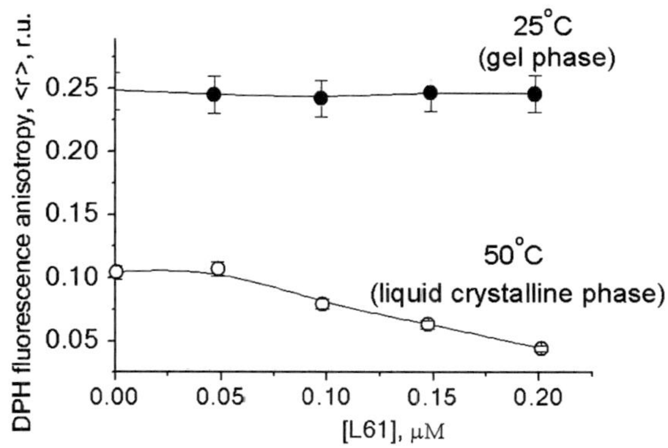

Fig. 22. Fluorescence anisotropy of DPH incorporated in the membrane of DPPC liposomes as a function of Pluronic L61 concentration. Lipid concentration was 0.1 mg/ml, lipid:DPH molar ratio was 1:2700. According to data published in Ref. [86].

in the bilayer, increases its fluidity, and accelerates the transmembrane migration of different substances.

3.2.2.3. The effect of Pluronics on transmembrane permeation of anti-tumor drug doxorubicin

In the previous section, we mentioned the remarkable ability of Pluronics to accelerate accumulation of DOX in tumor cells [3,12]. In an attempt to elucidate the mechanism of Pluronics action, their influence on DOX transport was studied using an artificial system based on pH-gradient vesicles described in Fig. 5. The addition of Pluronics to such vesicles resulted in acceleration of DOX transport, the effect being nearly proportional to the polymer concentration (Fig. 23) [61]. Note that nearly threefold acceleration of DOX permeation was caused by adsorption of only 20 molecules of Pluronic L61 per vesicle (Fig. 23, top axis). This effect is very significant, taking into consideration that each liposome was composed of about 90,000 lipid molecules, so 20 molecules of Pluronic with $M_w = 2000$ covered about 0.1% of the liposome surface. It follows that Pluronic incorporation into the bilayer strongly activates DOX translocation. This conclusion was confirmed in the direct experiments on planar bilayers by measuring the difference of surface potentials between two sides of planar lipid membrane [73].

The formation of void volume [87] in the bilayer was recently hypothesized for a branched fatty valproic acid, an anticonvulsant drug whose action was proposed to arise from membrane-disordering properties [88]. This conclusion was based on the results obtained by a molecular dynamics simulation [88].

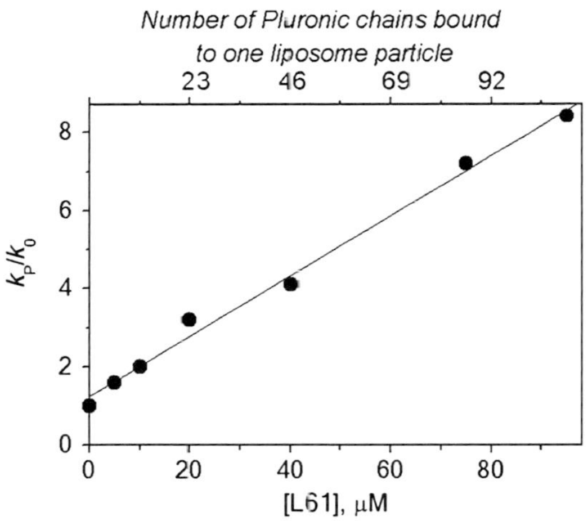

Fig. 23. Effect of Pluronic L61 on the rate of DOX permeation to liposomes as a function of the copolymer bulk concentration (bottom axis) and mean number of macromolecules bound to each liposome (N_P). Liposomes were prepared from EL, their concentration was 1.6 mg/ml; size about 100 nm; [DOX] = 50 μM; buffer solution inside liposomes contained 300 mM TRIS-citrate, external buffer contained 20 mM HEPES, 5 mM TRIS, 300 mM sucrose, pH 7.0, 30 °C. According to Ref. [61].

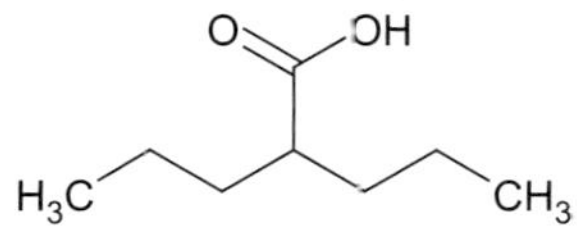

Valproic acid

The Pluronic effect on the rate of DOX permeation highly depended on the polymer structure. More hydrophobic Pluronics caused more pronounced effects on DOX permeation. An increase in the length of Pluronic hydrophobic block reinforced its ability to facilitate DOX permeation [86].

The relationship between the copolymer structure and its effect on the rate of DOX permeation was very similar to that observed for Pluronic-induced acceleration of flip-flop. A close correlation between the effects of copolymers on the rate of flip-flop ($\beta_{f\text{-}f}$) and DOX permeation (β_{DOX}) was observed (Fig. 24) [70]. It indicates that both processes are manifestations of the same phenomenon. However, the effects on flip-flop were about threefold higher than those on the DOX transport. The reasons of this quantitative difference remain unclear. It may be supposed that the copolymer-induced disturbance of the lipid packing may be assimilated to that induced by temperature. Formally, the quantitative measure of

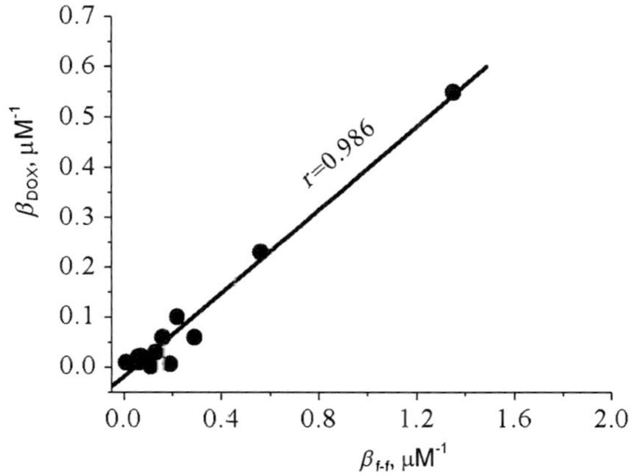

Fig. 24. Correlation between the copolymer influence on the rate of lipid flip-flop ($\beta_{f\text{-}f}$) and DOX transport (β_{DOX}). Each point corresponds to the effects caused by one copolymer on both processes. *r* is the correlation coefficient.

temperature-induced acceleration of a process is given by its activation energy, which is higher for the processes that are more sensitive to an increase in temperature. If the copolymer-induced disturbance is analogous to the temperature-induced disturbance, the effect of the copolymers on lipid flip-flop (activation energy about 42 kcal/mole [89]) should be considerably higher than that on the rate of DOX permeation (activation energy 28 kcal/mole [50]) (Fig. 24). Nevertheless, this hypothesis requires further investigation.

4. STRUCTURE REQUIREMENTS TO THE POLYMER THAT DETERMINE ITS MEMBRANE ACTIVITY

Many polymers are known to influence various properties of biological membranes [1–7]. The amount of new polymers and applications of water-soluble polymers in pharmacy and medicine grows intensively [3–12]. Therefore, it is important to understand the structural requirements and thermodynamic reasons for the copolymers membrane activity. In the present work, we summarized the effects induced by charged and amphiphilic polymers in lipid bilayers. It was shown that polycations induce lateral segregation of lipids with formation of domains enriched with anionic lipids and thus increase permeation of uncharged forms of bulky molecules, e.g. DOX. The amphiphiles, e.g. Pluronics, also increase membrane permeability. However, they do not form domains, their effect being achieved by another way. The hydrophobic mismatch of bulky amphiphile chains and fatty acid region of lipid bilayer results in the formation of void volume in the membrane [58] and stimulates the mobility of its components facilitating the transbilayer diffusion of a permeant [60,61,81,86]. Non-covalent interactions between the permeant and the polymer also contribute to the observed effect and favor overcoming the hydrophobic region of the membrane by the hydrophilic solute [73].

The experimental data overviewed in the present work demonstrate that the result of polymer–membrane interaction is determined by a superposition of a number of factors that depend on the properties of both the polymer and the membrane. To understand the regularities of such interaction, we should consider the thermodynamic changes induced by electrostatic adsorption of the copolymer on the bilayer (in case of polyelectrolytes) or its incorporation into the lipid bilayer (in case of amphiphilies).

In water solution, a polyelectrolyte is surrounded by a diffuse shell of counterions, while an amphiphile carries hydration water tightly bound to its hydrophobic block. Polymer binding to a lipid bilayer is accompanied by the release of these small molecules in the bulk. The increase in the degree of freedom of these small species determines the gain in entropy of the whole system and is the main driving force for polymer–membrane interaction. The latter can be virtually divided into several processes, and each of them may be associated with the corresponding free energy contribution: (i) release of water molecules or small counterions ($\Delta\Delta G_{\text{w/ci}}$); (ii) formation of a cavity in the lipid bilayer sufficient enough to accommodate the copolymer hydrophobic block (ΔG_{Cavity}); (iii) formation of electrostatic or Van der Waals bonds between the polymer and lipids ($\Delta\Delta G_{\text{Polymer–Lipid}}$); (iv) changes in the polymer conformational mobility due to its adsorption on or incorporation into the bilayer ($\Delta\Delta G_{\text{Polymer}}$); (v) changes in the bilayer structure due to polymer adsorption or incorporation ($\Delta\Delta G_{\text{Bilayer}}$). If we assume that the system is closed and the formation of the polymer–membrane complex has reached the equilibrium state, total changes of the system free energy equal zero:

$$\Delta\Delta G^0 = 0 = \Delta\Delta G_{\text{w/ci}} + \Delta\Delta G_{\text{Cavity}} + \Delta\Delta G_{\text{Polymer–Lipid}} + \Delta\Delta G_{\text{Polymer}} + \Delta\Delta G_{\text{Bilayer}} \quad (4)$$

The first term in this equation is always large and negative, and in the first approximation it adds up to the increase in the entropy of small molecules released into the bulk due to polymer adsorption. $\Delta\Delta G_{\text{w/ci}}$ is the main factor that determines the affinity of a polymer to the bilayer.

Electrostatic adsorption of polyelectrolytes is accompanied by formation of strong Coulomb interactions between charged groups and a simultaneous breaking of the same amount of electrostatic bonds between the counterions and charged groups of the polymer and lipids. So, the enthalpic gain due to formation of polymer–lipid electrostatic contacts is compensated by the enthalpic losses due to dissociation of the same amount of bonds with small counterions. Therefore, in the case of polyelectrolyte adsorption, $\Delta\Delta G_{\text{Polymer–Lipid}}$ can be neglected. Similar to this, neutral amphiphiles incorporated into the lipid bilayer should scarcely cause noticeable changes in the enthalpy of the system because the disruption of lipid–lipid interactions by the adsorbed copolymer is accompanied by formation of a similar amount of polymer–lipid contacts [19]. Therefore, in the case of adsorption of amphiphiles, the corresponding term $\Delta\Delta G_{\text{Polymer–Lipid}}$ may also be disregarded.

The term $\Delta\Delta G_{\text{Cavity}}$ corresponding to formation of a cavity in the bilayer is obviously positive and depends on the size of the polymer hydrophobic block and its conformation. This term should be taken into consideration only in case of amphiphiles, because hydrophilic polyelectrolytes usually are unable to penetrate between lipid chains.

Polymer adsorption onto the bilayer surface or its incorporation into the bilayer should obviously cause a restriction of the polymer conformational mobility. Therefore, the corresponding term $\Delta\Delta G_{\text{Polymer}}$ should be positive.

The last term in equation (4) ($\Delta\Delta G_{\text{Bilayer}}$) reflects the changes in the bilayer structure and mobility caused by the copolymer incorporation. If the polymer adsorption affects the density of lipid packing, it may be supposed that the main contribution to $\Delta\Delta G_{\text{Bilayer}}$ is made by changes in the bilayer entropy ($T\Delta\Delta S_{\text{Bilayer}}$). If the copolymer insertion increases the density of the bilayer, its entropy should decrease ($\Delta\Delta S_{\text{Bilayer}} < 0$) resulting in an increase of the bilayer free energy ($\Delta\Delta G_{\text{Bilayer}} > 0$). On the contrary, if the amphiphile disturbs lipid bilayer, its entropy would increase and $\Delta\Delta G_{\text{Bilayer}}$ would be negative. The more negative $\Delta\Delta G_{\text{Bilayer}}$ is, the more pronounced should be the copolymer effect on lipid flip-flop or DOX permeation. So, whether the copolymer influences the membrane properties or not, is determined by the sign of $\Delta\Delta G_{\text{Bilayer}}$, and the magnitude of its effect is determined by the value of $\Delta\Delta G_{\text{Bilayer}}$. An acceleration of flip-flop and an increase in the bilayer permeability would occur if $\Delta\Delta G_{\text{Bilayer}} < 0$.

Since we have assumed that $\Delta\Delta G^{\circ} = 0$, the sign and value of $\Delta\Delta G_{\text{Bilayer}}$ is determined by the balance between $\Delta\Delta G_{\text{w/ci}}$ that is negative and the positive sum $\Delta\Delta G_{\text{Cavity}} + \Delta\Delta G_{\text{Polymer}}$. Obviously, in case of polyelectrolyte adsorption, $\Delta\Delta G_{\text{Bilayer}}$ should be regarded as a sum of two contributions, corresponding to anionic lipid molecules bound to domains ($-T\Delta\Delta S_{\text{B}}^{\text{Lipid}}$) and free lipid molecules ($-T\Delta\Delta S_{\text{F}}^{\text{Lipid}}$). While the changes in entropy of the lipid bound to the domains is obviously negative (the free energy has a positive sign), the sign of the entropy of free lipid is determined by the ratio:

$$\delta = \frac{|\Delta\Delta G_{\text{polymer}} - T\Delta\Delta S_{\text{B}}^{\text{Lipid}}|}{|\Delta\Delta G_{\text{ci}}|} \tag{5}$$

If $\delta > 1$, i.e., the entropic losses due to the restriction of polymer and lipid mobility exceed the gain in the free energy of small counterions released in the bulk, the changes in the free energy of lipid molecules not bound to the polymer would be negative. It means that the $\Delta\Delta S_{\text{B}}^{\text{Lipid}}$ would be positive, indicating the increase in the mobility of membrane components and hence, its permeability. If the polymer forms loose domains with no match between charged groups, the numerator in the ratio (equation (5)) will decrease and δ will be <1. In this case, a stabilization of the lipid bilayer unoccupied by the polymer and formation of defects inside the domains should be expected.

The thermodynamic approach predicts that flexible polyelectrolytes would cause more disorder in the packing of surrounding bilayer than rigid ones. Indeed, if the polymer chain in water solution is flexible, its adsorption would cause more restrictions on the conformational mobility and hence would decrease $\Delta\Delta S_{\mathrm{Polymer}}$ This is in contrast to rigid polymers, whose mobility is originally frozen. This conclusion is confirmed by the above described experimental data (Fig. 7B) showing that the flexible carbon-chained PEVP caused a more pronounced effect on DOX permeation than rigid polypeptide PLL.

In case of hydrophobically driven insertion of amphiphiles into bilayer, the value and sign of $\Delta\Delta G_{\mathrm{Bilayer}}$ will be determined by the ratio similar to equation (5), but instead of $-T\Delta\Delta S_{\mathrm{B}}^{\mathrm{Lipid}}$, $\Delta\Delta G_{\mathrm{Cavity}}$ should be taken into consideration:

$$\delta = \frac{|\Delta\Delta G_{\mathrm{Polymer}} + \Delta\Delta G_{\mathrm{Cavity}}|}{|\Delta\Delta G_{\mathrm{Water}}|} \quad (6)$$

If a long flexible polymeric chain is incorporated into dense lipid bilayer, it is reasonable to expect a substantial restriction of its conformational mobility ($\Delta\Delta S_{\mathrm{Conf.}}<0$), so its configurational entropy would decrease giving the positive sign to $\Delta\Delta G_{\mathrm{polymer}}$. If the polymer has a bulky structure ($\Delta\Delta G_{\mathrm{Cavity}}\gg 0$), the numerator in the ratio (equation (6)) increases, indicating an enhancement of the bilayer mobility and permeability.

The thermodynamic consideration of the hydrophobically driven insertion of a copolymers leads to another important relationship. The decrease in the copolymer hydrophobicity ($\Delta\Delta G_{\mathrm{water}}$) would lower its affinity to the membrane (Fig. 14), leading to an increase of δ (equation (6)). The importance of the latter factor is confirmed by comparison of Pluronics and Brij surfactants.

Let us consider their effects from the point of view of the thermodynamic model. Their hydrophobicities are similar (Fig. 15). Hence, their $\Delta\Delta G_{\mathrm{water}}$ are similar also. At the same time, the hydrophobic block of Brij copolymers consists of only 12–16 methylene groups that is much shorter than the PPO block in Pluronics. Therefore, Brij copolymers should be regarded as short and rigid, so the changes in their configurational free energy ($\Delta\Delta G_{\mathrm{polymer}}$) would be small. The cavity in the bilayer sufficient to accommodate such short hydrophobic radical also should be small ($\Delta\Delta G_{\mathrm{Cavity}}\sim 0$). As a result, $\delta\leq 1$ (equation (6)) and $\Delta\Delta G_{\mathrm{Bilayer}}$ will be approaching zero indicating no disturbances in the lipid bilayer. This was confirmed experimentally by the extremely low ability of Brij surfactants to accelerate flip-flop and DOX permeation (Fig. 15). On the contrary, Pluronics whose hydrophobicity is rather low (Table 1) due to the presence of polar ether bonds in their hydrophobic block exhibit high membrane activity [70].

The thermodynamic consideration of the effects of the copolymer on lipid bilayer allowed us to establish several requirements to the copolymer structure that assure its ability to disturb lipid bilayer. Firstly, the polymer should be flexible (to increase $\Delta\Delta G_{\mathrm{Polymer}}$) and its hydrophobicity should not be very high ($|\Delta\Delta G_{\mathrm{water}} < |\Delta\Delta G_{\mathrm{Polymer}} + \Delta\Delta G_{\mathrm{Cavity}}|$), so that δ would be >1. Secondly, the presence of bulky and rigid motifs in the polymer structure would exert an additional disturbance in the membrane, since their insertion in the bilayer requires formation of a cavity (increase $\Delta\Delta G_{\mathrm{Cavity}}$).

From this point of view, the bulky dendrimers should be more effective than linear polymers. Indeed, polyamidoamine dendrimers were recently shown to be unusually active in facilitation of DNA permeation into eukaryotic cells [90] and permeabilization of prokaryotic cells [91].

Importantly, similar regularities govern the membrane activity of both polyelectrolytes and amphiphilic polymers, although the driving forces for their adsorption are quite distinct. The relationship between the copolymer structure and its ability to disturb lipid membranes demonstrated in the present chapter may be of an important value for the design of novel water-soluble polymers capable of influencing the properties of biological membranes. We believe that the requirements to the polymer structure formulated here may help in the development of synthetic strategies of novel pharmacologically active polymers.

ACKNOWLEDGEMENTS

This work was supported by VolkswagenStiftung (Az.:I/77 742) and the Russian Foundation for Basic Research (grant 03-03-32629). The authors express their gratitude to Dr. Sci. Irene Grozdova, Prof. Peter Pohl, Prof. Yuri Antonenko and Prof. Alexander Yaroslavov for invaluable discussion and to Mikhail Kozlov for his assistance in the preparation of manuscript.

REFERENCES

[1] C. Tribet, Hydrophobically driven attachments of synthetic polymers onto surfaces of biological interest: lipid bilayers and globular proteins, Biochimie 80 (1998) 461–473.

[2] A.F. Xie, S. Granick, Phospholipid membranes as substrates for polymer adsorption, Nat. Mater. 1 (2002) 129–133.

[3] A.V. Kabanov, T. Okano, in: H. Maeda, A. Kabanov, K. Kataoka, T. Okano (Eds.), Challenges in Polymer Therapeutics, in Polymer Drugs in Clinical Stage: Advantages and Prospects, Kluwer Academic/Plenum Publishers, New York, Boston, Dordrecht, London, Moscow, 2003, pp. 1–27.

[4] A.V. Kabanov, V.A. Kabanov, D.N.A. complexes, with polycations for the delivery of genetic material into cells, Bioconjug. Chem. 6 (1995) 7–20.

[5] J.P. Behr, B. Demenix, J.-P. Loeffer, J. Perez-Mutul, Efficient gene transfer into mammalian primary endocrine cells with lipopolyamine-coated DNA, Proc. Natl Acad. Sci. USA 86 (1989) 6982–6986.
[6] G.A. El'shina, M.A. Gorbunov, T.A. Bektimirov, N.I. Lonskaia, L.I. Pavlova, A.A. Nikul'shin, R.M. Khaitov, A.V. Nekrasov, A.S. Ivanova, M.N. Matrosovich, N.G. Puchkova, [The evaluation of the reactogenecity, harmlessness and prophylactic efficacy of Grippol trivalent polymer-subunit influenza vaccine administered to schoolchildren], Zh. Mikrobiol. Epidemiol. Immunobiol. 2000; 50–54.
[7] C.M. Peterson, J.G. Shiah, Y. Sun, P. Kopeckova, T. Minko, R.C. Straight, J. Kopecek, HPMA copolymer delivery of chemotherapy and photodynamic therapy in ovarian cancer, Adv. Exp. Med. Biol. 519 (2003) 101–123.
[8] R. Haag, Supramolecular drug-delivery systems based on polymeric core–shell architectures, Angew. Chem. Int. Ed. Engl 43 (2004) 278–282.
[9] R.J. Christie, D.W. Grainger, Design strategies to improve soluble macromolecular delivery constructs, Adv. Drug Deliv. Rev. 55 (2003) 421–437.
[10] M. Jones, J. Leroux, Polymeric micelles – a new generation of colloidal drug carriers, Eur. J. Pharm. Biopharm. 48 (1999) 101–111.
[11] S.M. Moghimi, A.C. Hunter, Poloxamers and poloxamines in nanoparticle engineering and experimental medicine, Trends Biotechnol. 18 (2000) 412–420.
[12] A.V. Kabanov, E.V. Batrakova, V.Yu. Alakhov, Pluronic block copolymers for overcoming drug resistance in cancer, Adv. Drug Deliv. Rev. 54 (2002) 759–779.
[13] G. Laroche, E.J. Dufourc, M. Pezolet, J. Dufourcq, Coupled changes between lipid order and polypeptide conformation at the membrane surface. A 2H-NMR and Raman study of polylysine-phosphatidic acid systems, Biochemistry 29 (1990) 6460–6465.
[14] T. Ikeda, H. Yamaguchi, S. Tazuke, Phase separation in phospholipid bilayers induced by biologically active polycations, Biochim. Biophys. Acta 1026 (1990) 105–112.
[15] A.A. Yaroslavov, A.A. Efimova, V.I. Lobyshev, V.A. Kabanov, Reversibility of structural rearrangements in the negative vesicular membrane upon electrostatic adsorption/desorption of the polycation, Biochim. Biophys. Acta 1560 (2002) 14–24.
[16] Y. Barenholz, D. Hirsch-Lerner, Probing DNA-cationic lipid interactions with the fluorophore trimethylammonium diphenyl-hexatriene TMADPH, Biochim. Biophys. Acta 1370 (1998) 17–30.
[17] C.M. Franzin, P.M. Macdonald, Polylysine-induced 2H NMR-observable domains in phosphatidylserine/phosphatidylcholine lipid bilayers, Biophys. J. 81 (2001) 3346–3362.
[18] P. Mitrakos, P.M. Macdonald, Polyelectrolyte molecular weight and electrostatically-induced domains in lipid bilayer membranes, Biomacromolecules 1 (2000) 365–376.
[19] P.M. Macdonald, K.J. Crowell, C.M. Franzin, P. Mitrakos, D.J. Semchyschyn, Polyelectrolyte-induced domains in lipid bilayer membranes: the deuterium NMR perspective, Biochem. Cell Biol. 76 (1998) 452–464.
[20] R. Welti, M. Glaser, Lipid domains in model and biological membranes, Chem. Phys. Lipids 73 (1994) 121–137.
[21] L. Yang, M. Glaser, Formation of membrane domains during the activation of protein kinase C, Biochemistry 35 (1996) 13966–13974.
[22] D. Hinderberger, H.W. Spiess, G. Jeschke, Dynamics, site binding, and distribution of counterions in polyelectrolyte solutions studied by electron paramagnetic resonance spectroscopy, J. Phys. Chem. B 2003; published in Web.
[23] G.S. Manning, J. Ray, Counterion condensation revisited, J. Biomol. Struct. Dyn. 16 (1998) 461–476.
[24] M. Glaser, S. Wanaski, C.A. Buser, V. Boguslavsky, W. Rashidzada, A. Morris, M. Rebecchi, M. Scarlata, L.W. Runnels, G.D. Prestwich, J. Chen, A. Aderemi, J. Ahni, S. McLaughlin, Myristoylated Alanine-rich C kinase substrate (MARCKS) produces reversible inhibition of phospholipase C by sequestering phosphatidylinositol 4,5-bisphosphate in lateral domains, J. Biol. Chem. 271 (1996) 26187–26193.

[25] M. Lafleur, I. Samson, M. Pezolet, Investigation of the interaction between melittin and dipalmitoylphosphatidylglycerol bilayers by vibrational spectroscopy, Chem. Phys. Lipids 59 (1991) 233–244.
[26] A. Nabet, J.M. Boggs, M. Pezolet, Study by infrared spectroscopy of the interaction of bovine myelin basic protein with phosphatidic acid, Biochemistry 33 (1994) 14792–14799.
[27] M. Lafleur, I. Samson, M. Pezolet, Investigation of the interaction between melittin and dipalmitoylphosphatidylglycerol bilayers by vibrational spectroscopy, Chem. Phys. Lipids 59 (1991) 233–244.
[28] A.A. Yaroslavov, O.Ye. Kuchenkova, I.B. Okuneva, N.S. Melik-Nubarov, N.O. Kozlova, V.I. Lobyshev, F.M. Menger, V.A. Kabanov, Effect of polylysine on transformations and permeability of negative vesicular membranes, Biochim. Biophys. Acta 1611 (2003) 44–54.
[29] A.A. Yaroslavov, V.E. Kul'kov, A.S. Polinsky, B.A. Baibakov, V.A. Kabanov, Polycation causes migration of negatively charged phospholipids from the inner to outer leaflet of the liposomal membrane, FEBS Lett. 340 (1994) 121–123.
[30] A.A. Yaroslavov, E.G. Yaroslavova, A.A. Rakhnyanskaya, F.M. Menger, V.A. Kabanov, Modulation of interaction of polycations with negative unilamellar lipid vesicles, Colloids Surf. B 16 (1999) 29–43.
[31] C. Madeira, L.M. Loura, M.R. Aires-Barros, A. Fedorov, M. Prieto, Characterization of DNA/lipid complexes by fluorescence resonance energy transfer, Biophys. J. 85 (2003) 3106–3119.
[32] F.M. Menger, J.S. Keiper, Chemistry and physics of giant vesicles as biomembrane models, Curr. Opin. Chem. Biol. 2 (1998) 726–732.
[33] F.M. Menger, V.M. Seredyuk, M.V. Kitaeva, A.A. Yaroslavov, N.S. Melik-Nubarov, Migration of poly-L-lysine through a lipid bilayer, J. Am. Chem. Soc. 125 (2003) 2846–2847.
[34] M.I. Angelova, I. Tsoneva, Interactions of DNA with giant liposomes, Chem. Phys. Lipids 101 (1999) 123–137.
[35] D. Marsh, Peptide models for membrane channels, Biochem. J. 315 (Pt 2) (1996) 345–361.
[36] T.I. Rokitskaya, Y.N. Antonenko, E.A. Kotova, Photodynamic inactivation of gramicidin channels: a flash-photolysis study, Biochim. Biophys. Acta 1275 (1996) 221–226.
[37] A.V. Krylov, Y.N. Antonenko, E.A. Kotova, T.I. Rokitskaya, A.A. Yaroslavov, Polylysine decelerates kinetics of negatively charged gramicidin channels as shown by sensitized photoinactivation, FEBS Lett. 440 (1998) 235–238.
[38] A.V. Krylov, E.A. Kotova, A.A. Yaroslavov, Y.N. Antonenko, Stabilization of *O*-pyromellitylgramicidin channels in bilayer lipid membranes through electrostatic interaction with polylysines of different chain lengths, Biochim. Biophys. Acta 1509 (2000) 373–384.
[39] Y.N. Antonenko, V. Borisenko, N.S. Melik-Nubarov, E.A. Kotova, G.A. Woolley, Polyanions decelerate the kinetics of positively charged gramicidin channels as shown by sensitized photoinactivation, Biophys. J. 82 (2002) 1308–1318.
[40] M. Colombini, C.L. Yeung, J. Tung, T. Konig, The mitochondrial outer membrane channel, VDAC, is regulated by a synthetic polyanion, Biochim. Biophys. Acta 905 (1987) 279–286.
[41] N.O. Kozlova, I.B. Bruskovskaya, N.S. Melik-Nubarov, A.A. Yaroslavov, V.A. Kabanov, Catalytic properties and conformation of hydrophobized alpha-chymotrypsin incorporated into a bilayer lipid membrane, FEBS Lett. 461 (1999) 141–144.
[42] N.O. Kozlova, I.B. Bruskovskaya, I.B. Okuneva, N.S. Melik-Nubarov, A.A. Yaroslavov, V.A. Kabanov, F.M. Menger, Interaction of a cationic polymer with negatively charged proteoliposomes, Biochim. Biophys. Acta 1514 (2001) 139–151.

[43] R. Bucki, F. Giraud, J.-C. Sulpice, Phosphatidylinositol 4,5-bisphosphate domain inducers promote phospholipid transverse redistribution in biological membranes, Biochemistry 39 (2000) 5838–5844.
[44] R.M. Epand, W. Lim, Mechanism of liposome destabilization by polycationic amino acids, Biosci. Rep. 15 (1995) 151–160.
[45] A.E. Gad, B.L. Silver, G.D. Eytan, Polycation-induced fusion of negatively-charged vesicles, Biochim. Biophys. Acta 690 (1982) 124–132.
[46] M. Vaara, The effect of oligolysines Lys-3, Lys-4, and Lys-5 on the outer membrane permeability of Pseudomonas aeruginosa, FEMS Microbiol. Lett. 55 (1990) 15–19.
[47] A.A. Yaroslavov, O.E. Kuchenkova, E.G. Yaroslavova, V.A. Kabanov, [On the cardinal difference in interaction of negatively charged liposomes with polylysine and poly(*N*-ethyl-4-vinylpyridinium bromides)], Dokl. Chem. (Engl.) 354 (1997) 350–352.
[48] E. Overton, Ueber die allgemeinen osmotischen Eigenschaften der Zelle, ihre vermutlichen Ursachen und ihre Bedeutung fur die Physiologie, Vjschr. Naturforsch. Ges. Zurich 44 (1899) 88–135.
[49] T.-X. Xiang, B.D. Anderson, Influence of chain ordering on the selectivity of dipalmitoylphosphatidylcholine bilayer membranes for permeant size and shape, Biophys. J. 75 (1998) 2658–2671.
[50] P.R. Harrigan, K.F. Wong, T.E. Redelmeier, J.J. Wheeler, P.R. Cullis, Accumulation of doxorubicin and other lipophilic amines into large unilamellar vesicles in response to transmembrane pH gradients, Biochim. Biophys. Acta 1149 (1993) 329–338.
[51] W. Hartmann, H.J. Galla, Binding of polylysine to charged bilayer membranes: molecular organization of a lipid–peptide complex, Biochim. Biophys. Acta 509 (1978) 474–490.
[52] G. Speelmans, R.W. Staffhorst, B. de Kruijff, F.A. de Wolf, Transport studies of doxorubicin in model membranes indicate a difference in passive diffusion across and binding at the outer and inner leaflets of the plasma membrane, Biochemistry 33 (1994) 13761–13768.
[53] E. Pantaler, D. Kamp, C.W. Haest, Acceleration of phospholipid flip-flop in the erythrocyte membrane by detergents differing in polar head group and alkyl chain length, Biochim. Biophys. Acta 1509 (2000) 397–408.
[54] M. Cócera, O. López, J.L. Coderch, J.L. Parra, A. de la Maza, Permeability investigations of phospholipid liposomes by adding cholesterol, Colloid. Surf. A: Phys. Eng. Asp. 221 (2003) 9–17.
[55] F.M. Goni, M.A. Urbaneja, J.L. Arrondo, A. Alonso, A.A. Durrani, D. Chapman, The interaction of phosphatidylcholine bilayers with Triton X-100, Eur. J. Biochem. 160 (1986) 659–665.
[56] A.V. Kabanov, V.P. Chekhonin, V.Yu. Alakhov, E.V. Batrakova, A.S. Lebedev, N.S. Melik-Nubarov, S.A. Arzhakov, A.V. Levashov, G.V. Morozov, E.S. Severin, The neuroleptic activity of haloperidol increases after its solubilization in surfactant micelles. Micelles as microcontainers for drug targeting, FEBS Lett. 258 (1989) 343–345.
[57] P. Anthony, K.C. Lowe, M.R. Davey, J.B. Power, Enhanced post-thaw viability of cryopreserved cells by oxygenated perfluorocarbon or Pluronic F-68, Artif. Cells Blood Substit. Immobil. Biotechnol. 26 (1998) 27–33.
[58] N.S. Melik-Nubarov, O.O. Pomaz, T.Yu. Dorodnych, G.A. Badun, A.L. Ksenofontov, O.B. Schemchukova, S.A. Arzhakov, Interaction of tumor and normal blood cells with ethylene oxide and propylene oxide block copolymers, FEBS Lett. 446 (1999) 194–198.
[59] M. Shinitzky, M. Inbar, Microviscosity parameters and protein mobility in biological membranes, Biochim. Biophys. Acta 433 (1976) 133–149.
[60] A.E. Zhirnov and N.S. Melik-Nubarov, Pluronic effect on the permeability of model lipid membranes of various composition. in Abstracts of III Russian Kargin Conference "Polymers-2004", Moscow, January 27–February 1, 2004, Pt 1, 246.

[61] O.O. Krylova, N.S. Melik-Nubarov, G.A. Badun, A.L. Ksenofontov, F.M. Menger, A.A. Yaroslavov, Pluronic L61 accelerates flip-flop and transbilayer doxorubicin permeation, Chemistry 16 (2003) 3930–3936.
[62] G. Karlstrom, A new model for upper and lower critical solution temperatures in poly(ethylene oxide) solutions, J. Phys. Chem. 89 (1985) 4962–4964.
[63] G. Howard, H.G. Schild, D.A. Tirrell, Microcalorimetric detection of lower critical solution temperatures in aqueous polymer solutions, J. Phys. Chem. 94 (1990) 4352–4356.
[64] A.F.M. Barton, CRC Handbook of Polymer–Liquid Interaction Parameters, CRC Press, Boca-Raton, FL, 1990. pp. 195–272.
[65] I.N. Topchieva, V.V. Volodin, Yu.A. Tyukavin, S.V. Osipova, V.N. Kulakov M.I. Banatskaya, The interaction of block copolymers of ethylene oxide and propylene oxide and their polymer–protein conjugates with lipids, Biomed. Sci. 2 (1991) 562–568.
[66] M.A. Firestone, A.C. Wolf, S. Seifert, Small-angle X-ray scattering study of the interaction of poly(ethylene oxide)–b-poly(propylene oxide)–b-poly(ethylene oxide) triblock copolymers with lipid bilayers, Biomacromolecules 4 (2003) 1539–1549.
[67] J.C. McIntyre, R.G. Sleight, Fluorescence assay for phospholipid membrane asymmetry, Biochemistry 30 (1991) 11819–11827.
[68] E.V. Batrakova, S. Lee, S. Li, A. Venne, V.Yu. Alakhov, A.V. Kabanov, Fundamental relationships between the composition of Pluronic block copolymers and their hypersensitisation effect in MDR cancer cells, Pharm. Res. 16 (1999) 1373–1379.
[69] A.V. Kabanov, I.R. Nazarova, L.V. Astafieva, E.V. Batrakova, V.Y. Alakhov, A.A. Yaroslavov, V.A. Kabanov, Micelle formation and solubilisation of fluorescent probes in poly(oxyethylene-*b*-oxypropylene-*b*-oxyethylene) solutions, Macromolecules 28 (1999) 2303–2308.
[70] T. Demina, I. Grozdova, O. Krylova, A. Zhinov, V. Istratov, H. Frey, H. Kautz, and N.S. Melik-Nubarov, Relationship between the structure of amphiphilic copolymers and their ability to disturb lipid layers, Biochemistry 44 (2005) 4042–4054.
[71] K. Matsuzaki, O. Murase, N. Fujii, K. Miyajima, An antimicrobial peptide, magainin 2, induced rapid flip-flop of phospholipids coupled with pore formation and peptide translocation, Biochemistry 35 (1996) 11361–11368.
[72] G. Boudet, S. Levy-Toledano, J. Maclouf, F. Rendu, R. Salesse, Change in the physical state of platelet plasma membranes upon ionophore A23187 activation. A fluorescence polarization study, Biochim. Biophys. Acta 812 (1985) 243–248.
[73] O.O. Krylova, P. Pohl, Ionophoric activity of pluronic block copolymers, Biochemistry 43 (2004) 3696–3703.
[74] M. Montal, P. Mueller, Formation of bimolecular membranes from lipid monolayers and a study of their electrical properties, Proc. Natl Acad. Sci. USA 69 (1972) 3561–3566.
[75] A.V. Krylov, P. Pohl, M.L. Zeidel, W.G. Hill, Water permeability of asymmetric planar lipid bilayers: leaflets of different composition offer independent and additive resistances to permeation, J. Gen. Physiol. 118 (2001) 333–340.
[76] B. Hille, Ionic Channels of Excitable Membranes, 2nd edition, Sinauer, Sunderlund MA, 1992.
[77] K.-J. Liu, Nuclear magnetic resonance studies of polymer solutions. V. Cooperative effects in the ion-dipole interaction between potassium iodide and poly(ethylene oxide), Macromolecules 1 (1968) 308–311.
[78] K. Ono', H. Honda, Proton NMR chemical shift induced by ionic association on a poly(ethy1ene oxide) chain, Macromolecules 25 (1992) 6368–6369.
[79] B.K. Annis, Y.S. Badyal, J.M. Simonson, Neutron-scattering determination of the Li^+ environment in an aqueous poly(ethylene oxide) solution, J. Phys. Chem. B 108 (2004) 2554–2556.

[80] Y.N. Antonenko, L.S. Yaguzhinsky, Generation of potential in lipid bilayer membranes as a result of proton-transfer reactions in the unstirred layers, J. Bioenerg. Biomembr. 14 (1982) 457–465.
[81] V.Yu. Erukova, O.O. Krylova, Yu.N. Antonenko, N.S. Melik-Nubarov, Effect of ethylene oxide and propylene oxide block copolymers on the permeability of bilayer lipid membranes to small solutes including doxorubicin, Biochim. Biophys. Acta 1468 (2000) 73–86.
[82] V.Yu. Evtodienko, D.I. Bondarenko, Yu.N. Antonenko, Permeation of dicarboxylic acids with different terminal position of two carboxylic groups through planar bilayer lipid membranes, Biochim. Biophys. Acta 1420 (1999) 95–103.
[83] Y.N. Antonenko, P. Pohl, G.A. Denisov, Permeation of ammonia across bilayer lipid membranes studied by ammonium ion selective microelectrodes, Biophys. J. 72 (1997) 2187–2195.
[84] R. Benz, Structural requirement for the rapid movement of charged molecules across membranes. Experiments with tetraphenylborate analogues, Biophys. J. 54 (1988) 25–33.
[85] A. Blume, Apparent molar heat capacities of phospholipids in aqueous dispersions. Effects of chain length and head group structure, Biochemistry 22 (1983) 5436–5442.
[86] O.O. Krylova, T.V. Demina, N.S. Melik-Nubarov, Effects of block copolymers of alkylene oxides on the permeability of lipid membranes: possible origins of biological activity, Dokl. Chem. (Engl.) 380 (2001) 267–271.
[87] V.P. Voloshin, M.G. Alinchenko, N.N. Medvedev, P. Zhedlovsky, A.L. Rabinovich, Study of structure of intermolecular cavities in lipid bilayer, Struct. Dyn. Mol. Syst. 10 (2003) 150–155.
[88] J. Ulander, A.D. Haymet, Permeation across hydrated DPPC lipid bilayers: simulation of the titrable amphiphilic drug valproic acid. Biophys. J. 85 (2003) 3475–3484.
[89] R. Homan, H.J. Pownall, Transbilayer diffusion of phospholipids: dependence on headgroup structure and acyl chain length, Biochim. Biophys. Acta 938 (1988) 155–166.
[90] J. Dennig, E. Duncan, Gene transfer into eukaryotic cells using activated polyamidoamine dendrimers, J. Biotechnol. 90 (2002) 339–347.
[91] Z.Y. Zhang, B.D. Smith, High-generation polycationic dendrimers are unusually effective at disrupting anionic vesicles: membrane bending model, Bioconjug. Chem. 11 (2000) 805–814.

CHAPTER 6

Hydration Pressure and Phase Transitions of Phospholipids

Helge Pfeiffer*

Laboratory of Chemical and Biological Dynamics, Department of Chemistry, Katholieke Universiteit Leuven, Celestijnenlaan 200D, 3001 Leuven, Belgium

Contents

Abstract

It is well known that dehydration enhances the main phase transitions temperature and it decreases the phase transition pressure in bilayers. Dehydration normally leads to a closer packing of such membrane aggregates and this rigidisation promotes the formation of a solid-like gel phase. On the other hand, the formation of solid-like phases is also favoured by high pressure. This chapter focuses on the relation between the piezotropic and thermotropic main phase transition of phospholipids and hydration pressure. The theoretical and experimental results from our laboratory show that the shift of the phase transition pressure is numerically equal to the hydration pressure, which follows directly from the Gibbs–Duhem equation. Using the Clausius–Clapeyron equation, the approach can be extended to thermotropic measurements. In this context, it is also interesting to consider the effects of non-aqueous polar solvents, such as propanediols, on the phase transition behaviour of phospholipids.

---

* Corresponding author. Tel.: (32) 16 32 7526; Fax: (32) 16 32 7982;
E-mail: helge. pfeiffer@fys.kuleuven.ac.be

ADVANCES IN PLANAR LIPID BILAYERS AND LIPOSOMES, VOLUME 2
ISSN 1554-4516 DOI: 10.1016/S1554-4516(05)02006-5

1. INTRODUCTION

Surface forces are of special importance in colloidal systems because of their high specific surface area. At low hydration, there is a dominating repulsive surface force which cannot be straightforwardly predicted by electrostatic, undulatory or van der Waals interactions [1]. This force is called hydration force and in the case of planar surfaces, it depends exponentially on the water layer distance, d_W (equation (1)).

$$F_h = F_{h,0} \exp\left(-\frac{d_w}{d_{w,0}}\right) \tag{1}$$

For phospholipids, the water layer thickness, d_w, is mostly calculated with the model of Luzzati [2] and it is determined from the water content and from the lamellar repeat distance which is usually obtained from X-ray scattering. It represents a simple parameter which should be carefully applied because the model makes the simplified *a priori* assumption of non-penetrating water and lipid layers [3]. The parameters $F_{h,0}$ and $d_{w,0}$ represent the hydration force at surface contact ($d_w = 0$) and the exponential decay constant, respectively.

Instead of hydration force, one may also use hydration pressure to describe the repulsive interaction at low hydration. Hydration pressure is defined as the hydrostatic pressure which is required to maintain the chemical equilibrium of enclosed surface water with a free water phase [4]. In the case of planar surfaces, hydration pressure depends exponentially on the amount of hydration water, R_w.

$$P_h = P_{h,0} \exp\left(-\frac{R_w}{R_{w,0}}\right) \tag{2}$$

$P_{h,0}$ and $R_{w,0}$ are the hydration pressure in the dehydrated state and the decay constant, respectively. In the case of phospholipid bilayers, the water content, R_w, is usually expressed by the molar ratio of water and phospholipid:

$$R_w = \frac{n_w}{n_{PL}} \tag{3}$$

Hydration pressure was found for lipid lamellae, but it seems also to play a role for the dehydration of surfactants, DNA, proteins, polyelectrolytes and polysaccharides [5]. It is of biological importance for the approach of biological surfaces, such as for cell fusion, stress on cartilage or osmotic dehydration, such as in cryobiology when biomembranes approach due to freezing induced dehydration [6].

One of the fundamental methods to determine hydration pressure is, beside direct surface force measurements, the 'osmotic stress method' (OSM). This method replaces piezotropic measurements by experiments at isopiestic conditions. The basic idea is the assumed equivalence of the Gibbs free energy of hydration under the conditions of piezotropic dehydration and isopiestic sorption [7]. According to the OSM, lipids are placed in an arrangement where the water activity, a_w, is adjusted by a defined water vapour or by an appropriate osmotic solution. If one determines an appropriate hydration parameter, such as the water content, R_w, and/or the water layer thickness, d_w, one obtains functional pairs of hydration pressure and hydration. The hydration pressure is given by:

$$P_h = -\frac{RT}{V_w} \ln a_w = -\Psi_w \quad (4)$$

where R is the gas constant, T, the absolute temperature and V_w, the molar volume of water [8]. Ψ_w is the water potential [9,10]. At progressing dehydration, the molar volume of water deviates more and more from the bulk property [11,12]. However, this fact is mostly not considered in the corresponding literature. Also in the present chapter, a further discussion is avoided and the molar volume of the bulk water phase is used as a fixed parameter. However, one should be conscious that a change of the molar volume of water influences the calculated hydration pressure, especially at very low hydration.

According to the principle of Le Chatelier, a shift of a phase equilibrium is always correlated to a shift of the magnitude of interactions. Therefore, hydration pressure is for example correlated to the main phase transition of phospholipids and to the liquid–solid phase transition of hydration water. It was for instance successfully proposed to use the melting temperature of hydration water to determine the magnitude of hydration pressure [13,14]. The basis is the well-known relationship between the water activity and the melting temperature of water. Other approaches relate hydration pressure also to the main phase transition of phospholipids. However, the prediction of the correlation between hydration pressure and the main phase transition temperature which is based on the polarisation theory gave only qualitatively correct results [15]. But it is also possible to apply pure thermodynamic relationships to obtain hydration pressure from the shift of the main phase transition of phospholipids. These approaches are applicable to the piezotropic main phase transition [16] and to the thermotropic main phase transitions as well [17]. The piezotropic approach is more straightforward because it relates hydration pressure to hydrostatic pressure at isothermal conditions. On the other hand, the thermotropic approach is more comfortable, i.e. thermotropic phase transitions are easier to measure and the Clausius–Clapeyron equation gives the possibility to link the shift of the main transitions temperature to a corresponding pressure shift. Furthermore, due to the pure thermodynamic character of the derivations, the approach is also applicable to other substances such as surfactants, DNA, proteins, polyelectrolytes and polysaccharides.

Because of the importance of water in biological systems, phospholipids are usually investigated in aqueous dispersions. However, it is also important to use non-aqueous polar solvents, especially for the verification of solvation theories [18]. It is for example possible to correlate packing density, dielectric constant and dipole moment of solvent molecules to solvation pressure parameters [19]. Therefore, it is adequate also to consider the piezotropic and thermotropic phase behaviour of phospholipids in non-aqueous solvents at different degrees of solvation.

2. PIEZOTROPIC MAIN PHASE TRANSITION

2.1. Theoretical background – application of the Gibbs–Duhem equation

In order to obtain a relationship between hydration pressure and the main phase transition pressure, let us consider the chemical potentials of the membrane domains of a two-state phase equilibrium between the liquid crystalline phase, L_α, and the gel phase, L_β (nomenclature according to Luzzati [2]). According to the equilibrium condition, the chemical potential of the liquid crystalline domain is equal to the chemical potential of the gel phase domain for all possible states of the equilibrium.

$$\mu_{L,liq} = \mu_{L,gel} \tag{5}$$

Therefore, one obtains for a shift of the phase equilibrium:

$$d\mu_{L,liq} = d\mu_{L,gel} \tag{6}$$

Water and phospholipid are considered as two components and the changes of the chemical potentials of lipid and water, $d\mu_L$ and $d\mu_w$ are related by the Gibbs–Duhem equation,

$$0 = n_L\, d\mu_L + n_w\, d\mu_w \tag{7}$$

which is applied at both the left and the right side of equation (6):

$$R_{w,liq}\, d\mu_{L,liq} = R_{w,gel}\, d\mu_{L,gel} \tag{8}$$

The change of the chemical potential, $d\mu$, can be related to the changes of pressure and activity by the second terms of the corresponding Taylor series.

$$d\mu = \left(\frac{\partial \mu}{\partial p}\right)_{a_w} dp + \left(\frac{\partial \mu}{\partial a}\right)_p da \tag{9}$$

The derivative of the chemical potential with respect to pressure is the partial molar volume, V_m, and the derivative with respect to the activity can be obtained from the well-known relationship, $\mu_w = \mu_{w,0} + RT \ln a_w$:

$$d\mu = V_w\, dp + RT\, d \ln a_w \tag{10}$$

A combination with equation (8) gives:

$$R_{\mathrm{w,liq}}(V_{\mathrm{w,liq}}\,\mathrm{d}p + V_{\mathrm{w,liq}}\,\mathrm{d}\Psi_{\mathrm{w}}) = R_{\mathrm{w,gel}}(V_{\mathrm{w,gel}}\,\mathrm{d}p + V_{\mathrm{w,gel}}\,\mathrm{d}\Psi_{\mathrm{w}}) \tag{11}$$

where Ψ_{w} is the isopiestic water potential which is given by equation (4). From this follows:

$$\mathrm{d}P = -\mathrm{d}\Psi_{\mathrm{w}} \tag{12}$$

Integration of equation (12) gives the relationship between the piezotropic phase transition pressure and the isopiestic water potential,

$$P_{\mathrm{tr}} - P_{\mathrm{tr,0}} = \Delta P_{\mathrm{tr}} = -\Psi_{\mathrm{w}} \tag{13}$$

where $P_{\mathrm{tr,0}}$ is the transition pressure at full hydration. From this follows that hydration pressure (equation (4)), P_{h}, is numerically equal to the dehydration-induced shift of the main phase transition pressure, ΔP_{tr}:

$$P_{\mathrm{h}} = \Delta P_{\mathrm{tr}} \tag{14}$$

2.2. Detection of piezotropic phase transitions at low hydration

The determination of the piezotropic main phase transition at low hydration is technically more challenging than comparable thermotropic measurements. One possibility for pressure generation is the use of the diamond anvil cell (DAC) which is also an appropriate tool for the investigation of the piezotropic behaviour of biomolecules [20–22]. The DAC requires a very small sample amount and enables a very wide pressure range. The pressure is usually determined with appropriate inert pressure sensors which are mixed with the sample, such as powdered quartz, ruby or $BaSO_4$ [23–25]. These pressure sensors show characteristic pressure-dependent vibration bands which can be detected by infrared or Raman spectroscopy. We found that $BaSO_4$ powder is inappropriate for samples at low hydration. The reasons are hygroscopic capillaries in $BaSO_4$ which were formed by the precipitation of tiny crystallites during $BaSO_4$ synthesis [26]. The phase transition can in principle be detected with all suitable methods which can be equipped with a DAC. A very appropriate method is infrared spectroscopy because spectral signals of the phospholipids in the mid-infrared region are very sensitive for changes of the phase state [27].

The tiny sample amount requires a careful filling of the sample holder of the DAC. Because of the paste-like nature of lipids at low hydration, the filling can be time-consuming leading to a change of the water content. An alternative procedure is the hydration inside of the DAC sample holder by a defined water vapour [16]. For the adjustment of the water vapour, we use a home-made humidity cell operating at controlled temperature. The constant water vapour is achieved by saturated salt solutions and the final water content can be obtained from the corresponding sorption isotherms of lipids. The kinetics of hydration can be determined by the intensity of the infrared water peak after different hydration times. With our equipment, the required equilibrium was reached after a few minutes at very low humidity up to one hour at RH=92%. After the hydration process, the DAC is immediately closed and mounted into the spectrometer.

2.3. Experiments – hydrostatic pressure *versus* hydration pressure

The piezotropic phase behaviour of POPC (1-palmitoyl-2-oleolyl-sn-glycero-3-phosphatidylcholine) determined by FTIR spectroscopy is shown in Fig. 1. The main phase transition appears as the clear sigmoidal discontinuity of the pressure-dependent symmetric stretching vibration, ν_S, of the CH-bonds. As expected, there is a strong dependence of the transition pressure on the hydration state of the lipid membrane, i.e. the transition pressure decreases with decreasing humidity. This can, on a molecular level, be explained by the dehydration-induced enhancement of the population of the all-trans conformation in the hydrocarbon chains reducing the number of gauche defects. The all-trans conformation occupies a smaller volume than gauche defects and according to the principle of Le Chatelier, all processes accompanied by volume reduction promote pressure-induced phase transitions.

A further characteristic is the broadening of the sigmoidal transition at decreasing hydration. This broadening is also observed for thermotropic scans (see below) and it can be explained by the reduction of the size of the co-operative units. For dry POPC at RH=0%, no transition appears, i.e. the membrane is already in a gel-phase at normal pressure conditions.

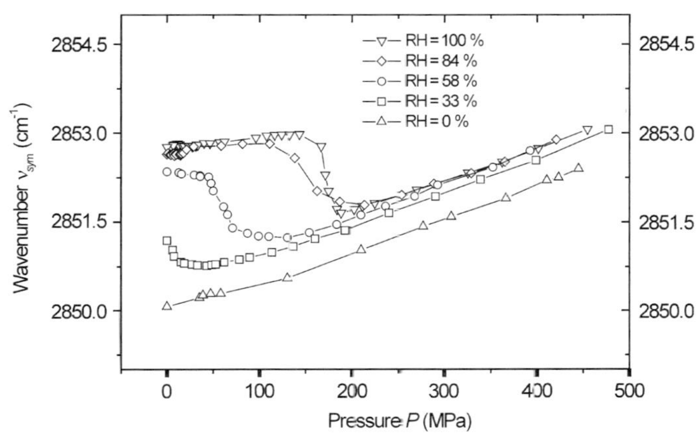

Fig. 1. Piezotropic main phase transition of POPC for selected humidities determined by the symmetric stretching vibration, ν_S, of the CH-bonds.

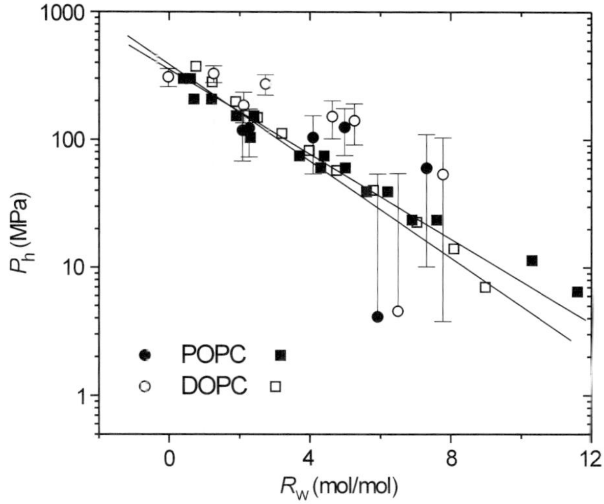

Fig. 2. Logarithmic representation of hydration pressure of POPC and DOPC at 25 °C determined by the piezotropic main phase transition (circles) and with the osmotic stress method (squares) as a function of the water content.

As an example for the principle suitability of the present approach, Fig. 2 shows the hydration pressure of POPC and DOPC (1,2-dioleoyl-sn-glycero-3-phosphocholine) as a function of water content. The hydration pressure was determined according to equation (14) and one can see that the measured values match very well with the values calculated by the isopiestic OSM.

On one hand, the present procedure establishes one of the most direct methods to determine hydration pressure because it is a piezotropic method and the model requires only a very few assumptions. On the other hand, the present method uses the parameters of an equilibrium between two phases and this is a principle difference compared to isothermal and isopiestic methods. Hydration pressure determined by isothermal and isopiestic methods is usually measured for one phase and the values obtained are more or less different depending on the phase state which is present [28]. However, this may not be considered as a limitation because there is indeed no intrinsic method for determining hydration pressure. Moreover, isopiestic and isothermal methods can also encounter phase equilibria, specifically lyotropic phase transitions. Other techniques are also based on phase equilibria, such as the dew-point method for the determination of vapour pressure which uses the equilibrium between the gas phase and the condensed phase.

The distribution of the data points in Fig. 2 clearly shows that the accuracy of the piezotropic method is poor compared to the OSM. The main reason is the limited accuracy of the spectroscopic determination of the hydrostatic pressure

($\pm$50 MPa in the present configuration) in the DAC. A solution could be the application of advanced pressure sensors or the use of different pressure devices. But other pressure devices require a remarkable enhancement of the sample amount and the pressure range could be decreased. Finally, also if the present approach can at the moment not be used for regular hydration pressure determination, it gives the experimental evidence that there is a direct correlation between the shift of the piezotropic main phase transition pressure and hydration pressure. Therefore, one should try to adapt the approach to thermotropic measurements, because it is easier to measure thermotropic instead of piezotropic phase transitions, as mentioned above.

3. THERMOTROPIC MAIN PHASE TRANSITION

The idea of using thermotropic phase transitions in phospholipid bilayers to obtain hydration pressure parameters is not new. The approach of Ulrich *et al.* [14], which was successfully demonstrated for DOPC is based on the freezing point depression of hydration water. The same approach was already proposed by Bach *et al.* [13], but instead of hydration pressure, the term ‘swelling pressure’ was used. The hydration pressure is given by Ref. [14]:

$$P_\mathrm{h} = \frac{\Delta H_{\mathrm{tr,w}}}{V_\mathrm{w}}\left(1 - \frac{T}{T_0}\right) \tag{15}$$

where $\Delta H_{\mathrm{tr,w}}$ is the molar melting enthalpy of ice, V_w, the molar volume of liquid water, T, the hydration-dependent melting temperature and T_0, the melting temperature of pure water.

Another approach predicts the hydration-induced shift of the main phase transition temperature of lipids, ΔT_m, on the basis of the water polarisation theory on hydration forces [29]:

$$\Delta T_\mathrm{m} = \Delta T_\mathrm{m}(0)\tanh\left(\frac{n_\mathrm{w} V_\mathrm{w}}{\lambda A}\right) \tag{16}$$

Here, $\Delta T_\mathrm{m}(0)$ is the transition temperature of the dehydrated state; n_w, the number of water molecules per lipid; λ, a decay length characteristic for the polarisation profile; and A, the area per lipid molecule. Unfortunately, the approach showed a qualitative, but not a quantitative agreement with experimental results, and it was concluded that calorimetric investigations of the thermotropic main phase transition could be inappropriate to give quantitative information on hydration processes [15].

3.1. Theoretical background – hydration pressure and the Clausius–Clapeyron equation

For a given constant temperature, T, the isothermal relationship between differential shifts of the piezotropic main phase transition, dP_{tr}, and the water potential, $d\Psi_w$, is given by: $dP_{tr} = -d\Psi_w$, (equation (12)). At a different temperature, T', there will be an analogous relationship, $dP'_{tr} = -d\Psi'_w$. The dependence of the transition pressure, P_{tr}, on temperature is given by the Clausius–Clapeyron equation (equation (17)), where $\Delta V_{L,tr}$ and $\Delta S_{L,tr}$ are the corresponding changes of volume and entropy.

$$\frac{dT}{dP} = \frac{\Delta V_{L,tr}}{\Delta S_{L,tr}} \tag{17}$$

There is abundant evidence that this equation is valid for the main phase transition of aqueous dispersions of phospholipids [30–32]. By using equation (12) the Clausius–Clapeyron equation can be adapted to the water potential, i.e., the ratio of the shifts of water potential and temperature is given by:

$$\frac{dT}{d\Psi_w} = -\frac{\Delta V_{L,tr}}{\Delta S_{L,tr}} \tag{18}$$

In order to simplify the notation, let us define the pressure-induced temperature increase, α, by:

$$\alpha = \frac{\Delta V_{L,tr}}{\Delta S_{L,tr}} \tag{19}$$

By a combination of equations (4), (18) and (19) as well as by subsequent integration over the temperature, one obtains the hydration pressure as a function of the dehydration-induced shift of the transition temperature, ΔT_{tr}:

$$P_h = \frac{1}{\alpha} \Delta T_{tr} \tag{20}$$

This is a similar equation compared to the piezotropic approach, $P_h = \Delta P_{tr}$, derived in the previous section (equation (14)). However, the non-isothermal nature of the thermotropic approach requires additional assumptions because of the appearance of the parameter α, i.e. the integration leading to equation (20) is only possible if α is a constant.

For the main phase transition, however, α is nearly a constant (about 0.2 K/MPa [30,32]) for about all phospholipids containing saturated hydrocarbon chains. Thus, if α is almost a constant for different phospholipids owing different transition temperatures at ambient condition, it should also be rather independent of the temperature at all. The pressure-induced temperature increase α is also independent of the use of different non-aqueous polar solvents. It is an inherent property of the CH_2-groups of aliphatic chains which are not in direct contact with the solvated bilayer surface (see Section 4.1). This leads to the conclusion that α is also independent of the hydration state.

3.2. Experiments – main phase transition temperature *versus* hydration pressure

The validity of the thermotropic approach can be shown by differential scanning calorimetry [17]. The transition temperatures (extrapolated onset temperatures) of POPC are given in Fig. 3 as a function of the water content, R_w. As in the case of piezotropic measurements (Section 2.3), there is a strong dependence of the transition temperature on the hydration of the lipid membrane, i.e. the transition temperature decreases with increasing water content. This can, on a molecular level, be explained by the hydration-induced enhancement of the population of gauche defects in the hydrocarbon chains. The liquid crystalline phase shows a higher population of gauche defects than the gel phase and according to the principle of Le Chatelier, all processes accompanied by an enhancement of the population of gauche defects promote temperature-induced main phase transitions.

The hydration pressure, P_h, calculated for POPC according to equation (20), is shown in Fig. 4. The logarithmic plot clearly shows the exponential behaviour which is typical for hydration pressure equation (2). The interesting comparison of those values with those obtained from the isothermal OSM could be complicated due to the non-isothermal nature of the present thermotropic approach. However, the striking agreement of both curves suggests that hydration pressure is

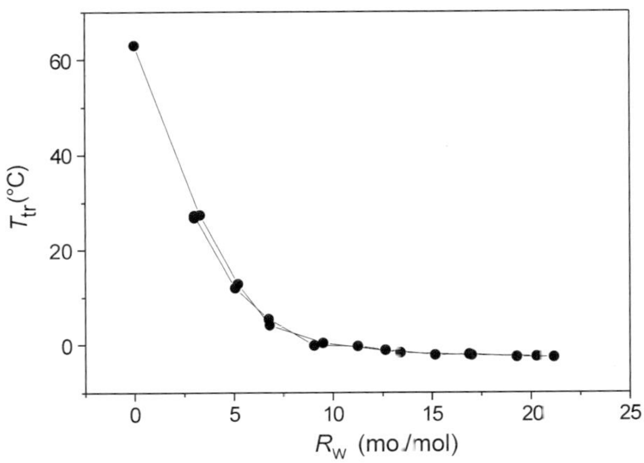

Fig. 3. Extrapolated onset temperature of the main phase transition of POPC as a function of the water content at normal pressure.

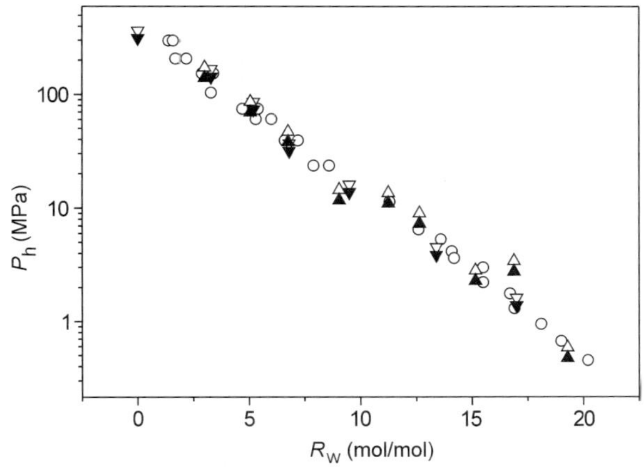

Fig. 4. Logarithmic representation of hydration pressure of POPC determined by the thermotropic main phase transition (triangles) and with the osmotic stress method (circles) as a function of the water content (open triangles: $\alpha=$ 0.18 K/MPa, solid triangles: $\alpha=0.21$ K/MPa).

essentially not a function of temperature, a fact which was also reported by Simon *et al.* [33].

The results for a group of lipids, POPC, DMPC (1,2-dimyristoyl-sn-glycero-3-phosphocholine monohydrate) and DPPC (1,2-dipalmitoyl-sn-glycero-3-phosphocholine), are summarised in a linear plot in Fig. 5. All curves show exponential behaviour and the corresponding parameters, $P_{h,0}$ and $R_{w,0}$ (see equation (2)), summarised in Table 1, are typical for hydration pressure; that values are

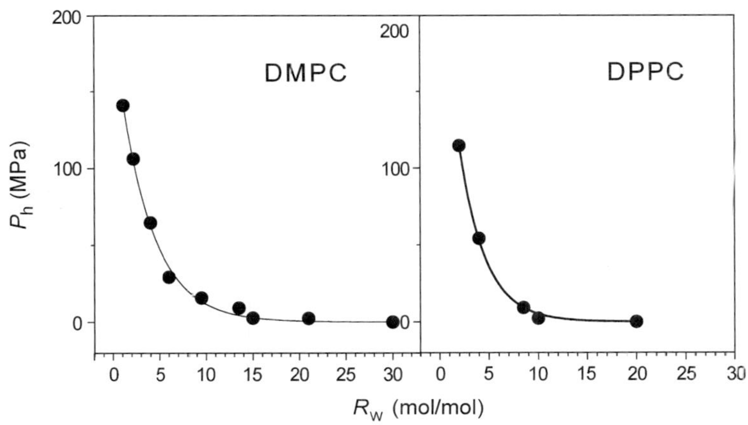

Fig. 5. Hydration pressure of DPPC and DMPC determined by the thermotropic main phase transition. The symbols are connected by the fitting line according to a single-order exponential decay.

Table 1. Parameter of hydration pressure determined by the thermotropic method (DSC) and by the osmotic stress method (OSM): (a)–(d) Refs. [36,28,37,14]

Lipid	ln $P_{h,0}$ (MPa)	$R_{w,0}$
POPC	$19.64_{(DSC)}$	$2.5 \pm 0.2_{(DSC)}$
	$19.56_{(OSM,\ a)}$	$2.5 \pm 0.2_{(L\alpha + L\beta',\ 25\ °C,\ a)}$
DMPC	$19.08_{(DSC)}$	$3.5 \pm 0.2_{(DSC)}$
	$19.62_{(OSM,\ 27\ °C,\ b)}$	$3.0_{(L\alpha',\ OSM,\ 27\ °C,\ b)}$
DPPC	$18.56_{(DSC)}$	$2.6 \pm 0.1_{(DSC)}$
	$19.87_{(OSM,\ 25\ °C,\ b)}$	$1.8_{(L\beta',\ OSM,\ 25\ °C,\ b)}$
	$20.21_{(OSM,\ 50\ °C,\ b)}$	$3.4_{(L\alpha,\ OSM,\ 50\ °C,\ b)}$
DOPC	$19.30_{(DSC)}$	$3.6 \pm 0.4_{(DSC)}$
	$19.52_{(OSM,\ c)}$	$3.3_{(OSM,\ c)}$
	$19.00_{(b)}$	$4.0_{(b)}$
	$19.61_{(DSC\text{-}H_2O,\ d)}$	$2.8 \pm 0.4_{(DSC\text{-}H_2O,\ d)}$

compared with those obtained from the literature and they show a satisfactory agreement considering the fact that no raw data were given (except of POPC). The decay constant of DPPC obtained by the thermotropic approach is exactly the average value ($R_{w,0} = 2.6$) of the decay constants of gel and liquid-crystalline phase ($R_{w,0} = 1.8$ and 3.4) obtained by an isothermal method. This seems to be a confirmation of the assumption that the phase transition method gives average parameters of the liquid-crystalline and the gel phase (Section 2.3). On the other hand, DMPC seems to show the opposite behaviour, but this can probably be explained by the limited hydration range for which the isothermal parameters were determined. More investigations can help to clarify such details.

Phospholipid phase transitions become broader during dehydration. The broadening is sometimes even accompanied by multiple overlapping transition-peaks [15]. However, this broadening is small compared to the range of the hydration-induced temperature shift. Multiple transition peaks can be explained by the Gibbs' phase rule, which allow the coexistence of at least three lipid phases at varied temperature. If this effect would be dominant, the phase equilibrium proposed by equation (1) must possibly be adapted.

An important test for the thermotropic approach is the agreement of the pressure-induced temperature increase α for hydrostatic pressure and hydration pressure. A direct comparison of the thermotropic phase transition as a function of hydrostatic and hydration pressure can be performed by spectroscopic and calorimetric measurements [17]. The results for POPC and DMPC are shown in Fig. 6. The transition shift due to hydration pressure was derived from lyotropic phase transitions performed at different temperatures with a FTIR spectrometer, and the transition temperature as a function of hydrostatic pressure at fully

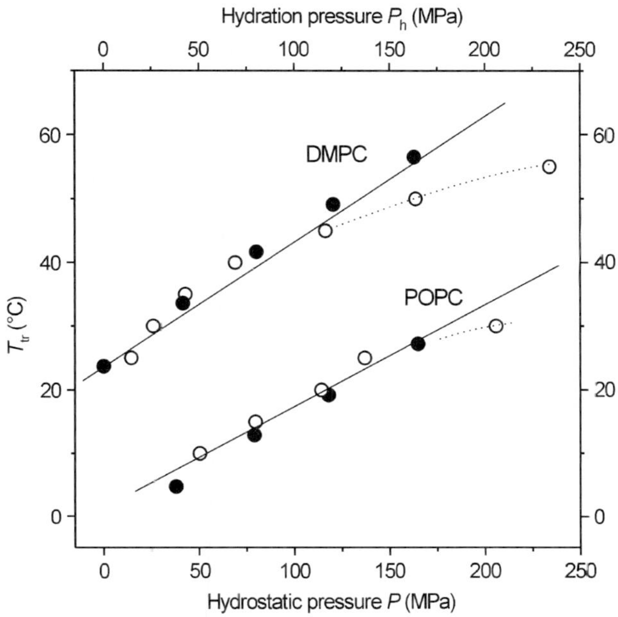

Fig. 6. Phase transition temperature of DMPC and POPC as a function of hydration pressure (open circles), determined by FTIR spectroscopy and hydrostatic pressure (solid circles), determined by DTA calorimetry.

hydration was determined by high-pressure differential thermoanalysis (DTA). The slope shows that the pressure-induced temperature increase α is indeed the same for hydrostatic and hydration pressure. The deviation for DMPC starting at about 120 MPa is most probably due to deviations of the water volume from the bulk property at very low water content (about $R_w = 1$, see Section 1). In this case, the hydration pressure calculated by equation (4) is overestimated. The same holds for POPC, but at higher hydration pressure (≈ 200 MPa $\leftrightarrow R_w \approx 1$).

4. PIEZOTROPIC AND THERMOTROPIC BEHAVIOUR WITH PROPANEDIOLS

The application of the thermotropic methods requires that the pressure-induced increase, $\alpha = dT/dP$, of the main phase transition temperature is essentially independent of the hydration state. An appropriate experiment to prove that condition is the investigation of the thermotropic and piezotropic phase behaviour of phospholipids in non-aqueous polar solvents, i.e. if α is invariant concerning the use of different polar solvents, it will also be invariant concerning different hydration states. Because of their structural similarities, the propanediols 1,2-PD and 1,3-PD are very appropriate polar solvents for investigating the solvation behaviour of phospholipids.

4.1. Main phase transition with propanediols at full solvation

The results for POPC and DMPC obtained from DTA scans are shown in Fig. 7. As expected, the transition temperatures increase in all cases linearly with increasing hydrostatic pressure. An exception might be DMPC which shows a slight curvature. But this is indeed a normal property which is also observed for other hydrocarbon chains, such as pure n-alkanes [34]. Interestingly, the shape and the area of the transition peaks show no systematic pressure dependence [32]. This indicates that the enthalpy change is essentially independent of pressure [34,35].

Figure 7 and Table 2 thus reveal that the linear slope, $dT/dP=\alpha$ is almost identical for all polar solvents investigated. This leads to the conclusion that the slope would also be essentially the same in the case of different hydration states. It is well known that this slope lies mostly in the range of 0.2 K/MPa [30], even in the case of pure n-alkanes [34]). This phenomenon can be discussed by the Clausius–Clapeyron equation (equation (17)). If the pressure-induced increase, dT/dP, is nearly constant for a large number of lipids and n-alkanes, i.e. if it does

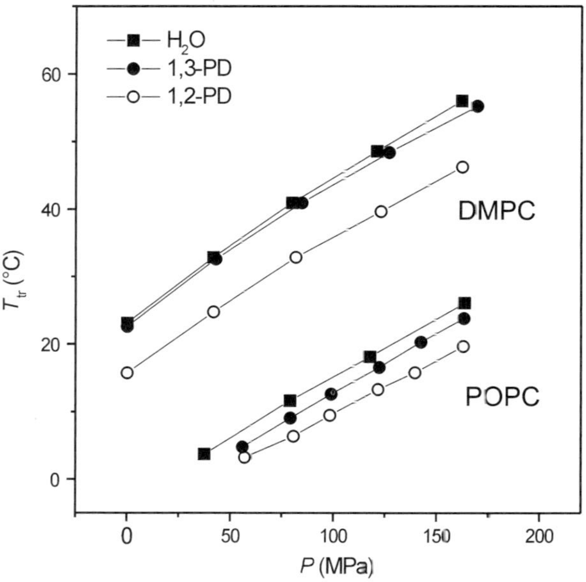

Fig. 7. Main phase transition temperatures of POPC and DMPC in different polar solvents as a function of hydrostatic pressure determined by DTA calorimetry.

Table 2. Pressure-induced temperature increase, $\alpha = dT/dP$ and the main phase transition temperature at normal pressure, T_0, for the POPC and DMPC in different polar solvents

Lipid	la (K/MPa)	T_0 (°C)
POPC/water	0.176 ± 0.003	-2.7 ± 0.4
POPC/1,2 PD	0.156 ± 0.003	-5.9 ± 0.3
POPC/1,3 PD	0.177 ± 0.001	-5.0 ± 0.2
DMPC/water	0.203 ± 0.007	23.9 ± 0.7
DMPC/1,2 PD	0.188 ± 0.007	16.5 ± 0.7
DMPC/1,3 PD	0.192 ± 0.009	23.7 ± 0.9

not depend on solvation state, headgroup structure, chain length or transition enthalpy, one must conclude that the quotient of volume and entropy change is an intrinsic property of a single aliphatic CH_2-group placed in a lattice of regular hydrocarbon chains. The slope will thus only deviate if the regular arrangement in the hydrocarbon chains is disturbed. This is the case for those lipids containing double bonds in the hydrocarbon chain, such as in the case of DOPC (two double bonds: $\alpha = 0.1$ K/MPa) and POPC (one double bond: $\alpha = 0.17$ K/MPa) [30].

Propanediols contain non-polar segments which give them partially the properties of an amphiphile, i.e., they have a tendency to mix with the lipids. The amphiphilic nature thus reduces the stability of the bilayer giving rise to a lowered transition temperature compared to water dispersions (Fig. 7 and Table 2). In the case of 1,2-PD, the destabilising effect even prevents the formation of a lamellar phase above the main phase transition. Below the main phase transition, there is a lamellar gel phase which could be shown by ^{31}P-NMR and FTIR spectroscopy [32]. Therefore, the main phase transition in lipid/1,2-PD dispersions is in fact a transition from a lamellar gel phase into an isotropic phase. The main phase transition of the lipid dispersions of 1,2-propanediol and 1,3-propanediol thus show principal differences concerning their conformational changes, but the pressure induced temperature increase, α, is nevertheless almost unaffected by these differences.

4.2. Phospholipids with propanediols at low solvation

The question is whether the thermotropic approach can also be applied to phospholipids which are dispersed in propanediols. The derivation of the corresponding equation, as given in Section 2.1, is quite general, i.e. it is not limited to water as solvent and the pressure induced temperature increase, α, is also known from piezotropic measurements (see previous Section). The corresponding main phase transition temperatures of DMPC and POPC as a

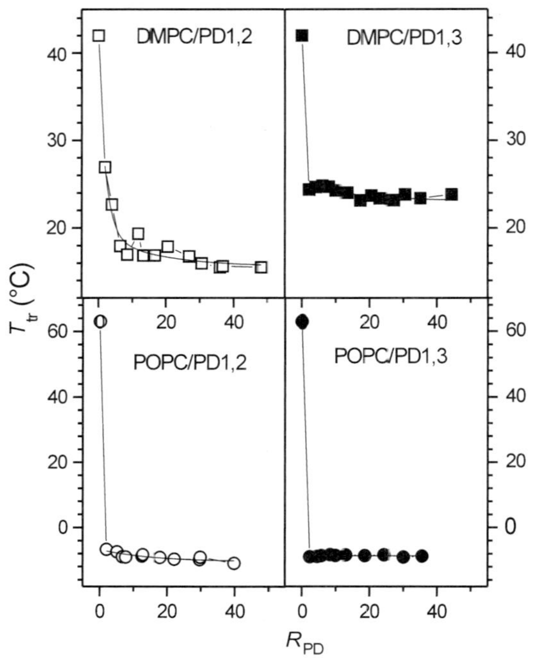

Fig. 8. Onset temperatures of the main phase transition of POPC and DMPC in 1,2-PD and 1,3-PD, determined by differential scanning calorimetry at normal pressure.

function of solvation are shown in Fig. 8. The solvation is, as in the case of hydration, given by the molar ratio of solvent molecules and lipids, $R_{PD}=n_{PD}/n_L$. The main phase transition temperatures indicate, that the solvation of the lipids is finished at very low degree of solvation. At about $R_{PD}=1\ldots2$, the system has already reached the plateau phase, i.e. that solvation range where the transition temperatures remain essentially constant. Such plateau phase is already known from lipid–water systems (see also Fig. 3).

By using of equation (20) and the pressure-induced temperature increase α given in Table 2, one can try to obtain an estimation of the corresponding solvation pressure parameters. The parameter, $P_{h,0}$, determined by the thermotropic approach for propanediol dispersions (DMPC/1,3-PD: $\ln(P_0)=18.4$ (MPa)) is comparable to the values which were obtained by the OSM for, e.g. egg–lecithin (EPC/PD1,3 $\ln(P_0)=18.5$ (MPa) [19]). It is also comparable to values that were measured for aqueous dispersions (e.g., DMPC/H_2O: $\ln(P_0)=19.6$ (MPa); (OSM [28]). The decay constant, $R_{RD,0}$, can only be estimated because of the lack of a sufficient number of points at very low solvation. Nevertheless, it seems clear that there is an inverse correlation between the size of the solvent molecules and the decay constant (DMPC/PD1,3 $R_{PD,0}=0.7\pm0.2$; DMPC/H_2O $R_{w,0}=3.0$ (OSM [28])). This could be explained by the fact that a propanediol molecule has about the fourfold size of a water molecule, and the decay constants show that a propanediol-molecules can replace about four water molecules in order to solvate the lipid with respect to the lowering of the transition temperature.

The question remains whether 1,2-PD dispersions can also be analysed with the thermotropic approach because of the decay of the bilayer in fully solvated dispersions mentioned above. However, X-ray measurements of the repeat distance (not shown) as a function of solvation indicate clear lamellar diffraction peaks up to a solvation of about $R_{PD}=8$ at 25 and 30 °C, respectively. Furthermore, Fig. 8 shows solvation curves for the thermotropic phase transition which are comparable to 1,3-PD dispersions. The interpretation of solvation pressure obtained under these circumstances will be similar, but not identical to the interpretation given for completely intact bilayers.

ACKNOWLEDGEMENTS

I want to express my gratitude to G. Klose, R. Winter, K. Heremans, H. Binder, B. Mädler, O. Reis and U. Dietrich. This work was supported by the Katholieke Universiteit Leuven, the Deutsche Forschungsgemeinschaft (SFB 294), the Universität Dortmund and the COST D10 Action of the European Community.

REFERENCES

[1] J.F. Nagle, S. Tristram-Nagle, Structure of lipid bilayers, Biochim. Biophys. Acta Rev. Biomembr. 1469 (2000) 159–195.

[2] V. Luzzati, X-ray diffraction studies of lipid–water systems, in: D. Chapman (Ed.), Biological Membranes, Physical Fact and Function, Academic Press, London, 1968, pp. 71–124.

[3] H. Schmiedel, P. Jorchel, M. Kiselev, G. Klose, Determination of structural parameters and hydration of unilamellar POPC/C12E4 vesicles at high water excess from neutron scattering curves using a novel method of evaluation, J. Phys. Chem. B 105 (2001) 111–117.

[4] D.M. Leneveu, R.P. Rand, V.A. Parsegian, Measurement of forces between lecithin bilayers, Nature 259 (1976) 601–603.

[5] V.A. Parsegian, R.P. Rand, D.C. Rau, Macromolecules and water: probing with osmotic stress, Methods Enzymol. 259 (1995) 43–94.

[6] J. Wolfe, Z.J. Yan, J.M. Pope, Hydration forces and membrane stresses—cryobiological implications and a new technique for measurement, Biophys. Chem. 49 (1994) 51–58.

[7] R.P. Rand, Interacting phospholipid-bilayers—measured forces and induced structural-changes, Annu. Rev. Biophys. Bioeng. 10 (1981) 277–314.

[8] D.M. Leneveu, R.P. Rand, V.A. Parsegian, D. Gingell, Measurement and modification of forces between lecithin bilayers, Biophys. J. 18 (1977) 209–230.

[9] J. Adam, P. Länger, G. Stark, Physikalische Chemie und Biophysik, Springer, Berlin, 1995.
[10] G. Bryant, J. Wolfe, Interfacial forces in cryobiology and anhydrobiology, Cryo Lett. 13 (1992) 23–36.
[11] J.R. Scherer, The partial molar volume of water in biological-membranes, Proc. Natl. Acad. Sci. USA 84 (1987) 7938–7942.
[12] S.H. White, R.E. Jacobs, G.I. King, Partial specific volumes of lipid and water in mixtures of egg lecithin and water, Biophys. J. 52 (1987) 663–665.
[13] D. Bach, B. Sela, C.R. Miller, Compositional aspects of lipid hydration, Chem. Phys. Lipids 31 (1982) 381–394.
[14] A.S. Ulrich, M. Sami, A. Watts, Hydration of DOPC bilayers by differential scanning calorimetry, Biochim. Biophys. Acta Biomembr. 1191 (1994) 225–230.
[15] S.A. Simon, C.A. Fink, A.K. Kenworthy, T.J. McIntosh, The hydration pressure between lipid bilayers—comparison of measurements using X-ray-diffraction and calorimetry, Biophys. J. 59 (1991) 538–546.
[16] H. Pfeiffer, H. Binder, G. Klose, K. Heremans, Hydration pressure and phase transitions of phospholipids—I. Piezotropic approach, Biochim. Biophys. Acta Biomembr. 1609 (2003) 144–147
[17] H. Pfeiffer, H. Binder, G. Klose, K. Heremans, Hydration pressure and phase transitions of phospholipids—II. Thermotropic approach, Biochim. Biophys. Acta Biomembr. 1609 (2003) 148–152
[18] S.A. Simon, T.J. McIntosh, Magnitude of the solvation pressure depends on dipole potential, Proc. Natl Acad. Sci. USA 86 (1989) 9263–9267.
[19] T.J. McIntosh, A.D. Magid, S.A. Simon, Range of the solvation pressure between lipid-membranes—dependence on the packing density of solvent molecules, Biochemistry 28 (1989) 7904–7912.
[20] K. Goossens, L. Smeller, J. Frank, K. Heremans, Pressure-tuning the conformation of bovine pancreatic trypsin inhibitor studied by Fourier-transform infrared spectroscopy, Eur. J. Biochem. 236 (1996) 254–262.
[21] M.C. Lin, P. Eid, P.T.T. Wong, R.B. Macgregor, High pressure Fourier transform infrared spectroscopy of poly(dA), poly(dT), poly(dA) and poly(dT), Biophys. Chem. 76 (1999) 87–94.
[22] O. Reis, R. Winter, T.W. Zerda. The effect of high external pressure on DPPC–cholesterol multilamellar vesicles: a pressure-tuning Fourier transform infrared spectroscopy study, Biochim. Biophys. Acta Biomembr. 1279 (1996) 5–16.
[23] M. Auger, H.C. Jarrell, I.C.P. Smith, D.J. Siminovitch, H.H. Mantsch, P.T.T. Wong, Effects of the local-anesthetic tetracaine on the structural and dynamic properties of lipids in model membranes—a high-pressure Fourier-transform infrared study, Biochemistry 27 (1988) 6086–6093.
[24] P.T.T. Wong, D.J. Moffatt, F.L. Baudais, Crystalline quartz as an internal-pressure calibrant for high-pressure infrared-spectroscopy, Appl. Spectrosc. 39 (1985) 733–735.
[25] P.T.T. Wong, D.J. Moffat, A new internal-pressure calibrant for high-pressure infrared-spectroscopy of aqueous systems, Appl. Spectrosc. 43 (1989) 1279–1281.
[26] Gmelins Handbuch der anorganischen Chemie, Ba: Barium, Ergänzungsband Nr. 60, Verlag Chemie, Weinheim/Bergstraße, 1960.
[27] A. Blume, Properties of lipid vesicles: FT-IR spectroscopy and fluorescence probe studies, Curr. Opin. Colloid Interface Sci. 1 (1996) 64–77.
[28] R.P. Rand, V.A. Parsegian, Hydration forces between phospholipid-bilayers, Biochim. Biophys. Acta 988 (1989) 351–376.
[29] G. Cevc, D. Marsh, Hydration of noncharged lipid bilayer-membranes—theory and experiments with phosphatidylethanolamines, Biophys. J. 47 (1985) 21–31.
[30] R. Winter, W.C. Pilgrim, A SANS study of high-pressure phase-transitions in model biomembranes, Ber. Bunsenges. Phys. Chem. 93 (1989) 708–717.

[31] H. Ichimori, T. Hata, H. Matsuki, S. Kaneshina, Effect of unsaturated acyl chains on the thermotropic and barotropic phase transitions of phospholipid bilayer membranes, Chem. Phys. Lipids 100 (1999) 151–164.
[32] H. Pfeiffer, R. Winter, G. Klose, K. Heremans, Thermotropic and piezotropic phase behaviour of phospholipids in propanediols and water, Chem. Phys. Lett. 367 (2003) 370–374.
[33] S.A. Simon, S. Advani, T.J. McIntosh, Temperature-dependence of the repulsive pressure between phosphatidylcholine bilayers, Biophys. J. 69 (1995) 1473–1483.
[34] A. Würflinger, G.M. Schneider, Differential thermal-analysis under high-pressures. 2. Investigation of rotational transition of several alkanes, Ber. Bunsenges. Phys. Chem. 77 (1973) 121–128.
[35] A. Landwehr, R. Winter, High-pressure differential thermal-analysis of lamellar to lamellar and lamellar to non-lamellar lipid phase-transitions, Ber. Bunsenges. Phys. Chem. 98 (1994) 214–218.
[36] B. König, Untersuchungen zum Hydratationsverhalten von mit nichtionischen Tensiden modifizierten Phospholipidmembranen, PhD thesis, Universität Leipzig, 1993.
[37] D. Marsh, Water-adsorption isotherms and hydration forces for lysolipids and diacyl phospholipids, Biophys. J. 55 (1989) 1093–1100.

CHAPTER 7

Modeling Protein–Lipid Interactions: Recent Advances

Monique Laberge,[1] István P. Sugár,[2] and Judit Fidy[1,*]

[1]*Department of Biophysics and Radiation Biology, Semmelweis University, Puskin u. 9, Budapest H-1088, Hungary*
[2]*Departments of Biomathematical Sciences and Physiology & Biophysics, Mount Sinai School of Medicine, One Gustave Levy Place, New York, NY 10029, USA*

Contents

Abstract

The lipid bilayer is the basic component of biological membranes. It functions as a relatively impermeable barrier for most water-soluble molecules. A large amount of proteins are intrinsic to the membrane or partially bound and mediate transport of specific molecules or act as receptors. The molecular modeling of lipid bilayers and proteins has now truly come of age with the widespread availability of parallel processing and advances in modeling methodologies.

This chapter will discuss recent advances in specifically modeling protein–lipid interactions focusing on two computational approaches: Molecular Dynamics (MD) and Monte Carlo (MC) simulation. MD is a choice method because the protein and lipid atoms are represented explicitly as well as solvation water and the whole system is followed in time, thus providing insights on both spatial organization and temporal dynamics. Even

* Corresponding author. Tel.: +36-1-267-6261; Fax: +36-1-266-6656; E-mail: judit@puskin.sote.hu

ADVANCES IN PLANAR LIPID BILAYERS AND LIPOSOMES, VOLUME 2
ISSN 1554-4516 DOI: 10.1016/S1554-4516(05)02007-7

though MC cannot provide time-dependent quantities, it represents a powerful method for sampling conformations and locations of a single peptide and also for simulating lateral diffusion of membrane lipids and proteins as well as the coupling of these distributions.

ABBREVIATIONS

A/H	area per headgroup
A/L	area per lipid
bR	bacteriorhodopsin
DMPC	1,2-dimyristoyl-*sn*-glycero-3- phosphatidylcholine
DMPG	1,2-dimyristoyl-*sn*-glycero-3-phosphatidylglycerol
DOPC	diolenoylphosphatidylcholine
DPPS	dipalmitoylphosphatidylserine
DPPC	dipalmitoylphosphatidylcholine
DPPE	dipalmitoylphosphatidylethanolamine
GA	gramicidin A
LJ	Lennard–Jones
MD	molecular dynamics
MC	Monte Carlo
OmpA	Outer membrane protein A
PB	Poisson Boltzmann
PBC	periodic boundary conditions
PC	phosphatidylcholine
PE	phosphatidylethanolamine
PME	particle-mesh Ewald summation
PMF	potential of mean force
POPC	palmitoyloleoyl phosphatidylcholine
POPS	palmitoyl-oleoyl phosphatidylserine
SDPC	1-stearoyl-2-docohexaencyl-*sn*-glycero-3-phosphocholine
TM	transmembrane

1. INTRODUCTION

This chapter reviews recent developments in the computational modeling of biological membrane systems with special emphasis on the challenges posed by simulations of protein–lipid complexes, reviewing contributions from both molecular dynamics (MDs) and Monte Carlo (MC) simulations, presently among the few methods capable of contributing atomic-level insights into the structure, properties and dynamics of lipid bilayers and membrane proteins. During the past decade, spectacular advances in computer hardware and parallel processing as well as new algorithmic developments in computational methodologies have allowed computational modeling methods to overcome their major limitations and contribute significantly to our knowledge of protein–lipid interactions. MD simulations have progressed from modeling very basic lipid molecules to that of complex bilayers containing physiological solutes and even transmembrane proteins [1]. Bilayer simulations using explicit lipid and solvent molecules are now routinely reported in the multinanosecond range. These advances have been coupled to significant technological progress in X-ray diffraction and NMR that yield increasing numbers of membrane protein structures, thus providing modelers with good initial configurations. The longer timescale of simulations also allow increasing correlation with the dynamical insights provided by experimental approaches that report on spin relaxations and bond mobilities, such as NMR, neutron scattering, spin-label EPR and FTIR [2].

The chapter is organized as follows: it starts with an introduction to membrane structure and protein–lipid interactions and proceeds to describe the MD method in the context of its application to lipid and membrane systems. Recent progress in methodological issues is surveyed as modeling efficiency critically depends on generating physically meaningful initial conditions and simulation protocols. We then review the recent contributions of MD simulations of pure bilayers, mixed bilayers and membrane protein and protein domain simulations. For the earlier literature, the reader is referred to the several reviews in print [3–12].

Finally, some insights obtained by MC simulations of lipid bilayer-protein systems are also discussed, and the chapter concludes with a brief outlook on future perspectives.

The first consequence of the recent computer hardware developments has been to allow significant increases in system size and simulation timescales, thus providing impetus to the modeling of membrane properties that are fast and localized – such as isomerizations, rotational relaxation and lateral diffusion in that they are now accessible to the average modeling laboratory. Aspects of membrane–protein interactions that are hard to access experimentally are now increasingly targeted by simulations. It should be noted, however, that methodological developments such as improved potential energy functions and parameterization, more accurate integrators, meaningful treatment of long-range electrostatics, more adequate statistical ensembles, standardization of periodic boundary conditions (PBC) in the case of MD simulations have been as important as the hardware developments.

Some of these methodological developments have also benefited all-atom MC simulations. Specific efforts have also been made to exploit the potential of MC methodology for efficiently sampling the conformational space of biomacromolecules, as demonstrated with the development of extension biasing, a sampling technique that allowed the sampling efficiency of MC to become comparable to that of MD [13].

Lipid membrane simulations have now truly come of age and we can expect them to yield significant new and important insights to our understanding of membrane structure, function, folding and dynamics at the atomic level in the near future. Besides being of interest because of their physiological importance, lipid membranes are also studied in relation to applications such as novel drug delivery techniques [14] and in many other biologically important topics such as the structure–function properties of lipoproteins, often found anchored to membranes.

2. MEMBRANE COMPLEXITY: LIPID BILAYER STRUCTURE AND PROTEIN–LIPID ASSOCIATION

Biological membranes are an essential component of living organisms and they are now recognized to play roles that far exceed their simple function as semipermeable physical barriers. Their characteristics have been extensively investigated for many decades, and experiments have provided substantial information about their physicochemical and structural properties [15–20].

They are highly complex and organized assemblies of lipids and proteins that contain ion pumps and channels that modulate signal transduction, energy conversion, storage and direct transport of small molecules in cells [21]. They are also supramolecular assembly sites and their role in organizing and folding membrane proteins into functional complexes has become the object of intense interest [22,23]. The complexity of their dynamics is also recognized [24]. The constituent lipids of a membrane bilayer are amphiphilic and conformationally very mobile, properties involved in the diffusion of other lipids, proteins and lipophilic molecules [20,25]. Integral membrane proteins, not covalently attached to membranes, also undergo conformational changes, but on a scale much slower than the rate of lipid diffusion, which is $\sim 10^3$ higher than that of proteins [26]. The classical fluid mosaic model has thus recently experienced significant refinement at the level of dynamics. For example, the role of membrane proteins that act as a barrier to the lateral diffusion of other proteins and lipids within the membrane has led to the increasingly sophisticated description of micro- and macro-domains within membranes. Diffusion is also better understood. Recent studies show that lateral diffusion of both integral and peripheral membrane proteins is controlled by their asymmetric orientation and distribution [27].

Lipid bilayers exist in a range of bilayer phases modulated by sometimes minute changes in hydration level or temperature. The crystal phase, referred to as L_c, is found at low temperature and hydration levels. The fatty acid chains are tightly packed and all *trans*. The degree of chain tilt, orientation and level of interdigitation of the two leaflets, have all been shown to vary significantly with lipid composition.

At high temperature and hydration level, the bilayer is in the liquid crystal, or L_α, phase and becomes highly fluid. Between L_c and L_α lies the phase usually referred to as the gel phase in which the fatty acid chains are packed less tightly than the crystal phase and are also mostly all *trans*. Intermediate phases have also been proposed, such as the subgel ($L_{c'}$) and ripple ($P_{\beta'}$) phases, the latter found at temperatures and hydration just below L_α. L_β is the phase found between subgel and ripple. Recent diffraction studies have also recently characterized two distinct L_β subphases, L_I and L_F with different fatty acid chain orientations. Membranes of biological interest are predominantly in the L_α phase with the result that most simulations are performed in this state [28].

Biological membranes come in a wide variety of composition and location within living organisms, with the consequence that their interactions with proteins are far from elucidated. This has to do with the wide diversity of proteins found to associate with membranes and that differ widely in their structural and dynamic properties. For instance, integral membrane proteins include proteins that span the entire membrane (TM proteins) or have a large section embedded within the lipid bilayer (Fig. 1) [29,30]. A large fraction also consist of transmembrane α-helices. Other membrane proteins are found in aqueous regions outside membranes and sometimes attached to membranes with covalent anchors, as for example, the glycosylphosphatidylinositol (GPI) – anchored proteins (e.g., acetylcholinesterase, amyloid protein Aβ). Finally, several peripheral membrane proteins have been identified that associate with membranes and anchored membrane proteins *via* noncovalent interactions.

Yet, in spite of this complex diversity, several of their characteristics have been identified, thus providing modelers with a good starting structural framework. It is now established that membrane proteins are amphipathic with polar regions found at the aqueous interface and nonpolar residues in the internal membrane regions (Fig. 2). Several recent 3D structures also show that secondary structure occurs in

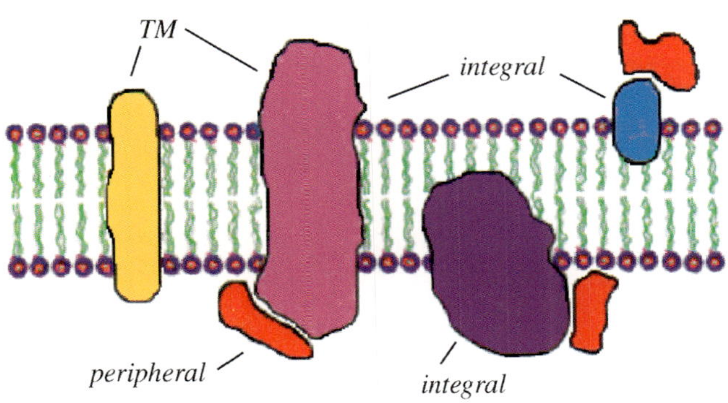

Fig. 1. Proteins associated with biological membranes.

Table 1. Lipid composition of eukaryotic membranes (% by weight)

Membrane type	Lipid type				Lipid (total)	Protein (total)
	Glycero-phos	Sphingo	Choles-terol	Other[a]		
Plasma membrane	45	10	10	35	50	50
Golgi complex	45	7	10	38	55	45
RER	80	2	3	15	48	52
Nuclear membrane	82	2	3	13	45	55
Outer mitochondrial	85	2	4	9	50	50
Inner mitochondrial	90	2	0	8	20	80

Source: http://www.kumc.edu/research/medicine/biochemistry/bioc800/mem01fra.htm.
[a] Various fatty acids, mono- and diacylglycerols.

both internal and external membrane regions (Fig. 3). Since they are amphipathic, they are not easily crystallized because they are not entirely soluble in aqueous or organic solvents. Thus the importance of molecular dynamics simulations in providing structural insights is evident.

Another feature of membrane complexity is that the overall lipid composition of biological membranes is far from homogeneous (c.f. Table 1) [20,31]. Not only do various lipids have different roles, such as supporting catalysis or facilitating protein sequestration/folding in the membrane, but they also contain various phospho- and glycolipid components that differ widely in fatty acid chain length,

Table 2. Composition of human erythrocyte membranes (% by weight)

Component	Weight (%)	Bilayer distribution (%) Outer	Inner
Proteins + glycoproteins			
Lipids	55.0	80	20
Phospholipids	45.0	75	25
spingomyelin	28.0	20	80
phosphatidylcholine	7.0	0	100
phosphatidylethanolamine	7.4		
phosphatidylserine	4.3		
phosphoinositides	1.0		
Cholesterol	13.0	50	50
Glycolipids	3.0	100	0
Fatty acids	1.0		

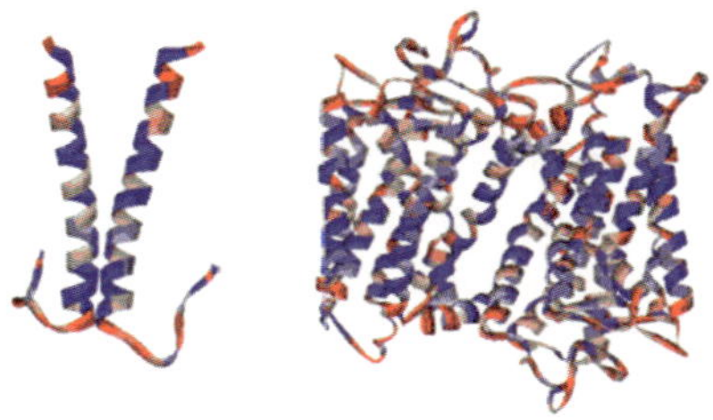

Fig. 2. Ribbon rendering of two membrane proteins colored according to the Eisenberg hydrophobicity scale: hydrophic residues in blue and polar residues in red. Left: glycophorin membrane-spanning domain (1afo.pdb); right: rhodopsin (1l9h.pdb).

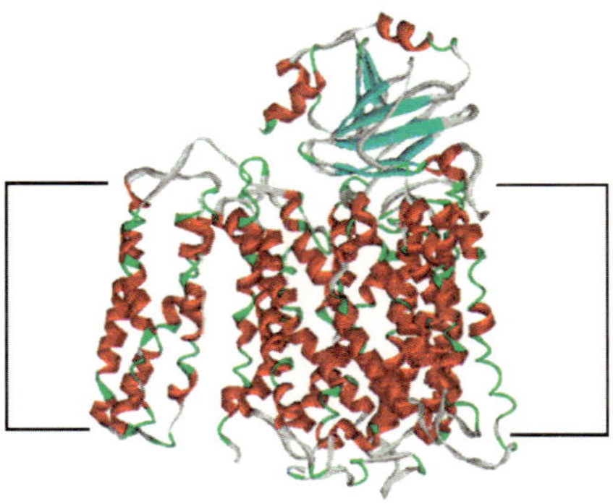

Fig. 3. Ribbon rendering of ubiquinol cytochrome *c* oxidase from *E. coli* (1fft.pdb). The oxidase helical structure (red) spans the membrane, note the random coil domain (blue) outside the bilayer with two small helices.

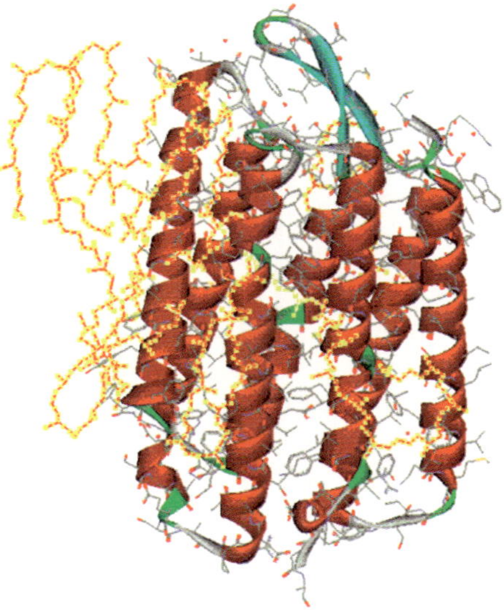

Fig. 4. Protein–lipid organization in bacteriorhodopsin at 1.90 Å resolution (1qhj.pdb). Helices red, coils blue and phytanyl lipid chains yellow.

saturation and headgroup structure. A membrane under physiological conditions thus consists of lipid molecules arranged as a bilayer which embeds various sterols, proteins and other lipids (c.f. Table 2). The polar headgroups (HG) and nonpolar fatty acid (FA) chains represent a highly flexible medium that allows these other lipids, proteins and solutes to diffuse into and out of the bilayer. Lipids and sterols amount to ~50% of most membranes by mass, the rest being accounted for by proteins. However, this ratio can vary in some specific membranes from 0.25 (purple membrane of BR) to 4 (lung surfactant) with some ~10% of lipids being charged. Phosphatidylethanolamines (PE) and phosphatidylcholines (PC) are the predominant species with PC amounting to some 50% of the HG. PC is known to be a bilayer-forming lipid and is found in most membranes (c.f. Table 2).

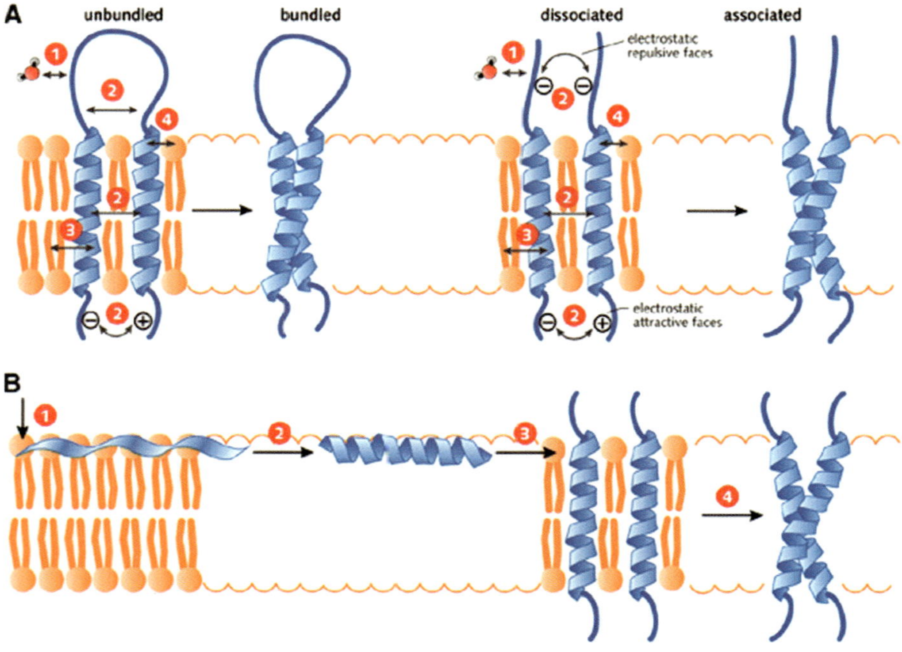

Fig. 5. Helix formation and bundling/association within membranes. (A) Schematic illustration of the forces that may lead to bundling of TM helices in proteins with multiple helices (left), and to association of TM helices belonging to different proteins (right). 1 represents protein–water, 2 protein–protein, 3 protein–bilayer core and 4 protein–bilayer interface interactions. (B) Depiction of the four-step model of membrane protein folding according to White *et al.* (JBC 276, 32395, 2001). 1 represents interfacial partitioning, 2 interfacial folding, 3 membrane insertion and 4-helix bundling. Proteins are blue, lipid molecules orange and water molecules red. (EMBO reports 3, 1133, 2002). Reproduction by kind permission from V. Helms, Bioinformatics Group, Saarland University, and the Nature Publishing Group.

Besides the bilayer lipids, membranes also incorporate lipid domains, defined as any area of a membrane that differs in lipid composition from the lipids that constitute the bilayer assembly. These domains can be transient, lasting for a few ns or very stable and their occurrence has been linked with protein–lipid interaction mechanisms. Nonbilayer-forming lipid components are often PE of monogalacto-syl/monoglycosyl which are neutral. Negative charge is usually contributed by PS, PG, CL and P acids. PS is the major anionic species in the plasma membrane of eukaryotes with PG and CL in prokaryote and mitochondrial membranes [31].

In the plasma membrane, most proteins have three different domains: two hydrophilic and one hydrophobic (Fig. 4). The latter threaded through the membrane and the former at the C- and N-terminals located in the cytoplasm and extracellular medium. Many are also glycosylated with glycans attached covalently to a hydrophilic domain. Another feature of integral membrane proteins is that they are known to associate with very specific membrane lipids (Fig. 5). For example, cytochrome *c* oxidase exhibits a strong specificity for cardiolipin. Another example is Ca^{2+} ATPase strongly associated to 12 lipid molecules [32] and bacteriorhodopsin (bR) is also tightly coupled to phytanyl chains (Fig. 6). The structural determinants of TM protein helix–helix packing are also known to be significantly different than that of soluble protein. For instance, the significance of the common

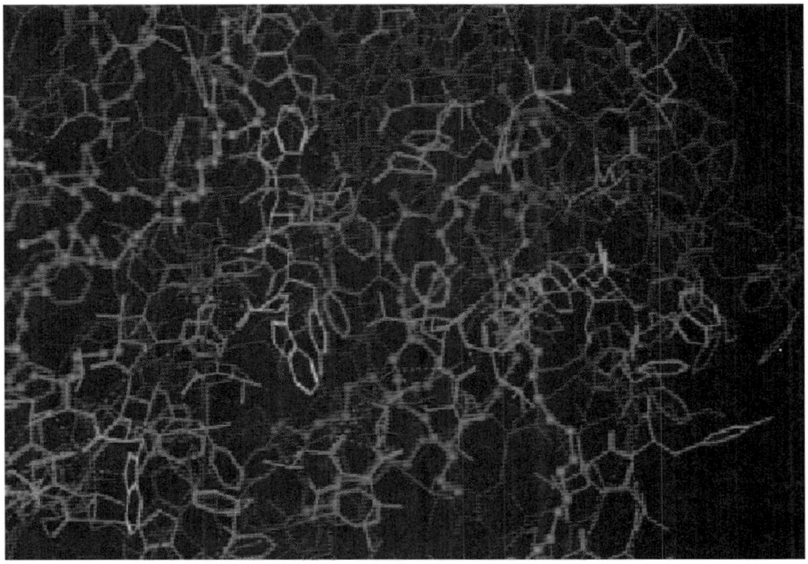

Fig. 6. Cytochrome *c* oxidase from *R. sphaeroides* structure (1m56.pdb) showing the distearoylphosphatidylethanolamine contact region. Protein residues are blue, hydrophobic residues purple and Trp–Tyr yellow. The lipids are rendered as ball and stick and H-bonds indicated with a white dashed line.

occurrence of glycine residues in TM helices has recently been elucidated in that it provides membrane helices with the flexibility required for the opening and closing of potassium-selective ion channels [33].

3. PROTEIN–LIPID INTERACTIONS

Protein–lipid interactions mediate the great diversity of membrane assembly and function. Understanding the molecular forces that govern the association of proteins and membranes as well as peptide–peptide interactions in a lipid environment at the atomic level, represents one of the major challenges of biophysics, to which simulations are one of the few methods capable of providing a unique contribution. For a discussion of lipid–protein interactions in biological membranes from a structural perspective, the reader is referred to Lee's [15] excellent review. Shai's [17] discussion of molecular recognition within the membrane milieu and Marsh's [34] review of lipid interactions with TM proteins are also indicated. For a specific discussion, Jones *et al.* [35] recently reviewed protein–lipid interactions in the purple bacterial reaction center.

The intensive research efforts of the past few years have yielded a deeper understanding of these interactions, but we are still very far from understanding them at the level required to make accurate predictions.

From a nonbonded interaction point of view, the first consideration is that some phases of a membrane system represent a highly charged medium with polar and charged lipid molecules interacting with each other and with small solute membrane molecules or proteins. Coulombic interactions are very strong between headgroups, headgroups and solvent, and solvent–solvent molecules. The second consideration is that the interior of a bilayer represents a hydrophobic environment of low dielectric constant in which membrane proteins or domains associate tightly with nonbilayer-forming lipids. In such an apolar environment, donor/acceptor groups are not in contact with solvent. Thus, the polar interactions that contribute significantly to the specificity of globular proteins in solution are replaced in the membrane interior by hydrogen-bonding interactions that have recently been shown to stabilize the helical conformation of membrane-spanning domains. In particular, the $C\alpha$–H$\cdots$O hydrogen bond has been recognized as an important determinant of membrane stability, packing and specificity [36].

Thus, the assembly of membrane components is energetically very different from the association of proteins in water, which is mainly driven by the hydrophobic driving force, not present in the case of TM proteins. The most recent model describing helix formation within membranes is based on a delicate interplay of electrostatic forces and hydrophobic interactions mediated by bilayer lipids that provide 'lipid entropy', the driving force for protein complex formation within membranes (Fig. 7). In this model, the specific lipids bound between

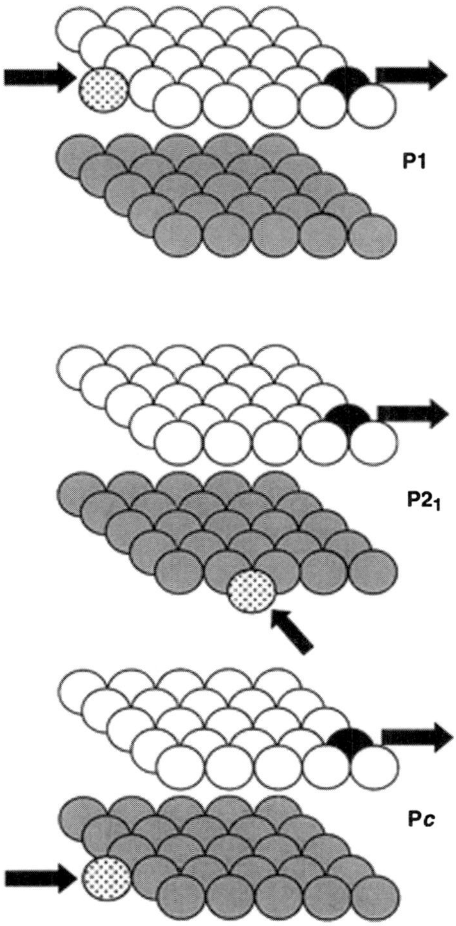

Fig. 7. Representation of a lipid bilayer under three different periodic boundary conditions. The commonly used *P*1 (top), and the recently developed *P*2$_1$ (middle), and *Pc* (bottom). The spheres represent lipid headgroups, with those in the top leaflet white and in the bottom leaflet gray. The black particle exits the primary cell, to be replaced by its image (dotted). The arrows show the direction of exit and entry for these particles for each PBC. Details in *Biophys. J.*, 2002, **82**, 2317–2325. Illustration reproduced by kind permission from B. Brooks, Laboratory of Biophysical Chemistry, National Institutes of Health and Biophysical Journal.

the subunits of supramolecular complexes are proposed to form a flexible interface between subunits, providing the binding energy required to stabilize electrostatic interactions as well as a flexible hydrophobic interface, thus increasing the entropy of interaction [26].

A description of protein–lipid interactions that goes beyond that of a hydrophobic bilayer core contained between polar boundaries exposed to

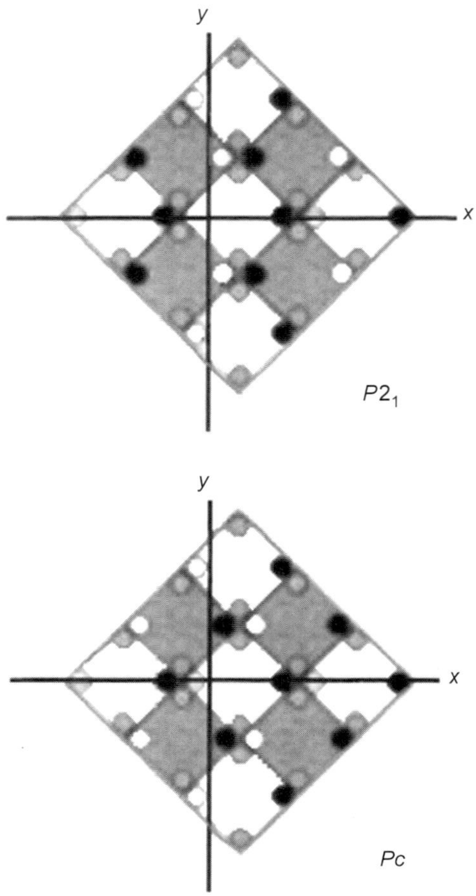

Fig. 8. The $P2_1(x, y+1/2, z)$ and $Pc(x+1/2, y+1/2, z)$ transformations, looking down the normal (z-axis) to the bilayer (xy plane). The primary unit cell is white, and located in the center; the four particles in the corners of the cell reside on the top lipid leaflet. In the adjacent cells (dark gray), the same four particles are in the bottom leaflet, are rotated with respect to the primary cell for $P2_1$, and inverted for *Pc*. The outer light gray cells are the result of a second transformation, and the four particles return to the top lipid leaflet. Detailed discussion in *Biophys. J.*, 2002, **82**, 2317–2325. Illustration reproduced by kind permission from B. Brooks, Laboratory of Biophysical Chemistry, National Institutes of Health and Biophysical Journal.

aqueous solvent on either side of the bilayer is now increasingly emerging. Convincing evidence [37–39] points to the occurrence of a ~10 Å water–lipid interface region composed of lipid headgroups and water molecules which interacts favorably with specific regions on the surface of membrane proteins. These regions can be polar, but also aromatic, as for example the aromatic 'surface belts' consisting mainly of Trp and Tyr sidechains believed to play an anchor role for membrane proteins. The presence of a shell of immobilized lipids located around membrane proteins has also been revealed by high-resolution structures (Fig. 8). Unlike that of soluble proteins, the molecular surface of a membrane protein is much rougher, consisting of nicks and grooves into which the fatty acid chains of the associated lipids can fit, thus providing the tight compactness required to ensure membrane nonpermeability.

4. SIMULATION METHODS

Numerical simulations of protein–lipid systems have contributed significantly to our understanding of several aspects of their interactions. For example, MD simulations have provided insights into bilayer lipid structure in different phases [28,39–42], the role of bilayer surface tension [43,44], the role of water and hydration [45,46], the interaction of peptides with bilayers [47–49], the interaction of peripheral and integral membrane proteins with bilayers [50–53], as well as aspects of transbilayer transport of water and ions [54–57]. MD is by far the preferred simulation method as it can provide an explicit description of the systems. But descriptions are limited by the number of molecules that can be included in the models, as well as by the timescale over which a given process can be studied. MC simulations applied to lipid membranes have avoided these limitations by introducing stochastic elements in the simulations and using simplified semiphenomenological models. However, these models also impose very severe restrictions. Nevertheless, MC methods have contributed most significantly to our understanding of the structure of pure lipid bilayers [13], of lipid bilayers containing cholesterol [58–60], and of the free energy and orientation of water molecules across the bilayer [61,62]. The interaction of a single peptide with a bilayer has been also investigated by using reduced representations of the peptide-bilayer system (see Section 4.2), and with the current advances in MC techniques, one can expect all-atom single peptide-bilayer MC simulations in the near future. By using MC methods and even further reduced peptide-bilayer representations, the lateral distribution of peptides and lipid molecules has been simulated. Neither MD nor MC is presently capable of elucidating the self-assembly process leading to a lipid membrane in aqueous solution, but both methods have addressed important 'subtopics' such as lipid membrane molecular organization and 3D conformational complexity [63].

4.1. Molecular dynamics simulation

The increasing number of available 3D membrane protein structures (Table 3) is yielding highly valuable insights into the molecular organization of these proteins.

Table 3. Some recent structures of membrane proteins

PDB ID	Protein	Res. (Å)	Year
1B9U	Membrane domain of subunit B of the *E. Coli* Fof1 ATP synthase	NMR	1999
1BM1	Bacteriorhodopsin in the light-adapted state	3.50	1999
1C3W	Bacteriorhodopsin	1.55	1999
1E54	Anion-selective porin from *C. Acidovorans*	2.10	2000
1E7P	Fumarate reductase from *W. Succinogenes*	3.10	2001
1EHK	Ba3-cytochrome *c* oxidase from *T. Thermophilus*	2.40	2000
1EIS	Photosynthetic reaction center from a thermophilic bacterium, *T. Tepidum*	2.20	2000
1EK9	Tolc integral outer membrane protein and efflux pump component from *E. Coli*	2.10	2000
1EOP	L intermediate of bacteriorhodopsin	2.10	2000
1EUL	Calcium ATPase with two bound calcium ions	2.60	2000
1EZV	Yeast cytochrome bc1 complex co-crystallized with an antibody Fv-fragment	2.30	2000
1FFT	Ubiquinol oxidase from *E. Coli*	3.50	2000
1FX8	*E. Coli* glycerol facilitator (Glpf) with glycerol	2.20	2000
1G90	Outer membrane protein A TM domain	NMR	2000
1H68	Sensory rhodopsin II	2.10	2001
1H9E	Lem-like domain of human membrane protein Lap2	NMR	2001
1IIE	Trimeric domain of the Mhc class II-associated chaperonin and targeting protein II	NMR	1999
1I78	Outer membrane protease Ompt from *E. Coli*	2.60	2001
1IJD	LH3 complex from *R. Acidophila*	3.00	2001
1ILX	PS II	3.80	2001
1J95	Kcsa potassium channel	2.80	2001
1JBO	PS I	2.50	2001
1INM	Intimin-190 (Int190) from enteropathogenic *E. Coli*	NMR	1999
1K4C	Potassium channel Kcsa-Fab complex I	2.80	2001
1LDF	*E. Coli* glycerol facilitator (Glpf)	2.10	2002

(*continued*)

Table 3. *continued*

PDB ID	Protein	Res. (Å)	Year
1M56	Cytochrome *c* oxidase from *R. Sphaeroides*	2.30	2002
1MG1	Htlv-1 Gp21 ectodomain/maltose-binding chimera	2.50	1999
1MHS	*N. Crassa* proton ATPase	8.00	2002
1OSM	Ompk36, the osmoporin of *K. Pneumoniae*	3.20	1999
1PF4	Msba from *V. Cholera*	3.80	2003
1QD5	Outer membrane phospholipase A from *E. Coli*	2.17	1999
1QHJ	Bacteriorhodopsin with phytanyl segments	1.90	1999
1QJ8	Outer membrane protein Ompx from *E. Coli*	1.90	1999
2SQC	Membrane protein squalene-hopene cyclase	2.00	1999
1UUN	Main porin from *M. Smegmatis* (Mspa)	2.50	2004

But X-ray diffraction detects only those lipids that are highly ordered, providing a static picture and little insight into the dynamics of protein–lipid interactions. Heteronuclear NOE can measure spin relaxation, reflecting bond vector mobility on a ps–ns timescale, but as with all spectroscopies, resolution in NMR depends on obtaining sharp and intense resonance bands, which is related to how fast the molecule is tumbling. The larger the molecule, the more slowly it tumbles, and the longer the rotational correlation time which results in band broadening and loss of intensity. This imposes an upper limit for the protein size, a clear drawback for membrane proteins embedded in large lipid bilayers [66].

During the past decade, however, MD simulations have been able to complete this experimentally derived structural knowledge on the timescale of fast and localized dynamics and we can expect more contributions [65].

4.1.1. Overview of the method

MD is a powerful theoretical approach that can now study the structure and dynamics of biological macromolecules as complex as lipid bilayers and membrane proteins in the picosecond to the nanosecond range. Originally developed in the early 1960s to simulate atomic motion in simple liquids [64], MD simulations started to be applied to biological macromolecules in the mid-1970s, with the very first protein simulations reported on bovine pancreatic trypsin inhibitor by MacCammon *et al.* [67].

MD uses the equations of classical mechanics to describe a potential energy function for a system and to predict the time evolution of its structure and properties. In all MD approaches, a molecular system is a collection of atoms that interact *via* simple analytical functions of the atomic coordinates. In this approach, the potential energy function is commonly the sum of the individual components of the potential energy, which accounts for all interactions, both intra- and intermolecular. A typical energy function has the form:

$$U(\vec{R}) = \sum_{\text{bonds}} k_b(r-r_0)^2 + \sum_{\text{UB}} k_{\text{UB}}(s-s_0)^2 + \sum_{\text{angles}} k_\theta(\theta-\theta_0)^2 + \sum_{\text{pr.dihe}} |\phi| - k_\phi\cos(n\phi) + \sum_{\text{impr.dihe}} k_\omega(\omega-\omega_0)^2 + \sum_{\text{nb.pairs}, i\neq j}\left[\frac{A_{ij}}{r_{ij}^{12}} - \frac{B_{ij}}{r_{ij}^{6}}\right] + \sum_{\text{nb.pairs}, i\neq j}\left[\frac{q_i q_j}{\varepsilon r_{ij}}\right] \quad (1)$$

where k_b, k_{UB} k_θ, k_ϕ, k_ω are the bond, Urey–Bradley, angle, dihedral angle, improper dihedral angle force constants, respectively; r, S, θ, ϕ and ω are the bond length, Urey–Bradley 1–3 distance, bond angle, dihedral angle and improper torsion angle with the subscript zero representing the equilibrium values. The bonds between atoms are modeled with harmonic expressions while the Coulomb and Lennard–Jones (LJ) terms describe the nonbonded interactions. The sixth term is the LJ potential and I expresses the van der Waals interaction energy using the constants A and B that depend on the atom type, and for which values are determined by a variety of methods, such as from nonbonding distances in crystals or from scattering measurements. The electrostatic interaction is represented by the last term, a Coulomb potential in which q is the partial atomic charge, ε the effective dielectric constant of the medium, and r_{ij} the distance between atoms i and j. Given $\check{R}$, the vector of the atom coordinates, the distances and angles required to evaluate equation (1) can readily be determined. In this form of the potential energy function, nonbonded interaction terms are evaluated for all atoms separated by three or more covalent bonds. For some specific atoms (aliphatic carbons, amide nitrogens, oxygens), there is scaling of the 1–4 Lennard–Jones term. The function does not contain an explicit term for H-bonding because the nonbonded term can accurately represent the hydrogen-bonding interaction. This potential energy function – and the parameters characterizing the interactions described by the terms is commonly referred to as the force field, whose accuracy and validity is tested against experimental data. The function is also differentiable with respect to the atomic coordinates; this gives the value and the direction of the force acting on an atom according to Newton's second law:

$$F_i = m_i a_i = -\frac{dV}{dr_i} = m_i \frac{d^2 r_i}{dt^2} \tag{2}$$

where F_i is the force exerted on particle i, m_i is the mass of particle i and a_i is the acceleration of particle i. Knowing the force on each atom, it is then possible to also know the acceleration of each atom in the system. The force can also be expressed as the gradient of the potential energy V.

Newton's law can then relate the derivative of V to the changes in position as a function of time. In a MD simulation, the goal is to acquire 'trajectories' which are the atomic positions (generalized coordinates) as a function of time. With a 1D description of the trajectory, equation (2) can be written as:

$$x = at^2 + v_0 t + x_0 \tag{3}$$

relating the value of x at time t as a function of the acceleration a, the initial position x_0 and initial velocity v_0. Thus, a trajectory can be generated provided that the initial coordinates of the atoms of the system are available, as well as an initial distribution of velocities and the acceleration.

Most MD codes approximate the many-atom system by a simple kinetic gas model, that is, they generate the initial distribution of velocities (v_i) from a random Gaussian or Maxwell–Boltzmann distribution at a given temperature:

$$p(v_{ix}) = \left(\frac{m_i}{2\pi k_B T}\right)^{1/2} \exp\left[-\frac{1}{2}\frac{m_i v_{ix}^2}{k_B T}\right] \tag{4}$$

Temperature is determined from the velocities using:

$$T = \frac{1}{3N}\sum_{i=1}^{N}\frac{|p_i|}{2m_i} \tag{5}$$

After calculating v_0 and a in equation (3), the equations of motion are integrated to generate a trajectory consisting of a set of frames of coordinates and velocities that describe the motion of the atoms over time. In this way, the average structure of a molecular system can be calculated as well as fluctuations of geometric parameters, properties, and time-dependent processes of the system. Numerous properties computed from a simulation can then be compared to experimentally determined quantities. Several algorithms are available to perform the numerical integration, the Verlet algorithm being the 'historical' one [68]. In its basic implementation, two third-order Taylor expansions are written for the positions of the coordinates $r(t)$, the first for time $t+\delta t$ and the second for $t-\delta t$.

With r the velocities, i.e., the first derivative of r with respect to time, a the accelerations, i.e., the second derivative, and b the third derivative, we have at timestep $t+\delta$:

$$r(t+\delta t) = r(t) + v(t)\delta t + \frac{1}{2}a(t)\delta t^2 + \frac{1}{6}b(t)\delta t^3 + O(\delta t^4) \quad (6)$$

And at timestep $t-\delta$:

$$r(t-\delta t) = r(t) - v(t)\delta t - \frac{1}{2}a(t)\delta t^2 - \frac{1}{6}b(t)\delta t^3 - O(\delta t^4) \quad (7)$$

Adding both expressions yields:

$$r(t+\delta t) = 2r(t) - r(t-\delta t) + a(t)\delta t^2 + O(\delta t^4) \quad (8)$$

The positions and accelerations at time t and the positions from time $t-\delta t$ are used to calculate new positions at time $t+\delta t$. For a detailed description of MD and other numerical methods for integrating differential equations (the simple Verlet algorithm described above is not used as such, variants have been developed), the reader is referred to more comprehensive monographs [69,70]. All MD methods follow this basic scheme, they differ in the equations of motions that are being solved (e.g., Newton's equations, Langevin equations, etc.) and in the computational procedures used to account for the timescale associated with a given process.

4.1.2. Methodological challenges and advances

All-atom MD simulations of lipid–protein systems are performed using various computational protocols and several groups are actively engaged in designing and optimizing MD simulations conditions that best fit the experimental results on membrane systems. These include PBC that maintain constant volume, thermodynamic ensemble definitions such as constant pressure or constant surface tension, and treatment of electrostatic interactions. No unique 'best' protocol has yet emerged, but the refinements of the past decade have been very substantial [71].

4.1.2.1. Sampling efficiency

An important concern inherent to MD simulations has always been the ergodicity – or sampling problem that affects any multiminimum system not being able to cross barriers frequently enough so as to achieve equilibrium within the timescale of the simulation. The consequence is that large systems such as membrane-protein models are usually nonergodic on a given timescale and at a given temperature because the time average of several properties (e.g., equipartition of kinetic energy, relaxation of nonbonded interactions, atomic fluctuations…) depends on initial conditions. Even with frequent barrier crossings, the bias introduced by the starting conditions is not likely to disappear within the simulation time. Thus, increased CPU performance is important. In the case of bilayer simulations, it has been shown that an hydrated 128 lipid molecule-system requires between 10 and 20 ns of dynamics before reaching equilibration, due to the occurrence of large membrane fluctuations that have long correlation times. This long timescale is also required to adequately follow lateral lipid diffusion and membrane undulations [72].

To reduce the CPU time required for equilibration and increase the conformational sampling time, Pastor *et al.* [73] developed approaches to start lipid simulations from a structural model that corresponds as closely as possible to an equilibrium state. Woolf and Roux further developed this approach to assemble protein-membrane complexes from pre-equilibrated and pre-hydrated phospholipid molecules. To embed a protein in a lipid bilayer, lipid molecules are randomly selected from a pre-equilibrated and pre-hydrated set, and placed around the protein, reducing the number of core–core overlaps between heavy atoms through systematic rotations about the *Z*-axis and translations in the *XY* plane of the lipids and protein. The construction of such systems is consistent with the concept of a membrane being a partly ordered, partly disordered liquid-crystalline phase. This initial configuration is then equilibrated and subjected to MD, significantly reducing the equilibration CPU requirement. Woolf and Roux applied the approach to the simulation of the gramicidin A (GA) channel [74], pf1 coat protein [75], melittin [76] and the KcsA potassium channel [77].

4.1.2.2. Force fields and parameterization

Initial MD efforts aimed at simulating bilayers were severely hampered by the lack of adequate force fields and parameters. They also produced structures that were highly distorted due to an inadequate balance between the physical description of the forces responsible for some aspects of lipid structure and the algorithms used to run the dynamics. Recently, the development of force fields and parameters specifically designed – or improved for the MD simulation of phospholipid systems has been the object of considerable effort by several groups and is an on-going process [11]. Feller and MacKerrell [78] recently reported improvements to the CHARMM all-atom force field that includes substantial adjustments to the LJ hydrocarbon and torsional parameters and to the partial atomic charges and torsional parameters of the phosphate group. The resulting energy function was also validated by an 11 ns MD simulation of a hydrated phospholipid bilayer. CHARMM was also improved by the recent inclusion of new parameters for polyunsaturated lipids, tested on a 16-ns MD simulation of a SDPC bilayer [79].

Another MD simulation program, GROMACS – claimed to be the fastest code in the world and very popular for coarse-grained simulations of chemical systems, has also been optimized for lipid simulations. Berger *et al.* [80] parameterized the united-atom LJ parameters of hydrocarbon tails using a cutoff of 1.0 nm for the LJ interactions and an isotropic long-range dispersion correction which were incorporated in GROMACS along with other new parameters that allow the modeling of gel-phase models Marrink *et al.* further developed the force field, performing simulations of lipid adhesion forces using both GROMACS and GROMOS [72], another force field available to model lipid–protein interactions [81].

Recently, the performance of the GROMOS96 parameter set developed for aliphatic alkanes was validated on a DPPC solvated in water in the liquid-crystalline L_α phase. The area per lipid (*A*/*L*), as a measure of structure, the order parameters of the chain carbons, as a measure of membrane fluidity, and the translational diffusion of the lipids in the plane of the bilayer were calculated and compared with experimental data. An optimal set of headgroup area, chain order parameters and a lateral diffusion coefficient in accordance with experimental data were obtained [82].

Similarly, AMBER is continuously improved for simulating lipid systems. Hydrocarbon nonbonded parameters were recently added for use in lipid simulations [83] and Smondyrev and Berkowitz [84] refined the united atom AMBER force field for the simulation of phospholipid membranes. An accurate set of torsional parameters corresponding to a set of atomic partial charges calculated for DPPC and DMPC molecules was derived and a Ryckaert–Bellemans potential was used for the torsion potentials for the united atoms in lipid chains. The force field was validated by performing a constant pressure simulation on a DPPC bilayer in water. The average area per headgroup (*A*/*H*), peak-to-peak distance in the electron density, and order parameter profile for the hydrocarbon chains agreed with the experimental values and with the results of a simulation using a model with explicit hydrogens [84,85].

4.1.2.3. Treatment of long-range electrostatics

The most time-consuming part of a MD simulation is the calculation of the electrostatic forces, described by the coulombic term in the potential energy function. The term is evaluated between every pair of atoms. To speed up calculations, the interactions between two atoms separated by a distance greater than a pre-defined distance, i.e., the cutoff distance, are ignored as the artifacts due to truncation, if any, are expected to become smaller as the cutoff distance is increased, and that for reasonably large cutoffs, the system should not be influenced by truncation. Molecular dynamics programs propose several different truncation methods to truncate the interaction that have proven adequate for systems close to overall neutrality. However, membrane systems are highly charged, and the treatment of long-range electrostatic interactions becomes an important issue. This is because the summation in the coulombic term formally includes all atoms. In contrast to the Lennard–Jones term which rapidly converges as r^6/r^{12}, truncation of the coulombic term has been shown to lead to over or underestimation of the electrostatic forces acting on the system and to severe artifacts, especially when electrostatic interactions are truncated between dipoles or whole charges [3,86]. It has been shown that properties at the surfaces of lipid monolayers are indeed affected by truncation [87]. Thus, the obvious solution, which is to ignore long-range electrostatic interactions, and the next obvious, which is to use truncation, are not easily applicable to membrane simulations.

To address this issue, lipid bilayer system simulations are often performed with cylindrical or spherical cutoffs. In this approach, all interactions within a cylinder/sphere perpendicular to the lipid–water interface with a given radius are calculated explicitly and the remainder is accounted for by solving the Poisson equation based on the average charge distribution in a direction perpendicular to the interface. But the most accurate electrostatic treatment is to use PBC and solve the Poisson equation for the entire system using some form of Ewald treatment, commonly the particle-mesh Ewald (PME) summation [88].

The electrostatic energy of a periodic system can be expressed by a lattice sum over all pair interactions and over all lattice vectors excluding the $I=j$ term in the primary box. Under certain conditions, the summation can converge. Ewald's method transforms this summation into two more complex but rapidly convergent sums, the first carried out in reciprocal space and the second in real space. The lattice sum is thus reformulated to yield absolutely convergent summations, which define the electrostatic potential. Given the periodicity present in both crystal calculations and in MD simulations performed under PBC, PME and its variants have become increasingly popular to describe electrostatic energy and force.

Recent $L_{\beta'}$ bilayer MD simulations performed by Venable *et al.* clearly show that PME summation of electrostatics is superior to spherical cutoff methods and that summation corrections have to be included [28]. However, the approach remains to be validated for L_{α} bilayer simulations which have to model a phase in which lipids are much more disordered and liquid-like with the risk of introducing unwanted periodicities when using Ewald summations [11].

For all-atom simulations, the current trend is definitely to use PME summation, the reaction field approach that corrects the pair-wise electrostatic interactions to account for the polarizable medium outside a given cutoff being computationally far too expensive [70].

4.1.2.4. Use of implicit solvation models

Since the inclusion of explicit solvent considerably increases CPU requirements, implicit solvation approaches are continuously refined, the generalized Born solvent model being considered one of the most promising. For simulations that are not specifically focused on assessing the role of solvation, these models provide modelers with a tool that significantly decreases computational requirements. Roux and Simonson [89] recently reviewed the theoretical implications of implicit solvation. They provide a thorough description of the quantity of interest – the solute potential of mean force (PMF), and of the models presently available for an accurate description. Since the total solvation energy can be described by the sum of two contributions – nonpolar and electrostatic contributions, two basic models can be used to approximate them, either solvent-exposed area models or macroscopic continuum models, the latter based on the Poisson Boltzmann (PB) equation for macroscopic media and becoming increasingly efficient due to recent advances in refining semi-analytical approximations. In the context of membrane systems, Roux *et al.* have shown how the influence of the transmembrane potential can easily be incorporated into the theory using a modified form of the PB equation, validating the approach by modeling the voltage-dependent membrane insertion of a polyalanine α-helix and by computing the electrostatic potential across a 60-Å-diameter sphere used to represent a large intrinsic protein [90,91]. In a recent review, Tobias assesses the performance of implicit solvation models and discusses the new approaches for the treatment of electrostatics in the context of membrane system simulations [92].

4.1.2.5. Periodic boundary conditions

PBC enable MD simulations to be performed using a relatively small number of particles in such a way that the particles experience forces as though they were in a bulk solution. The last decade has seen steady progress in the application of PBC to MD simulations of solvated protein systems. The advantages are that the boundary effects experienced with solvated shells and layers are minimized and that PME/Ewald summations can be used to describe the electrostatics. PBC also allows larger systems to be modeled because only the coordinates of the primary cell are stored. In the case of lipid bilayer modeling, however, a severe PBC limitation is involved when the relative number of lipids in the two leaflets of a bilayer requires adjustment during a simulation. In the usual implementation of PBC, primary cells are replicated by translation only, as illustrated in Fig. 7, which shows a particle in the primary cell (black) and its image (dotted) under $P1$ symmetry. When the particle exits the primary cell from the right side of the top leaflet, the image particle enters the cell at the left. A microscopic patch of as few as 50–100 lipids can plausibly model the short timescale dynamics of a much larger bilayer. However, adequate modeling of a lipid bilayer requires more than the $P1$ translation described in Fig. 7. Brooks *et al.* [93] recently developed two alternative transformations in PBC designed to allow exchange of lipids between leaflets ($P2_1$ and Pc symmetry). Figure 8 illustrates these new transformations. In $P2_1$, the primary cell can be rotated about the y-axis and translated along the y-axis by one-half a cell length; the rotation has the effect of transforming x to x and z to z. In a simulation of a lipid bilayer, the effect of the $P2_1$ boundary condition is that a lipid exiting either leaflet can re-enter the opposite leaflet *via* an orthogonal face. In Pc, the cell is inverted through a mirror plane at $z=0$, another way to overcome the limitations of the translationally invariant $P1$.

For other important methodological developments, the reader is referred to the comprehensive overview published by a team of expert investigators leading five different research groups on the algorithmic work involved in MD long-time integration, crystallographic refinement approaches, fast electrostatic approaches and the implementation of computation-intensive programs on the newly available computing architectures [94].

4.1.3. Survey of recent MD simulations

In this section, we briefly survey the major and most recent MD simulation reports of lipid–protein bilayer systems. Several excellent earlier reviews discuss MD applications in this context with a remarkable depth of insight and almost all include discussion of significant contributions at the time of writing. The reader is referred particularly to the reviews by Feller [95], Tieleman *et al.* [3], Saiz and Klein [6], Sansom *et al.* [9], Pastor *et al.* [96] and Scott [11].

4.1.3.1. Pure bilayers

The first simulations performed on pure bilayer systems containing only one type of lipid molecule historically provided the initial insights into bilayer properties. Pure lipid bilayer MD simulations represent the systems that benefit the most from the recently available computer power because of their chemical simplicity, thus the hardware advances immediately translate into both qualitative and quantitative improvements. Also, very few lipid molecule X-ray structures are available. In those that are, crystal packing constraints force them to adopt conformations different from those of membrane lipids; thus the special relevance of pure bilayer simulations.

The influence of truncating the electrostatic interactions was investigated in a fully hydrated DPPC bilayer through 20 ns MD simulations. Trajectories in which the electrostatic interactions were truncated were compared to similar simulations acquired using PME. All examined truncation distances (1.8–2.5 nm) led to major new effects on the bilayer properties, such as enhanced order of acyl chains together with decreased *A*/*L*. The results obtained using PME, on the other hand, were shown to be consistent with experiments [86]. Tieleman *et al.* [97] report a 40-ns simulation of a palmitoyl-oleoyl phosphatidylserine (POPS) bilayer in the liquid crystalline phase with Na^+ counterions and NaCl that investigates the structure of the negatively charged lipid bilayer and the effect of salt on its structure. The authors show that Na^+ counterions play an important role in determining the properties of the bilayer in that counterions can condense close to the headgroup region or intercalate within the bilayer, leading to a significant role in membrane electrostatics [97].

Another simulation reporting on a dipalmitoylphosphatidylserine (DPPS) bilayer confirms the importance of including counterions in simulations [98]. The authors report that H-bonding between the NH_3^+ group and the phosphate group leads to a reduction in the *A*/*H* when compared to the area of a dipalmitoylphosphatidylcholine (DPPC) bilayer. The Na+ ions bind to the oxygen in the carboxyl group of serine, thus giving rise to a dipolar bilayer similar to a dipalmitoylphosphatidylethanolamine (DPPE) bilayer. The results of the simulation clearly show that counterions play a crucial role in determining the structural and electrostatic properties of the DPPS bilayer.

A bilayer system with 128 POPC lipids was studied by Böckman *et al.* [99] with all simulations performed under PBC conditions with a total $>21{,}000$ atoms, comparing the effect of counterion addition. Increasing NaCl concentration decreased the self-diffusion of POPC lipids within the bilayer. Self-diffusion coefficients calculated from 100-ns simulations suggest that most of the relaxation processes relevant for lipid diffusion are faster than the simulation timescale. The MD simulations also revealed a tight binding of Na^+ to the carbonyl oxygens of, on an average, three lipids leading to larger complexes with reduced mobility. Pandit *et al.* [100] also report on the effect of NaCl on a DPPC bilayer containing 128 lipid molecules and subjected to 10-ns NPT dynamics with and without salt. Ion binding to the membrane surface was characterized by the loss of ion hydration. The binding of Cl^- was slightly weaker than that of Na^+. Excess binding of Na^+ yielded a net positive surface charge density just outside the bilayer producing a positive electrostatic potential in this region, whereas the system without salt showed an electrostatic potential of zero.

Marrink *et al.* [101] report most interesting DPPC simulations designed to investigate hydrophobic pores. The system was simulated in two different states; first, as a bilayer containing a meta-stable pore and second, as an equilibrated bilayer without a pore with surface tension applied in both cases to study the formation and stability of hydrophilic pores inside the bilayers. The flux of water molecules through these stabilized pores was analyzed, and the structure and size of the pores characterized. Marrink and Mark also studied the effect of membrane undulations on surface tension glycerolmonoolein (GMO) bilayers [102]. Up to 40 ns MD simulations were performed on systems ranging from 200 to 1800 GMO molecules. As the system size increased, long wavelength undulatory modes appeared with a concomitant increase in the area compressibility due to the coupling of the undulation modes to area fluctuations.

4.1.3.2. Mixed bilayer simulations

Biological membranes are much more complex than pure bilayer systems and MD simulations have now reached a point where more realistic models can be studied, incorporating more than one type of lipid (Fig. 9) or small physiologically relevant molecules. For example, the properties of mixed dioleoylphosphatidylcholine (PC)/dioleoylphosphastidylethanolamine (PE) bilayers were studied as a function of PC/PE headgroup composition. Adding small amounts of PE to PC bilayers caused a strong reduction of the area of the bilayer and adding small amounts of PE to PC bilayers increased the hydration of the PC headgroups compared to PC hydration in pure PC bilayers [103]. Another recent study reports on simulations of two mixed bilayers containing dipalmitoyl-PC and dipalmitoyl-PS including

different concentrations of NaCl electrolyte [98]. The authors conclude that dipalmitoyl-PC alone is less likely to form interlipid complexes than in the presence of ions or dipalmitoyl-PS.

The importance of the work on mixed bilayers lies in that the significant effects reported as a result of mixing different lipids in bilayers raises the question of the validity and of the caution that must be exercised in performing/evaluating simulations performed on more complex systems (e.g., incorporating peptides or proteins) with bilayers consisting of only one type of lipid.

Besides simulations mixing two different lipid types, more complex membrane sections incorporating lipophilic molecules have also become increasingly sophisticated with respect to system size and trajectory length.

For example, the role of cholesterol (Chol) was recently investigated by Edholm *et al.* in a series of 10-ns simulations of 1024 hydrated DPPC molecules varying the cholesterol contents from 0 to 40%. The main effect of cholesterol was to reduce the *A*/*L*, increase the ordering of the fatty acid chains, in good agreement with experiment. The simulations provided further insights: the effect of Chol is dependent on the depth in the bilayer and also reduces membrane undulations [65].

A simulation of this size represents an increase of a factor of 5 compared to previous simulations. The benefit derived from the greater number of molecules and longer simulation is that the conclusions reached have more statistical validity while decreasing the sampling bias, especially important since lipid molecules can adopt a wide range of conformations in the presence of Chol.

Another recent study reports on the effect of epicholesterol (Echol) on a PC bilayer comparing the bilayer properties of a pure DMPC bilayer and that of a mixed DMPC–Echol bilayer [104]. Like Chol, Echol was found to change the organization of the bilayer/water interface and to increase membrane order and condensation. Smondyrev and Berkowitz [85] report five MD simulations on different phospholipid/sterol membrane systems, using Chol, ergosterol and lanosterol. The simulations showed that the sterol structure also significantly affects bilayer properties. Ubiquinone (UQ) was also simulated in a dipalmitoyl-PC bilayer. Multinanosecond trajectories were acquired, allowing the authors to study UQ mobility and its preferred positions within the bilayer, while monitoring its diffusion rate, found to be strongly dependent on the UQ position within the bilayer [105].

Thus mixed bilayer MD simulations suggest that lipid and small lipophilic molecule stoichiometry cannot be ignored as they significantly affect bilayer properties, such as lateral diffusion and headgroup electrostatic interactions.

4.1.3.3. Bilayer-peptide simulations

MD simulations of peptides – often α-helices – embedded in bilayers are an area of great interest not only because the membrane-spanning portions of TM proteins often consist of α-helix bundles, but also because several types of peptides are of high biomedical interest, for example, the small peptides that are known to bind membranes and destroy invasive microorganisms, affecting membrane fluidity in the process [106]. Recently, several MD simulations of polypeptides embedded in bilayers have been reported, investigating their stability and dynamics as well as the role of hydrophilic and hydrophobic membrane domains in stabilizing protein structures. Other simulations have reported on how a bilayer restricts polypeptide flexibility, while others monitored the effect of peptide insertion on bilayer properties.

In this context, two 6-ns simulations of the somatostatin analog sandostatin were recently reported. In the first simulation, the analog was placed in solvent and allowed to bind the bilayer surface. In the second, the N-terminus of the bound species was deprotonated to model experimental observations and the simulation proved consistent with experiment, in that MD yielded increased peptide-bilayer interaction. Thus, details of peptide-bilayer complexes properties could be described, such as peptide orientation, percent buried surface area, and lipid headgroup orientation, again in agreement with experimental data [48].

Another study reports on simulating the structural properties of an opioid peptide, dynorphin A (1–17), a potent analgesic, embedded in DMPC bilayers. NMR data was used to correctly insert the peptide in the bilayer. The resulting model encompassed some $\sim$26,400 atoms including solvent and $>$2-ns simulations were performed. Specific interactions of the opioid with lipid headgroups, fatty acid chains and water were identified with the simulations also predicting a tilted orientation for the helical region of the peptide [107].

Another κ-opioid receptor docking study reports on modeling synthetic benzomorphan-based ligands docked to dynorphin A in a DPPC bilayer using MD to refine the docking results. The interactions of the protein with its ligands was characterized with results in agreement with site-directed mutagenesis data showing that all TM regions are important modulators of ligand affinity [108].

In the context of simulating TM α-helix bundle formation, Petrache *et al.* performed MD simulations on both dimeric and monomeric forms of glycophorin A (GpA) in four different fluid phase lipid bilayers (DMPC, DPPC, DOPC, POPC). Trajectory analysis showed how lipid chain length and saturation modulate the structural and energetic properties of TM helices [109]. Specifically, the simulations showed that the four types of bilayer environment did not change the structure of GpA significantly. However, the range of fluctuations about the average were modulated by lipid properties, thus supporting the concept that membrane proteins at physiological temperatures are best described by fluctuations in the range of their accessible structures. This is important in the context of understanding how protein stability correlates with structural fluctuations.

Melittin, a small amphipathic peptide found in bee venom has also recently been simulated in a DMPC bilayer to study the microscopic details of protein–lipid interactions. Calculated density profiles and order parameters of the lipid acyl chains extracted from the trajectories showed that melittin insertion affects both layers of the membrane. The lipid chains also closely pack around melittin, suggesting their importance in stabilizing the helix in the bilayer [76].

4.1.3.4. Bilayer-protein simulations

The past 4 years have seen substantial progress in the MD simulations of membrane-channels and proteins [4,9,106,110–114]. Given the experimental difficulties in studying these systems, we expect that MD simulations based on more accurate starting systems will play an increasingly larger role in the elucidation of membrane protein structure and function. Considering that the translocation of a single ion across a channel is a μs process, MD simulations attempting to provide insight into the fluctuations of ion channels have always been limited by severe timescale considerations. Nevertheless, the field of ion channel simulation has seen major advances, and is finally starting to bridge the gap between relating the function of a channel to its underlying molecular structure. A consequence of the timescale limitations was that a description of ion fluxes represented a major problem for simulating channels. Recently, Roux *et al.* developed a PMF approach that successfully describes the free energy surface governing ion conduction and allows to calculate conductance from channel MD simulations [115]. This contribution illustrates the spectacular advances in MD simulations if one considers that conductance was a property beyond the reach of the method only 3 years ago. Bernèche and Roux applied their PMF approach to investigating the energetics of ion conduction through the KcsA K^+ channel, a wide, nonpolar pore of some 8 Å in radius on the intracellular side that narrows on the extracellular side to a pore of some 12 Å in diameter – the so-called 'selectivity filter'. They were able to elucidate the mechanism of K^+ fluxes through this filter at the atomic level and simulate how they are diffusion-limited [114].

The same authors applied the PMF method to the GA channel, the hydrophobic pentadecapeptide which, as a dimer, forms cation-selective membrane pores of some ~4 Å in diameter. MD had already allowed to clarify several aspects of GA function, such as ion binding, ion selectivity, multiple occupancy and gating, properties successfully simulated and found entirely consistent with experimental data [110]. The PMF approach allowed to investigate ion conduction through the GA channel, simulated for 0.1 μs with explicit solvent and membrane [116]. On the basis of the simulations, single channel conductance was determined for K^+ ions (0.8 pS), in good agreement with experiment. Single ion dissociation constants were calculated, also within the range of experimental determinations.

The KcsA channel was also simulated by Domene and Sansom [117] who investigated interactions of Na^+, K^+, Rb^+ and Cs^+ ions within the selectivity filter in 48-ns simulations of KcsA embedded in a POPC bilayer. Results showed that buried waters help to stabilize the filter and described the concerted motions of ions and water within the filter.

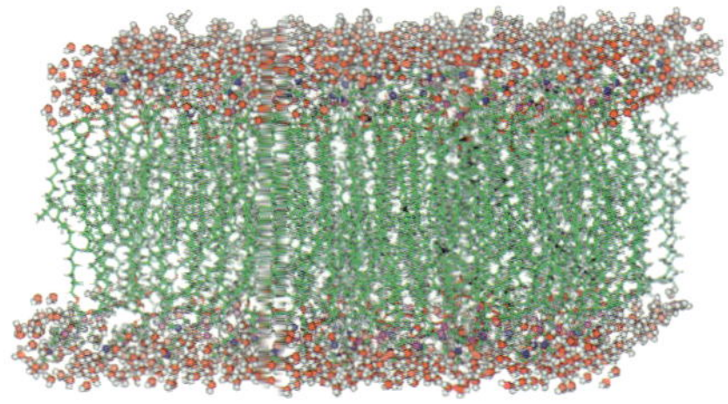

Fig. 9. Mixed DMPC/DMPG bilayer: the model contains 174 lipid molecules for a total of 19020 atoms. Gel phase simulation after 200 ps in the context of monitoring HSA-lipid interactions (see J. Fidy *et al.*, *Langmuir*, 2003, **19**, 146–153).

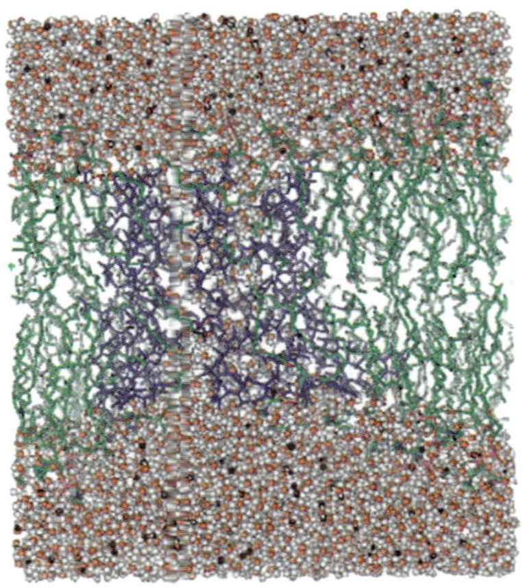

Fig. 10. 8-helix alamethicin bundle structure after a 4 ns-simulation in a POPC bilayer. The coordinates were obtained from Peter Tieleman, University of Calgary, simulation details in: D.P. Tieleman, B. Hess and M.S.P. Sansom, Analysis and evaluation of channel models: simulations of alamethicin, *Biophys. J.*, 2002, **83**, 2393–2407.

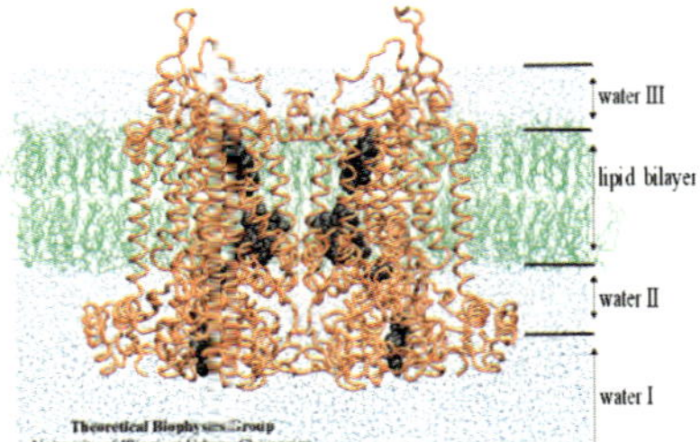

Fig. 11. Model of the structure of the cytochrome bc1 complex from chicken heart mitochondria with the iron–sulfur protein inserted in the membrane bilayer. Backbone represented by orange tubes; the Fe_2S_2 cluster and the heme groups are rendered as gray spheres; the lipid molecules are rendered as green lines, and the solvent water molecules are shown as blue lines and dots. The model contains a total of 206,720 atoms and a subset of 91,061 atoms was simulated with 45,131 moving atoms. Details in *Biophys. J.*, 1999, **77**, 1753–1768. Illustration reproduced by kind permission from K. Schulten, Theoretical and Computational Biophysics Group, University of Illinois at Urbana-Champaign and Biophysical Journal.

Tieleman *et al.* recently evaluated channel models by simulating alamethicin (Fig. 10), the antimicrobial channel-forming peptide [112]. Alamethicin bundles consisting of 4, 5, 6, 7 and 8 helices were simulated in POPC bilayers for 20 ns. The main motions were ascribed more to individual helices rather than to bundles as a whole. A result of methodological significance is that the authors performed the simulations under different electrostatic treatments, comparing cutoff methods and PME. Water ordering inside the channels showed a dependence on the algorithm used.

Recent simulations by Sansom *et al.* [118] report on the bacterial outer membrane protein OmpA. OmpA was simulated in a DPC detergent micelle and in a DMPC bilayer for 10 ns. Enhanced flexibility was observed in the micellar environment, allowing changes in the sidechain torsion angles that resulted in the formation of a pore in the middle of the OmpA molecule. These interesting results offer an explanation for the experimentally observed channel formation by OmpA. The authors also report OmpA simulations that describe other features of the OmpA transient pore formation mechanism [119].

The high level of molecular characterization achieved by MD methods on ion channel systems has also started to be reproduced on protein-membrane systems.

Steered MD simulations of the cytochrome bc1 complex (Fig. 11) were recently reported with the structure inserted in a phospholipid bilayer (206,720 atoms). A subset of 91,061 atoms was simulated with 45,131 moving atoms for 1 ns. Point charge distributions for the force field parameterization of heme groups and the Fe2S_2 cluster of the Rieske protein included in the simulated complex were determined. The simulations showed that rotation of the soluble domain of the iron–sulfur protein is actually feasible. Several metastable conformations of the ISP during its rotation were identified and the interactions stabilizing the initial, final and intermediate positions of the soluble head of the ISP domain were characterized [1].

Bovine rhodopsin, the prototypical G-protein coupled receptor (GPCR), has also been recently simulated in a DOPC fully hydrated bilayer (41,623 atoms) for 40 ns [120]. The simulation was undertaken to investigate the retinal conformational change that triggers the rhodopsin conformational change, which activates G-proteins and signal amplification of the visual response. The simulations provided insights into H-bonding changes that occur near retinal within $\sim$10 ns and that lead to a shift in helical tilt. This result is interesting because it provides a rare glimpse into the coupling of local and larger scale protein motions.

Another GPCR MD simulation was performed with rhodopsin embedded in a POPC bilayer with acquisition of 15 ns trajectories that revealed a conformation significantly different from that of the recent crystal structures. Hydrophobic matching occurred between the lipids and the receptor surface within the MD timescale in good agreement with experiment [121].

The dynamics of water molecules in the bR trimer were recently simulated in an explicit lipid/water environment. The trimer was inserted in a fully solvated POPC bilayer (84,000-atoms) and 5-ns trajectories were acquired. Two large areas of higher water density were identified in bR, located between the central proton binding site, the Schiff base, and the extracellular proton release site. Residues were identified – H-bonded to the water clusters and proposed to facilitate proton conduction. The author suggest that this may explain why several site-directed mutations alter the proton release kinetics without a direct involvement in proton transfer [122].

The role of dynamic water networks was also investigated in *P. denitrificans* cytochrome *c* oxidase MD simulations. The oxidase was embedded in a DMPC bilayer and 1.125 ns trajectories acquired. Most water molecules had residence times shorter than 100 ps, but a few water molecules were fixed inside the protein for up to 1.125 ns. The H-bond network in the oxidase was not uniformly distributed, and the degree of water arrangement was also variable. The average number of solvent sites in the proton-conducting K- and D-pathways were determined. The highly fluctuating H-bond networks, combined with the significant diffusion of individual water molecules, provide a basis for the transfer of protons in cytochrome *c* oxidase, therefore, leading to a better understanding of the mechanism of proton pumping [123].

4.2. Monte Carlo simulations

4.2.1. MD versus MC simulations of protein–bilayer systems

Molecular dynamics provides a powerful means to investigate the conformation and dynamics on all-atom models of one-component lipid bilayers involving about 100–1000 lipid molecules in nanosecond time regimes or one-component lipid bilayers with an initially positioned single integral or peripheral protein. MC simulations have been demonstrated in recent studies to be also an important tool in the investigation of membrane interaction with a single protein or peptide. The stochastic nature of the MC method excludes the calculation of time-dependent properties. This limitation can also give an edge to the MC methodology over MD: being uncoupled from the time evolution of the system the size of the configuration space exploring steps is not limited by any fundamental reasons, only by the practical requirement of obtaining a sufficient number of accepted moves and it can be increased by introducing appropriate biasing.

Typically, the MC methods are based on reduced representation of the protein-membrane system, in contrast to the all-atom representation used in MD simulations. The reduced representation enables comprehensive sampling of peptide conformations and locations in the membrane in an accelerated manner. In most of these

MC studies the membrane is considered as a continuous hydrophobic profile, while the peptide is represented in atomic or residue resolution [124–132].

The technical problems of simulating protein-membrane system become even more complex when one goes from a single species lipid bilayer to a system of mixed lipids or of lipids mixed with other types of molecules such as peptides. The relaxation times for different components of a heterogeneous membrane to sample conformations and orientations relative to each other are orders of magnitude longer than the nanosecond time-scale sampled by molecular dynamics [133]. Similarly, the distance-scales over which membrane domains organize themselves are often larger than 100 nm^2 surface area typical of the largest molecular dynamics simulation to date [134]. In order to model these heterogeneous systems one has to use further reduced peptide-membrane models such as the lattice models [135–137]. In spite of the simplicity of the lattice models, reviewed in the next sections, they provide results in quantitative agreement with a broad range of experimental data [138–143]. It is particularly important that only these lattice models can simulate lateral distribution of membrane lipids and membrane proteins as well as their coupling.

4.2.2. One-component two-state lipid bilayer

Before reviewing the lattice models of the protein-membrane systems it is useful to give a relatively detailed description of the simplest lattice model of one-component phospholipid membranes. This will help to understand lattice models *per se* and the abbreviated descriptions of the more complicated lattice models of protein-membrane systems.

The simplest lattice model has been originally developed by Sugar *et al.* [144] to simulate the gel-to-fluid transition in dipalmitoylphosphatidylcholine small unilamellar vesicles. It was further developed to study gel/fluid and compositional clusters in two-component phospholipid membranes [138–141] and also to study the effect of lipid–protein interaction on the lateral distribution of membrane proteins [135,136,142,145].

4.2.2.1. Lattice geometry, states, and configuration

In the simplest lattice model one layer of the lipid bilayer is modeled on a two-dimensional triangular lattice, of N lattice points [146]. Each lattice point represents a lipid molecule. Each lipid molecule exists in either gel or fluid state. In gel state the hydrocarbon chains of the molecule are in all-*trans* configuration, while in fluid state the chains contain a given number of rotational isomers. The actual membrane configuration can be characterized by specifying the state of the lipid molecule at every lattice point. Each configuration can be roughly characterized by the number of gel and fluid molecules, i.e., by N_g and N_f.

respectively, where $N_g + N_f = N$, and by the number of pairs of nearest neighbor molecules (N_{ij}), where one is in state *i* and the other is in state *j*.

4.2.2.2. Configuration energy and degeneracy

The energy of a configuration is:

$$E = E_g N_g + E_f N_f + E_{gg} N_{gg} + E_{gf} N_{gf} + E_{ff} N_{ff} \tag{9}$$

where the first two terms give the intra-molecule energy of the configuration and the last three terms is the inter-molecule interaction energy of the configuration. E_i and E_{ij} is the intra-molecule energy in state *i*, and the interaction energy between a molecule in state *i* and a nearest neighbor molecule in state *j*, respectively. $E_f - E_g > 0$ because each rotational isomer in the fluid molecule has about 2.1 kJ/mol energy. Only nearest neighbor interactions are considered because short-range van der Waals interactions are dominating between the hydrocarbon chains.

In the case of PBC the following relationships hold [144]:

$$zN_i = 2N_{ii} + N_{ij} \quad \text{for every } i \text{ state} \tag{10}$$

where z is the coordination number of the lattice, and $i \neq j$. After substituting equation (10) into equation (9), we get

$$E = \Delta E N_f + w_{gf} N_{gf} \tag{11}$$

where $\Delta E = [E_f + (z/2)E_{ff}] - [E_f + (z/2)E_{ff}]$ is the energy difference between the all-fluid and all-gel configuration, and

$$w_{gf} = E_{gf} - \frac{E_g + E_f}{2} \tag{12}$$

is the cooperativity energy, associated with the half-width of the gel-to-fluid transition, and strongly affects the lateral distribution of the gel/fluid state molecules in the membrane.

The rotational isomers can be arranged along the hydrocarbon chains of the fluid molecule in many ways, while the intra-molecule energy remains the same. The number of possible arrangements, Ω_f is the degeneracy of the fluid state molecule. The degeneracy of the gel state molecule is $\Omega_g = 1$, because the all-*trans* state of the hydrocarbon chains is the only micro-state of the molecule. The degeneracy of a membrane configuration is:

$$\Omega = (\Omega_f)^{N_f} (\Omega_g)^{N_g} \tag{13}$$

4.2.2.3. Simulation of equilibrium fluctuations

The membrane is in thermal contact with the surrounding (canonical ensemble). It can take energy from and give up energy to the surrounding. As a result the membrane energy fluctuates around an average, while higher and lower energy configurations form. In order to simulate these fluctuations one can use Monte Carlo methods. Each simulation starts from either an all-gel or all-fluid state. During the equilibration, trial configurations are generated by changing the state of a randomly selected lipid molecule from gel to fluid or from fluid to gel. Each trial configuration is accepted according to the following criterion [147]:

$$\mathrm{RAN} < \frac{\Omega_{\mathrm{trial}}}{\Omega_{\mathrm{orig}}} \mathrm{e}^{-(E_{\mathrm{trial}} - E_{\mathrm{orig}})/RT} \tag{14}$$

where RAN is a random number generated in the (0,1) interval, Ω_{trial} and Ω_{orig} is the degeneracy of the trial and the original configuration, respectively, E_{trial} and E_{orig} is the energy of the trial and original configuration, respectively, T is the absolute temperature, and $R=8.3$ J/mol/deg is the Boltzmann constant. If the inequality in equation (14) is not fulfilled the trial configuration is rejected. This way one can generate a chain of accepted membrane configurations. The specificity of the above criterion is that it drives the system from any initial configuration to the thermodynamic equilibrium, i.e., the energy of the equilibrium configurations follows a Boltzmann distribution. Along the chain of the accepted configurations the membrane energy is monitored. The thermodynamic equilibrium is attained when the membrane energy starts to fluctuate around a constant value, i.e., around the equilibrium average energy.

4.2.2.4. MC cycle and analysis of simulated configurations

After the attainment of the equilibrium one may start analyzing the configurations. The subsequent trial configurations are strongly correlated. However, the correlation decreases with the increasing distance of the trial configurations. During N subsequent trial state generations, where N is the number of lattice points in the membrane model, the system has the opportunity of realizing all of its configurations one time. In this series of trial state generations, called Monte Carlo cycle, the first and last accepted configuration will be uncorrelated. Thus by analyzing the configuration only at the end of each Monte Carlo cycle, one analyzes uncorrelated configurations. For example, by determining the membrane energy at the end of many Monte Carlo cycles one can calculate the average energy of the membrane at a given temperature. By running the simulations at several temperatures one can get the average energy at different temperatures, $\langle E(T)\rangle$. The numerical derivative of the average energy according to the temperature provides the calculated heat capacity curve of the membrane.

4.2.2.5. Determination of model parameters

In order to complete our membrane model for a particular lipid membrane, e.g., dipalmitoylphosphatidylcholine large unilamellar vesicle [136], we have to determine the model parameters ΔE, ΔS $(=R \ln \Omega_f - R \ln \Omega_g)$ and w_{gf}. This can be done by fitting the simulated heat capacity curve to the experimental one. The experimental heat capacity curve has one peak characterizing the gel-to-fluid transition of the membrane. The maximum of the peak is at the melting temperature, $T_m = 314.2$ K of the membrane, the integral of the curve gives the transition enthalpy, $\Delta H_{tr} = 36.4$ kJ/*mol* and the transition entropy is $\Delta S_{tr} = \Delta H_{tr}/T_m = 115.87$ J/mol K. Since the volume work is much smaller than the transition energy $\Delta E \approx \Delta H_{tr} = 36.4$ kJ/*mol*. The measured transition entropy $N\Delta S_{tr}$ can be identified with the calculated transition entropy $N\Delta S = R \ln(\Omega_f)^N - R \ln(\Omega_g)^N = NR \ln(\Omega_f/\Omega_g)$. Note that $\Omega_f/\Omega_g = \Omega_{trial}/\Omega_{orig}$ if in the trial state generation the randomly picked molecule is in gel state and it is changed into fluid state, otherwise $\Omega_f/\Omega_g = \Omega_{orig}/\Omega_{trial}$. The remaining parameter w_{gf} can be determined by fitting the heat capacity calculated at T_m to the respective experimental value, and thus $w_{gf} \approx 1326$ J/mol.

It is important to note that models using more lipid states [148] do not alter the overall physical picture. It is also not crucial whether lipids are located on triangular lattices or whether lipid chains are located on such lattices [149].

4.2.2.6. Lateral distribution of gel/fluid lipid molecules

The values of the model parameters have profound effects on the type of the gel-to-fluid transition and on the lateral distribution of the gel/fluid lipid molecules in the membrane (see Table 4).

When $w_{gf} = 0$ the gel and fluid molecules are randomly distributed in the membrane, i.e., the average number of nearest neighbor fluid state molecules around a gel or fluid state molecule, $\langle N_f^{nn} \rangle = z\langle N_f \rangle / N$, is proportional to the concentration of the fluid state molecules. If $w_{gf} > 0$ and the transition temperature is above the critical temperature $RT_m > 1.82 w_{gf} = RT_c$ [150] fluid and gel state molecules coexist at the region of the transition temperature, T_m. In this coexistence region fluid state molecules are surrounded preferentially by fluid state molecules, i.e., $\langle N_f^{nn} \rangle > z\langle N_f \rangle / N$, and gel molecules are surrounded preferentially by gel state molecules. The gel (or fluid) state molecules form islands or clusters. The size and number of the clusters in a membrane configuration can be determined by using a cluster counting algorithm, labeling each cluster by different numbers [151]. During the gel-to-fluid transition equilibrium can be attained at every temperature, i.e., the transition is continuous, equilibrium transition.

If $w_{gf} > 0$ and the transition temperature is below the critical temperature $RT_m < 1.82 w_{gf} = RT_c$ the equilibrium distribution is either attainable or unattainable during the transition. When it is attainable, the system spontaneously flip-flops between mostly gel and mostly fluid configurations at around the transition

Table 4. Effect of cooperativity parameter on the lateral distribution of gel/fluid lipids

Parameter	Distribution	Heat capacity curve	Comments
$w_{gf}=0$	Random	Broad, symmetric	System of independent two-state molecules
$w_{gf}>0$ $RT_m>1.82w_{gf}$	Clustering of similar state molecules	Sharp, symmetric There is no hysteresis	Equilibrium system. Gel–fluid clusters coexist. Continuous transition
$w_{gf}>0$ $RT_m<1.82w_{gf}$	In the transition region flip-flop between mostly-gel and mostly-fluid configurations	Sharp, symmetric. There is no hysteresis at slow scan rate	Attainable thermodynamic equilibrium. First-order equilibrium transition
	Depending on thermal history either mostly-gel or mostly-fluid configuration	Sharp. There is hysteresis	Unattainable thermodynamic equilibrium. First order non-equilibrium transition
$w_{gf}>0$	Preference for gel–fluid nearest neighbors	Broad symmetric	

temperature. In this case the transition is still an equilibrium transition, and it is called first order phase transition. This type of transition may take place primarily in small systems where the barrier between the two sets of configurations is not high enough to prevent jumps induced by the thermal fluctuations. In the case of certain large systems, however, the equilibrium distribution is not attainable, and the system is confined into either the mostly gel configurations or the mostly fluid configurations. The change from mostly gel configurations to mostly fluid configurations take place vehemently at temperatures above T_m and the change from mostly fluid configurations to mostly gel configurations take place vehemently at temperatures below T_m, i.e., the transition has a hysteresis, and it is a nonequilibrium process in both directions.

4.2.2.7. Model building

In the lattice models, one tries to keep the number of model parameters to an absolute minimum otherwise the parameter fitting is not manageable. The complexity of the model can be increased step by step, where each step introduces one or maximum two new model parameters. This strategy assumes that experimental data are available at every level of complexity. The next model of integral proteins is a good example of this strategy.

4.2.3. One-component two-state lipid bilayer with integral proteins

Integral peptides and proteins are embedded into the lipid matrix of the membrane. Their lateral distribution can be understood by the concept of 'hydrophobic matching', which implies that the interaction between integral proteins and matrix lipids depends on the relative length of their hydrophobic cores [152–154]. The integral proteins tend to aggregate, when the 'hydrophobic matching' is poor in order to minimize the lipid–protein interface. The lattice model described in Section 4.2.2 has been extended to a model of one-component lipid bilayers which also contain integral peptides (or proteins) [136]. In the simplest version of the model, it is assumed that each peptide occupies one lattice point. The number of peptides, fluid and gel-state lipid molecules is N_p, N_f and N_g, respectively, where $N_g + N_f + N_p = N$. The configuration energy of this two-component membrane is:

$$E = \Delta E N_f + w_{gf} N_{gf} + w_{gp} N_{gp} + w_{fp} N_{fp} \tag{15}$$

where N_{gp} and N_{fp} is the number of gel-state lipid–peptide and fluid-state lipid–peptide contacts, respectively. The gel-to-fluid transition energy, ΔE is defined by equation (11), the cooperativity parameters, w_{gp} and w_{fp} are related to the peptide–lipid interaction energies as it is defined by equation (12). Equation (15) can be derived by assuming PBC, and *assuming that the microstates of a peptide do not affect its interaction with the lipid molecules*. The degeneracy of the configuration is:

$$\Omega = (\Omega_f)^{N_f} (\Omega_g)^{N_g} (\Omega_p)^{N_p} \tag{16}$$

where Ω_p is the number of microstates of a peptide.

The steps of the Monte Carlo simulation are rather similar to the steps described in Section 4.2.2. Initially the peptides are located at the first N_p lattice points, while gel-state lipid molecules are situated on the rest of the lattice points. Besides the state alteration there is an additional way of trial configuration generation: one lipid and one peptide are picked randomly and their positions are exchanged. This step is necessary to get the equilibrium lateral distribution of the peptides. Note that this is a nonphysical step, which may accelerate the attainment of the equilibrium distribution [155]. The number of model parameters increases by the two cooperativity parameters: w_{gp} and w_{fp}, while the degeneracy of the peptide, Ω_p cancels out in the Metropolis criterion, equation (14). The values of the cooperativity parameters depend on the actual lipid and peptide (or protein) compounds. Table 5 lists the effect of the values of the cooperativity parameters on the peptide lateral distribution and on the heat capacity curve.

Note that the model in its current form is unable to explain the heat capacity profile of DMPC-gramicidin S [156], where the peak slightly shifts to lower

Table 5. Effect of cooperativity parameters on the peptides' lateral distribution

Cooperativity parameters	Peptides lateral distribution	Heat capacity curve	Comments and examples
$w_{gp} = w_{fp} = 0$	Random in both phases	Peak's symmetric broadening	Peptides at the gel–fluid interface lower the line tension and the co operativity of the melting
$w_{gp} = w_{fp} > 0$	Aggregation in both phases	Peak's small symmetric broadening	
$w_{gp} = 0$, $w_{fp} > 0$	Aggregation in fluid phase only	Peak shifts to higher temperature and has a wing at the high-temperature end	At high temperature accumulation of the peptides in the remaining gel phase domains. See example at Ref. [33]
$w_{gp} > 0$, $w_{fp} = 0$	Aggregation in gel phase only	Peak shifts to lower temperature and has a wing at the low-temperature end	At low temperature accumulation of the peptides in the remaining fluid phase domains. See examples at Ref. [35]

temperature with increasing peptide concentration, while the wing appears on the high-temperature end. We think that by releasing the assumption that the microstates of a peptide do not affect its interaction with the lipid molecules one would be able to model this effect too.

Usually the cross-sectional area of a peptide (or protein) is larger than that of a lipid molecule. Lattice models have been developed where a peptide occupies more than one lattice point [135,136]. In these models the diffusion of a randomly picked peptide was performed by moving the peptide in one of the possible z directions of the lattice, and moving the lipids from one side to the opposite side of the peptide. This refinement of the model does not change the goodness of the fit to the experimental data, although the value of the model parameters will change.

4.2.4. One- and two-component lipid bilayers with peripheral proteins

The simplest lattice model of one-component lipid bilayers has been modified by Heimburg and Biltonen [145] and also by Hinderliter *et al.* [135] to simulate the lateral distribution of peripheral proteins attached to a one-component two-state and to a two-component one-state lipid bilayer, respectively. In the model of Heimburg and Biltonen [145] the binding of peripheral proteins to the membrane surface is characterized with the binding enthalpy, ΔH_p (J/mol lipid) and binding entropy, ΔS_p (J/mol lipid/K). The binding enthalpy, measurable by titration calorimetry, depends on the type of the protein, on the type of the lipid and also on the state of the lipid membrane. For example, at saturating amount of cytochrome c the binding enthalpy to dimyristoylphosphatidylglycerol bilayer decreases strongly from $\Delta H_p = +28.33$ kJ/mol lipid in the gel phase to $\Delta H_p = +11.66$ kJ/mol lipid in the fluid phase [157], i.e., the change is $\delta\Delta H_p = -16.67$ kJ/mol lipid. At the same time the binding entropy changes too. Its change, $\delta\Delta S_p$ can be estimated by means of the following equation:

$$T_m^{LP} = \frac{\Delta H_{tr} + \delta\Delta H_p}{\Delta S_{tr} + \delta\Delta S_p} \tag{17}$$

where ΔH_{tr} and ΔS_{tr} is the transition enthalpy and transition entropy of the one-component lipid membrane without peripheral protein, T_m^{LP} is the transition temperature when peripheral protein is bound to the membrane at saturating amount. T_m^{LP} can be determined from the excess heat capacity curve of the protein/lipid bilayer system. Note, that the peak is broad and asymmetric, and thus the estimation of $\delta\Delta S_p$ from equation (17) is not precise.

The lattice model of one-component two-state bilayer with peripheral proteins contains two similar parallel triangular lattices: the lattice of the lipid matrix and the lattice of the peripheral proteins. The lattice of the lipid matrix is similar to the lattice model of one-component two-state membranes. On the protein lattice a constant number of N_p proteins are situated, each centered at a lattice point. One peripheral protein usually covers more than one lipid molecules, and thus in the lattice more than one lattice point belong to one peripheral protein, e.g., 19, hexagonally ordered, lattice points in the case of cytochrome c [145]. The configuration of the lattice of the peripheral proteins is defined by the location of each protein, where the proteins must not overlap each other. The configuration of the lattice of lipid matrix is defined by the state of each lattice point. The overall configuration of the membrane/protein system is defined by the configuration on both lattices. Through the protein–lipid interaction these lattices are coupled to each other. One can use Monte Carlo simulation methods to study the inter-relationships between the lateral distribution of gel/fluid state lipid molecules and peripheral proteins.

There are two ways of trial configuration generation: (1) a lipid molecule is picked randomly and its state is changed from gel to fluid, or from fluid to gel, (2) a peripheral protein is picked randomly and moved in one of the randomly picked directions by one lattice unit. The number of possible directions is equal with the coordination number of the lattice, z.

The decision of keeping or rejecting the trial configuration is made according to the Metropolis criterion, equation (14). The calculation of $E_{trial} - E_{orig}$ and $\Omega_{trial}/\Omega_{orig}$ is different for different trial configuration generations.

In the case of trial configuration generation (1) the calculations depend on the association of the picked lipid molecule with peripheral protein. If the molecule does not interact with a peripheral protein then

$$E_{\text{trial}} - E_{\text{orig}} = \pm\Delta E + w_{\text{gf}}\Delta N_{\text{gf}} \quad \Omega_{\text{trial}}/\Omega_{\text{orig}} = (\Omega_{\text{f}}/\Omega_{\text{g}})^{\pm 1} = e^{\pm\Delta S_{\text{tr}}/R}. \tag{18}$$

where the positive sign belongs to gel-to-fluid transition and the negative to fluid-to-gel transition, and $\Delta N_{\text{gf}} = N_{\text{gf}}^{\text{trial}} - N_{\text{gf}}^{\text{orig}}$. If however, the picked lipid molecule interacts with a peripheral protein then

$$E_{\text{trial}} - E_{\text{orig}} = \pm(\Delta E + \delta\Delta H_{\text{p}}) + w_{\text{gf}}\Delta N_{\text{gf}} \quad \Omega_{\text{trial}}/\Omega_{\text{orig}} = e^{\pm(\Delta S_{\text{tr}} + \delta\Delta S_{\text{p}})/R} \tag{19}$$

where again the positive sign belongs to gel-to-fluid transition and the negative to fluid-to-gel transition. Note, it is assumed that protein binding does not change the cooperativity parameter, w_{gf}.

In the case of trial configuration generation (2) the protein is moved by one lattice unit. If the move results in overlap between proteins the trial configuration is rejected, otherwise the Metropolis criterion, equation (14) is used in the decision making. The energy difference between the trial and original configuration and the ratio of the degeneracy of the trial and original configuration can be calculated as follows:

$$E_{\text{trial}} - E_{\text{orig}} = (n_{\text{trial}} - n_{\text{orig}})\delta\Delta H_{\text{p}} \quad \Omega_{\text{trial}}/\Omega_{\text{orig}} = e^{(n_{\text{trial}} - n_{\text{orig}})\delta\Delta S_{\text{p}}/R} \tag{20}$$

where the number of fluid state molecules covered by the protein in the original and trial configuration is n_{orig} and n_{trial}, respectively.

The distribution of the peripheral protein on the membrane surface depends on the actual value of the model parameter, $\delta\Delta S_{\text{p}}$ (see Table 6).

The lattice model of Hinderliter *et al.* [135] describing the interaction between peripheral proteins (C2 protein motif) and two-component (phosphatidylcholine and phosphatidylserine) one-state lipid bilayers is more complicated. It permits the association and dissociation of the peripheral proteins to and from the protein lattice. Each protein, when bound to the protein lattice, is represented by 19-site hexagon. This hexagon is further divided into a core of inner 7-site hexagon, plus a 12-site border. The core binds similarly to each lipid component, while the border preferentially binds negatively charged phosphatidylserine molecules. The simulations showed that lipid domain formation and protein domain formation are coupled through the preferential interaction of the protein with the anionic lipid. Since no attractive protein–protein interaction was incorporated into the model, it is clear that large protein and lipid clusters can be formed in a process mediated by preferential protein–lipid and lipid–lipid interactions. These interactions can impart dynamic control on the spatial and temporal organization of signal transduction and trafficking machinery.

Table 6. Effect of model parameters on the lateral distribution of peripheral proteins

Model parameters	Protein's lateral distribution	Heat capacity curve	Examples
$\delta\Delta H_p - T_m\delta\Delta S_p > 0$	Preferential aggregation on the gel clusters	Peak's up-shift and asymmetric broadening	Dimyristoylphosphatidyl-glycerol with cytochrome *c* [22]
$\delta\Delta H_p - T_m\delta\Delta S_p < 0$	Preferential aggregation on the fluid clusters	Peak's down-shift and asymmetric broadening	Dimyristoylphosphatidylserine with myelin basic protein [37]
$\delta\Delta H_p - T_m\delta\Delta S_p = 0$	Random distribution. It is not coupled to the gel/fluid clustering	No change	

T_m is the transition temperature of the one-component membrane without protein.

5. PERSPECTIVES

In this chapter, we have attempted to provide an overview of the recent advances in modeling protein–lipid interactions and progress in MD/MC methods primarily resulting from the recent availability of increasingly powerful computing. These developments have contributed a wealth of new insights in membrane protein structure–dynamics–function relationships. A survey of the recent modeling literature has allowed us to show that MD and MC simulations are very appropriate tools for this purpose. From a methodological point of view, MD has significantly benefited from parameterization developments, improved treatments to account for electrostatic interactions (PME), increased refinements of PBC protocols – not to mention improved conformational sampling time. Further, the PMF approach that allows successful simulation of ion fluxes is a remarkable achievement.

As for all-atom MC simulations, the application of new sampling techniques, such as extension biasing was particularly beneficial.

However, both methods still have unjustified approximations and limitations that need to be addressed in future work. In our view, the accuracy of the force fields used in MD simulations, still need to be improved by including polarization effects. This could be achieved by using *ab initio* MD [111]. It should be noted, however, that the good agreement increasingly reported between simulations and experiment, point to the validity of the approximations included in the present force fields to account for polarization effects. The results of pure bilayer simulations also show that the simulation of a protein in a fully solvated explicit phospholipid bilayer still represents a major MD challenge in that proteins need to be modeled in more complex bilayers systems. Future MD and MC simulation developments will also depend on the success achieved in resolving structures of membrane proteins. Because MD simulations rely on atomic-detail structure, such information is essential for using MD in structure–function studies.

To conclude, membrane protein simulations represent the next frontier in computational biophysics. Some 30% of proteins coded by the human genome are membrane proteins and so are some 60% of drug targets in the pharmaceutical industry because of the important physiological functions that occur inside cells or on their membranes. The complexity of the membrane protein environment requires more accurate models, but the rapid pace of processor performance developments lead us to state that this now represents a realistic goal for the very near future, as well as direct comparison of simulation results with experimental data.

ACKNOWLEDGEMENTS

Fidy and Laberge acknowledge financial support from OTKA TS-044730 (J.F). Sugar acknowledges T.E. Thompson and R.L. Biltonen for initiating and supporting his research in membrane lattice models and financial support from NIH GM 59205 (R.L.B.) and Pfizer Inc. (I.P.S.). Sugar also acknowledges Mrs. Lawrence Garner's generous support. The authors also thank Peter Tieleman (Department of Biological Sciences, University of Calgary) for the MD simulation coordinates of the alamethicin bundle, Volkhard Helms (Bioinformatics Group, Saarland University), B. Brooks (Laboratory of Biophysical Chemistry, National Institutes of Health), Klaus Schulten (Theoretical and Computational Biophysics Group, University of Illinois at Urbana-Champaign) for permission to reproduce illustrations from their work as well as the Nature Publishing group and Biophysical Journal.

REFERENCES

[1] S. Izrailev, A.R. Crofts, E.A. Berry, K. Schulten, Steered molecular dynamics simulation of the Rieske subunit motion in the cytochrome bc1 complex, Biophys. J. 77 (1999) 1753–1768.

[2] A. Arora, L.K. Tamm, Biophysical approaches to membrane protein structure determination, Curr. Opin. Struct. Biol. 11 (2001) 540–547.

[3] D.P. Tieleman, S.J. Marrink, H.J.C. Berendsen, A computer perspective of membranes: molecular dynamics studies of lipid bilayer systems, Biochim. Biophys. Acta 1331 (1997) 235–270.

[4] L.R. Forrest, M.S.P. Sansom, Membrane simulations: bigger and better?, Curr. Opin. Struct. Biol. 10 (2000) 174–181.

[5] S.E. Feller, Molecular dynamics simulations of lipid bilayers, Curr. Opin. Colloid Interface Sci. 5 (2000) 217–223.

[6] L. Saiz, M.L. Klein, Computer simulation studies of model biological membranes, Acc. Chem. Res. 35 (2002) 482–489.
[7] R.W. Pastor, Molecular dynamics and Monte Carlo simulations of lipid bilayers, Curr. Opin. Struct. Biol. 4 (1994) 486–492.
[8] D.J. Tobias, K. Tu, M.L. Klein, Atomic-scale molecular dynamics simulations of lipid membranes, Curr. Opin. Colloid Interface Sci. 2 (1997) 15–26.
[9] M.S.P. Sansom, I.H. Shrivastava, J.N. Bright, J. Tate, C.E. Capener, P.C. Biggin, Potassium channels: structures, models, simulations, Biochim. Biophys. Acta Biomembr. 1565 (2002) 294–307.
[10] M.S.P. Sansom, Models and simulations of ion channels and related membrane proteins, Curr. Opin. Struct. Biol. 8 (1998) 237–244.
[11] L.H. Scott, Modeling the lipid component of membranes, Curr. Opin. Struct. Biol. 12 (2003) 495–502.
[12] K.M. Merz Jr., Molecular dynamics simulations of lipid bilayers, Curr. Opin. Struct. Biol. 7 (1997) 511–517.
[13] P. Jedlovszky, M. Mezei, Grand canonical ensemble Monte Carlo simulation of lipid bilayer using extension biased rotations, J. Chem. Phys. 111 (1999) 10770–10773.
[14] R. Langer, Drug delivery and targeting, Nature 392 (Suppl. 6679) (1998) 5–10.
[15] A.G. Lee, Lipid–protein interactions in biological membranes: a structural perspective, Biochim. Biophys. Acta Biomembr. 1612 (2003) 1–40.
[16] G. von Heijne, Principles of membrane protein assembly and structure, Prog. Biophys. Mol. Biol. 66 (1996) 113–139.
[17] Y. Shai, Molecular recognition within the membrane milieu: implications for the structure and function of membrane proteins, J. Membr. Biol. 182 (2001) 91–106.
[18] S. Tristram-Nagle, J.F. Nagle, Lipid bilayers: thermodynamics, structure, fluctuations, and interactions, Chem. Phys. Lipids 127 (2004) 3–14.
[19] H.T. Tien, A. Ottova-Leitmannova (Eds.), Planar Lipid Bilayers (BLMs) and Their Applications, Elsevier, Amsterdam, 2003.
[20] H.T. Tien, A. Ottova-Leitmannova, Membrane Biophysics as Viewed from Experimental Lipid Bilayer Membranes, Membrane Science and Technology Series 5, Elsevier, Amsterdam, 2000.
[21] G. Panyi, G. Vamosi, Z. Bacso, M. Bagdany, A. Bodnar, Z. Varga, R. Gaspar, L. Matyus, S. Damjanovich, Kv1.3 potassium channels are localized in the immunological synapse formed between cytotoxic and target cells, Proc. Natl Acad. Sci. USA 1001 (2004) 1285–1290.
[22] T. Gil, J.H. Ipsen, O.G. Mouritsen, M.C. Sabra, M.M. Speroto, M.J. Zuckermann, Theoretical analysis of protein organization in lipid membranes, Biochim. Biophys. Acta 1376 (1998) 245–266.
[23] L.K. Tamm, J. Crane, V. Kiessling, Membrane fusion: a structural perspective on the interplay of lipids and proteins, Curr. Opin. Struct. Biol. 12 (2003) 239–243.
[24] R.B. Gennis, Biomembrane Molecular Structure and Functions, Springer, New York, 1989.
[25] U. Essmann, M.L. Berkowitz, Dynamical Properties of Phospholipid Bilayers from Computer Simulation, Biophys. J 76 (1999) 2081–2089.
[26] V. Helms, Attraction within the membrane: forces behind transmembrane protein folding and supramolecular complex assembly, EMBO Rep. 3 (2002) 1133–1138.
[27] J.F. Tocanne, L. Cezanne, A. Lopez, B. Piknova, V. Schram, J.F. Tournier, M. Welby, Lipid domains and lipid/protein interactions in biological membranes, Chem. Phys. Lipids 73 (1994) 139–158.
[28] R.M. Venable, B.R. Brooks, R.W. Pastor, Molecular dynamics simulations of gel (LI) phase lipid bilayers in constant pressure and constant surface area ensembles, J. Chem. Phys. 112 (2000) 4822–4832.
[29] I. Ubarretxena, D.L. Engelman, Helical membrane proteins: diversity of functions in the context of simple architecture, Curr. Opin. Struct. Biol. 11 (2001) 370–376.

[30] L.K. Tamm, A. Arora, J.H. Kleinschmidt, Strucrure and assembly of beta-barrel membrane proteins, J. Biol. Chem. 276 (2001) 32399–32402.
[31] M. Opekarova, W. Tanner, Specific lipid requirements of membrane proteins – a putative bottleneck in heterologous expression, Biochim. Biophys. Acta 1610 (2003) 11–22.
[32] A.G. Lee, How lipids interact with an intrinsic membrane protein: the case of the calcium pump, Biochim. Biophys. Acta 1376 (1998) 381–390.
[33] E. Gouaux, S.H. White, Membranes: proteins and membranes – a fusion of new ideas, Curr. Opin. Struct. Biol. 13 (2003) 401–403.
[34] D. Marsh, Lipid interactions with transmembrane proteins, Cell. Mol. Life Sci. 60 (2003) 1575–1580.
[35] M.R. Jones, P.K. Fyfe, A.W. Roszak, N.W. Isaacs, R.J. Cogdell, Protein–lipid interactions in the purple bacterial reaction centre, Biochim. Biophys. Acta – Biomembr. 1565 (2002) 206–214.
[36] A. Senes, I. Ubarretxena-Belandia, D.M. Engelman, The C–H···O hydrogen bond: a determinant of stability and specificity in transmembrane helix interactions, Proc. Natl Acad. Sci. USA 98 (2001) 9056–9061.
[37] M.C. Wiener, S.H. White, Structure of a fluid dioleoylphosphatidylcholine bilayer determined by joint refinement of X-ray and neutron diffraction data. III. Complete structure, Biophys. J. 61 (1992) 437–447.
[38] E. Lindahl, O. Edholm, Molecular dynamics simulation of NMR relaxation rates and slow dynamics in lipid bilayers, J. Chem. Phys. 115 (2001) 4938–4950.
[39] C. Domene, P.J. Bond, S.S. Deol, M.S.P. Sansom, Lipid/protein interactions and the membrane/water interfacial region, J. Am. Chem. Soc. 125 (2003) 14966–14967.
[40] K. Tu, D.J. Tobias, M.L. Klein, Constant pressure and temperature molecular dynamics simulation of a fully hydrated liquid crystal phase dipalmitoylphosphatidylcholine bilayer, Biophys. J. 69 (1995) 2558–2562.
[41] K. Tu, D.J. Tobias, K. Blasie, M.L. Klein, Molecular dynamics investigation of the structure of a fully hydrated gel-phase dipalmitoylphosphatidylcholine bilayer, Biophys. J. 70 (1996) 595–608.
[42] D.P. Tieleman, H.J.C. Berendsen, Molecular dynamics simulation of a fully hydrated dipalmitoylphosphatidylcholine bilayer with different macroscopic boundary conditions and parameters, J. Chem. Phys. 105 (1996) 4871–4880.
[43] S.E. Feller, R.W. Pastor, Constant surface tension simulations of lipid bilayers: the sensitivity of surface areas and compressibilities, J. Chem. Phys. 111 (1999) 1281–1287.
[44] B. Roux, Commentary: surface tension of biomembranes, Biophys. J. 71 (1996) 1346–1347.
[45] L. Perera, U. Essmann, M.L. Berkowitz, Role of water in hydration force acting between lipid bilayers, Langmuir 12 (1996) 2625–2629.
[46] K. Aman, E. Lindahl, O. Edholm, P. Hakansson, P.O. Westlund, Structure and dynamics of interfacial water in an *L* alpha phase lipid bilayer from molecular dynamics simulations, Biophys. J. 84 (2003) 102–115.
[47] K.M. Merz Jr., Molecular dynamics simulations of lipid bilayers, Curr. Opin. Struct. Biol. 7 (1997) 511–517.
[48] C.M. Shepherd, K.A. Schaus, H.J. Vogel, A.H. Juffer, Molecular dynamics study of peptide-bilayer adsorption, Biophys. J. 80 (2001) 579–596.
[49] C.M. Shepherd, H.J. Vogel, D.P. Tieleman, Interactions of the designed antimicrobial peptide MB21 and truncated dermaseptin S3 with lipid bilayers: molecular-dynamics simulations, Biochem. J. 370 (2003) 233–243.
[50] F. Zhou, K. Schulten, Molecular dynamics study of phospholipase A2 on a membrane surface, Proteins 25 (1996) 12–27.

[51] A. Kessel, K. Schulten, N. Ben-Tal, Calculations suggest a pathway for the transverse diffusion of a hydrophobic peptide across a lipid bilayer, Biophys. J. 2000; 2322–2330.
[52] S.J. Marrink, E. Lindahl, O. Edholm, A.E. Mark, Simulation of the spontaneous aggregation of phospholipids into bilayers, J. Am. Chem. Soc. 123 (2001) 8638–8639.
[53] O. Edholm, O. Berger, F. Jahnig, Structure and fluctuations of bacteriorhodopsin in the purple membrane: a molecular dynamics study, J. Mol. Biol. 250 (1995) 94–111.
[54] B. Roux, M. Karplus, Molecular dynamics simulations of the gramicidin channel, Annu. Rev. Biomol. Struct. Dyn. 23 (1994) 731–761.
[55] T.B. Woolf, B. Roux, Molecular dynamics simulation of the gramicidin channel in a phospholipid bilayer, Proc. Natl Acad. Sci. USA 91 (1994) 11631–11635.
[56] D.P. Tieleman, H. Leontiadou, A.E. Mark, S.J. Marrink, Simulation of pore formation in lipid bilayers by mechanical stress and electric fields, J. Am. Chem. Soc. 125 (2003) 6382–6383.
[57] S.J. Marrink, F. Jahnig, H.J. Berendsen, Proton transport across transient single-file water pores in a lipid membrane studied by molecular dynamics simulations, Biophys. J. 71 (1996) 632–647.
[58] P. Jedlovszky, M. Mezei, Effect of cholesterol on the properties of phospholipid membranes. 1. Structural features, J. Phys. Chem. B 107 (2003) 5311–5321.
[59] P. Jedlovszky, M. Mezei, Effect of cholesterol on the properties of phospholipid membranes. 2. Free energy profile of small molecules, J. Phys. Chem. B 107 (2003) 5322–5332.
[60] P. Jedlovszky, M. Mezei, Effect of cholesterol on the properties of phospholipid membranes. 3. Local lateral structure, J. Phys. Chem. B 108 (2003) 465–472.
[61] P. Jedlovszky, M. Mezei, Calculation of the free energy profile of H_2O, O_2, CO, CO_2, NO, and $CHCl_3$ in a lipid bilayer with cavity insertion variant of the Widom method, J. Am. Chem. Soc. 122 (2000) 5125–5131.
[62] P. Jedlovszky, M. Mezei, Orientational order of the water molecules across a fully hydrated DMPC bilayer: a Monte Carlo simulation study, J. Phys. Chem. B 105 (2001) 3614–3623.
[63] O.G. Mouritsen, K. Jørgensen, Small-scale lipid-membrane structure: simulation versus experiment, Curr. Opin. Struct. Biol. 7 (1997) 518–527.
[64] M.P. Allen, D.J. Tildesley, Computer Simulation of Liquids, Clarendon Press, Oxford, 1987. p. 385.
[65] C. Hofsass, E. Lindahl, O. Edholm, Molecular dynamics simulations of phospholipid bilayers with cholesterol, Biophys. J. 84 (2003) 2192–2206.
[66] J. Torres, T.J. Stevens, M. Samso, Membrane proteins: the “Wild West” of structural biology, Trends Biochem. Sci. 28 (2003) 137–144.
[67] J.A. McCammon, S.C. Harvey, Dynamics of proteins and nucleic acids, Cambridge University Press, Cambridge, 1987. p. 234.
[68] L. Verlet, Computer experiments on classical fluids. I. Thermodynamical properties of Lennard–Jones molecules, Phys. Rev. 159 (1967) 98–103.
[69] I.P. Omelyan, I.M. Mryglod, R. Folk, Optimized Verlet-like algorithms for molecular dynamics simulations, Phys. Rev. E Stat. Nonlinear Soft Matter Phys. 65 (2002) 056706-1-5.
[70] T. Hansson, C. Oostenbrink, W. van Gunsteren, Molecular dynamics simulations, Curr. Opin. Struct. Biol. 12 (2002) 190–196.
[71] T. Schlick, Computational molecular biophysics today: a confluence of methodological advances and complex biomolecular applications, J. Comp. Phys. 151 (1999) 1–8.
[72] C. Anézo, A.H. de Vries, H.-D. Höltje, D.P. Tieleman, S.J. Marrink, Methodological issues in lipid bilayer simulations, J. Phys. Chem. B 107 (2003) 9424–9433.
[73] R.W. Pastor, R.M. Venable, M. Karplus, Model for the structure of the lipid bilayer, Proc. Natl Acad. Sci. USA 88 (1991) 892–896.

[74] T.B. Woolf, B. Roux, Molecular dynamics simulation of the gramicidin channel in a phospholipid bilayer, Proc. Natl Acad. Sci. USA 91 (1994) 11631–11635.
[75] B. Roux, T.B. Woolf, Molecular dynamics simulation of Pf1 coat protein in a phospholipid bilayer, in: K. Merz, B. Roux (Eds.), Biological Membranes: A Molecular Perspective from Computation and Experiments, Birkhauser, Cambridge, MA, 1996, pp. 555–587.
[76] S. Bernèche, M. Nina, B. Roux, Molecular dynamics simulation of melittin in a dimyristoylphosphatidylcholine bilayer membrane, Biophys. J. 75 (1998) 1603–1618.
[77] S. Bernèche, B. Roux, Molecular dynamics of the KcsA K+ channel in a bilayer membrane, Biophys. J. 78 (2000) 2900–2917.
[78] S.E. Feller, A.D. MacKerrell Jr., An improved empirical potential energy function for molecular simulations of phospholipids, J. Phys. Chem. B 104 (2000) 7510–7515.
[79] S.E. Feller, K. Gawrisch, A.D. MacKerell Jr., Polyunsaturated fatty acids in lipid bilayers: intrinsic and environmental contributions to their unique physical properties, J. Am. Chem. Soc. 124 (2002) 318–326.
[80] O. Berger, O. Edholm, F. Jahnig, Molecular dynamics simulations of a fluid bilayer of dipalmitoylphosphatidylcholine at full hydration, constant pressure, and constant temperature, Biophys. J. 72 (1997) 2002–2013.
[81] S.J. Marrink, O. Berger, P. Tieleman, F. Jähnig, Adhesion forces of lipids in a phospholipid membrane studied by molecular dynamics simulations, Biophys. J. 74 (1998) 931–943.
[82] I. Chandrasekhar, M. Kastenholz, R.D. Lins, C. Oostenbrink, L.D. Schuler, D.P. Tieleman, W.F. van Gunsteren, A consistent potential energy parameter set for lipids – dipalmitoylphosphatidylcholine as a benchmark for the GROMOS96 45A3 force field, Eur. Biophys. J. 32 (2003) 67–77.
[83] S.W. Chiu, M.M. Clark, E. Jakobsson, S. Subramaniam, L.H. Scott, Optimization of hydrocarbon chain interaction parameters: application to the simulation of fluid phase lipid bilayers, J. Phys. Chem. B 103 (1999) 6323–6327.
[84] A.M. Smondyrev, M.L. Berkowitz, United atom force field for phospholipid membranes: constant pressure molecular dynamics simulation of dipalmitoylphosphatidylcholine/water system, J. Comput. Chem. 20 (1999) 531–545.
[85] A.M. Smondyrev, M.L. Berkowitz, Molecular dynamics simulation of the structure of dimyristoylphosphatidylcholine bilayers with cholesterol, ergosterol, and lanosterol, Biophys. J. 80 (2001) 1649–1658.
[86] M. Patra, M. Karttunen, M. Hyvönen, E. Falck, P. Lindqvist, I. Vattulainen, Molecular dynamics simulations of lipid bilayers: major artifacts due to truncating electrostatic interactions, Biophys. J. 84 (2003) 3636–3645.
[87] S.E. Feller, R.W. Pastor, A. Rojnuckarin, S. Bogusz, B.R. Brooks, Effect of electrostatic force truncation on interfacial and transport properties of water, J. Phys. Chem. 100 (1996) 17011–17020.
[88] U. Essmann, L. Perera, M.L. Berkowitz, T. Darden, H. Lee, L.G. Pedersen, A smooth particle mesh Ewald method, J. Chem. Phys. 103 (1995) 8577–8593.
[89] B. Roux, T. Simonson, Implicit solvent models, Biophys. Chem. 78 (1999) 1–20.
[90] W.I.D. Beglov, B. Roux, Continuum solvation model: electrostatic forces from numerical solutions to the Poisson–Boltzmann equation, Comp. Phys. Comm. 111 (1998) 59–75.
[91] B. Roux, Influence of the membrane potential on the free energy of an intrinsic protein, Biophys. J. 73 (1997) 2980–2989.
[92] D.J. Tobias, Electrostatic calculations: recent methodological advances and applications to membranes, Curr. Opin. Struct. Biol. 11 (2001) 253–261.
[93] E.A. Dolan, R.M. Venable, R.W. Pastor, B. Brooks, Simulations of membranes and other interfacial systems using $P2_1$ and *Pc* periodic boundary conditions, Biophys. J. 82 (2002) 2317–2325.

[94] T. Schlick, R.D. Skeel, A.T. Brunger, L.V. Kalé, J.A. Board Jr., J. Hermans, K. Schulten, Algorithmic challenges in computational molecular biophysics, J. Comp. Phys. 151 (1999) 9–48.
[95] S.E. Feller, Molecular dynamics simulation of phospholipid bilayers, in: J. Katsaras, T. Gutberlet (Eds.), Lipid Bilayers, Springer, New York, 2000, pp. 89–107.
[96] R.W. Pastor, R.M. Venable, S.E. Feller, Lipid bilayers, NMR relaxation, and computer simulations, Acc. Chem. Res. 35 (2002) 438–446.
[97] P. Mukhopadhyay, L. Monticelli, D.P. Tieleman, Molecular dynamics simulation of a palmitoyl-oleoyl phosphatidylserine bilayer with Na^+ counterions and NaCl, Biophys. J. 86 (2004) 1601–1609.
[98] S.A. Pandit, D. Bostick, M.L. Berkowitz, Mixed bilayer containing dipalmitoylphosphatidylcholine and dipalmitoylphosphatidylserine: lipid complexation, ion binding, and electrostatics, Biophys. J. 85 (2003) 3120–3131.
[99] R.A. Bockmann, A. Hac, T. Heimburg, H. Grubmuller, Effect of sodium chloride on a lipid bilayer, Biophys. J. 85 (2003) 1647–1655.
[100] S.A. Pandit, M.L. Berkowitz, Molecular dynamics simulation of dipalmitoylphosphatidylserine bilayer with Na^+ counterions, Biophys. J. 82 (2002) 1818–1827.
[101] H. Leontiadou, A.E. Mark, S.J. Marrink, Molecular dynamics simulations of hydrophilic pores in lipid bilayers, Biophys. J. 86 (2004) 2156–2164.
[102] S.J. Marrink, A.E. Mark, Effect of undulations on surface tension in simulated bilayers, J. Phys. Chem. B 105 (2001) 6122–6127.
[103] A.H. de Vries, A.E. Mark, S.J. Marrink, The binary mixing behavior of phospholipids in a bilayer: a molecular dynamics study, J. Phys. Chem. B 108 (2004) 2454–2463.
[104] T. Rog, M. Pasenkiewicz-Gierula, Effects of epicholesterol on the phosphatidylcholine bilayer: a molecular simulation study, Biophys. J. 84 (2003) 1818–1826.
[105] J.A. Soderhall, A. Laaksonen, Molecular dynamics simulations of ubiquinone inside a lipid bilayer, J. Phys. Chem. B 105 (2001) 9308–9315.
[106] T.H. Duong, E.L. Mehler, H. Weinstein, Molecular dynamics simulation of membranes and a transmembrane helix, J. Comp. Phys. 151 (1999) 358–387.
[107] R. Sankararamakrishnan, H. Weinstein, Molecular dynamics simulations predict a tilted orientation for the helical region of dynorphin A(1–17) in dimyristoylphosphatidylcholine bilayers, Biophys. J. 79 (2002) 2331–2344.
[108] M. Iadanza, M. Holtje, G. Ronsisvalle, H.-D. Holtje, k-Opioid receptor model in a phospholipid bilayer: molecular dynamics simulation, J. Med. Chem. 45 (2002) 4838–4846.
[109] H.I. Petrache, A. Grossfield, K.R. MacKenzie, D.M. Engelman, T.B. Woolf, Modulation of glycophorin A transmembrane helix interactions by lipid bilayers: molecular dynamics calculations, J. Mol. Biol. 302 (2000) 727–746.
[110] B. Roux, Computational studies of the gramicidin channel, Acc. Chem. Res. 35 (2002) 366–375.
[111] S.-H. Chunga, S. Kuyucak, Recent advances in ion channel research, Biochim. Biophys. Acta (BBA) – Biomembr. 1565 (2002) 267–286.
[112] D.P. Tieleman, B. Hess, M.S.P. Sansom, Analysis and evaluation of channel models: simulations of alamethicin, Biophys. J. 83 (2002) 2393–2407.
[113] A. Baumgartner, S. Grudinin, J.-F. Gwan, J.-H. Lin, Molecular dynamics simulation of membrane proteins, in: D. Wolf, G. Muenster, M. Kremer (Eds.), NIC Symposium 2004, Proceedings NIC Series 20, John von Neuman Institute for Computing, Juelich, 2003, pp. 365–374.
[114] S. Berneche, B. Roux, Energetics of ion conduction through the K^+ channel, Nature 414 (2001) 73–77.
[115] S. Bernèche, B. Roux, A microscopic view of ion conduction through the K^+ channel, Proc. Natl Acad. Sci. USA 100 (2003) 8644–8648.

[116] T.W. Allen, O.S. Andersen, B. Roux, Energetics of ion conduction through the gramicidin channel, Proc. Natl Acad. Sci. USA 101 (2004) 117–122.
[117] C. Domene, M.S.P. Sansom, Potassium channel ions, and water: simulation studies based on the high resolution X-ray structure of KcsA, Biophys. J. 85 (2003) 2787–2800.
[118] P.J. Bond, M.S.P. Sansom, Membrane protein dynamics *versus* environment: simulations of OmpA in a micelle and in a bilayer, J. Mol. Biol. 329 (2003) 1035–1053.
[119] P.J. Bond, J.D. Faraldo-Gomez, M.S.P. Sansom, OmpA: a pore or not a pore? Simulation and modeling studies, Biophys. J. 83 (2002) 763–775.
[120] P.S. Crozier, M.J. Stevens, L.R. Forrest, T.B. Woolf, Molecular dynamics simulation of dark-adapted rhodopsin in an explicit membrane bilayer: coupling between local retinal and larger scale conformational change, J. Mol. Biol. 333 (2003) 493–514.
[121] T. Huber, A.V. Botelho, K. Beyer, M.F. Brown, Membrane model for the G-protein-coupled receptor rhodopsin: hydrophobic interface and dynamical structure, Biophys. J. 86 (2004) 2078–2100.
[122] C. Kandt, J. Schlitter, K. Gerwert, Dynamics of water molecules in the bacteriorhodopsin trimer in explicit lipid/water environment, Biophys. J. 86 (2004) 705–717.
[123] E. Olkhova, M.C. Hutter, M.A. Lill, V. Helms, H. Michel, Dynamic water networks in cytochrome *c* oxidase from *Paracoccus denitrificans* investigated by molecular dynamics simulations, Biophys. J. 86 (2004) 1873–1889.
[124] M. Milik, J. Skolnick, Insertion of peptide chains into lipid membranes: an off-lattice Monte Carlo dynamics model, Proteins 15 (1993) 10–25.
[125] M. Milik, J. Skolnick, Monte Carlo model of FD and PF1 coat proteins in lipid membranes, Biophys. J. 69 (1995) 1382–1386.
[126] R.G. Efemerov, D.E. Nolde, G. Vergoten, A.S. Arseniev, A solvent model for simulations of peptides in bilayers. I. Membrane-promoting α-helix formation, Biophys. J. 76 (1999) 2448–2459.
[127] R.G. Efemerov, D.E. Nolde, G. Vergoten, A.S. Arseniev, A solvent model for simulations of peptides in bilayers. I. Membrane-spanning α-helices, Biophys. J. 76 (1999) 2460–2471.
[128] R.G. Efemerov, P.E. Volinsky, D.E. Nolde, P.V. Dubovsky, A.S. Arseniev, Interaction of cardiotoxins with membranes: a molecular modeling study, Biophys. J. 83 (2002) 144–153.
[129] R.G. Efemerov, P.E. Volinsky, D.E. Nolde, A. van Dalen, B. de Kruijff, A.S. Arseniev, Monte Carlo simulation of voltage-driven translocation of a signal sequence, FEBS Lett. 526 (2002) 97–100.
[130] L. Lins, B. Charloteaoux, A. Thomas, R. Brasseur, Computational study of lipid-destabilizing protein fragments: towards a comprehensive view of tilted peptides, Proteins 44 (2001) 435–447.
[131] M.W. Maddox, M.L. Longo, A Monte Carlo study of peptide insertion into lipid bilayers: equilibrium conformations and insertion mechanisms, Biophys. J. 82 (2002) 244–263.
[132] A. Kessel, D. Shental-Bechor, T. Haliloglu, N. Ben-Tal, Interactions of hydrophobic peptides with lipid bilayers: Monte Carlo simulations with $M2\delta$, Biophys. J. 85 (2003) 3431–3444.
[133] W.L.C. Vaz, E.C.C. Melo, T.E. Thompson, Translational diffusion and fluid domain connectivity in two-component, two-phase phospholipid bilayers, Biophys. J. 56 (1989) 869–876.
[134] H.L. Scott, Modeling the lipid component of membranes, Curr. Opin. Struct. Biol. 12 (2002) 495–502.
[135] A. Hinderliter, P.F.F. Almeida, C.E. Creutz, R.L. Biltonen, Domain formation in a fluid mixed lipid bilayer modulated through binding of C2 protein motif, Biochemistry 40 (2001) 4181–4191.

[136] V.P. Ivanova, I.M. Makarov, T.E. Schaffer, T. Heimburg, Analyzing heat capacity profiles of peptide-containing membranes: cluster formation of gramicidin A, Biophys. J. 84 (2003) 2427–2439.
[137] P. Hoyrup, K. Jorgensen, O.G. Mouritsen, Nano-scale structure in membranes in relation to enzyme action – computer simulation vs. experiment, Comp. Phys. Comm. 147 (2002) 313–320.
[138] I.P. Sugar, T.E. Thompson, R.L. Biltonen, Monte Carlo simulation of two-component bilayers: DMPC/DSPC mixtures, Biophys. J. 76 (1999) 2099–2110.
[139] I.P. Sugar, R.L. Biltonen, Structure–function relationships in two-component phospholipid bilayers. A Monte Carlo simulation approach using a two state model, Methods Enzymol. 323 (2000) 340–372.
[140] I.P. Sugar, E. Michonova-Alexova, P.L.-G. Chong, Geometrical properties of gel and fluid clusters in DMPC/DSPC bilayers: Monte Carlo simulation approach using a two-state model, Biophys. J. 81 (2001) 2425–2441.
[141] E.I. Michonova-Alexova, I.P. Sugar, Component and state separation in DMPC/DSPC lipid bilayers: a Monte Carlo simulation study, Biophys. J. 83 (2002) 1820–1833.
[142] T. Heimburg, D. Marsh, Thermodynamics of the interaction of proteins with lipid membranes, in: K.M. Merz, B. Roux (Eds.), Biological Membranes. A Molecular Perspective from Computation and Experiment, Birkhauser, Boston, 1996, pp. 405–462.
[143] O.G. Mouritsen, K. Jorgensen, Small-scale lipid-membrane structure: simulation versus experiment, Curr. Opin. Struct. Biol. 7 (1997) 518–527.
[144] I.P. Sugar, R.L. Biltonen, N. Michard, Monte Carlo simulations of membranes: the phase transition of small unilamellar DPPC vesicles, Methods Enzymol. 240 (1994) 569–593.
[145] T. Heimburg, R.L. Biltonen, A Monte Carlo study of protein-induced heat capacity changes and lipid-induced protein clustering, Biophys. J. 70 (1996) 84–96.
[146] T.L. Hill, An Introduction to Statistical Thermodynamics, Dover Publications, Inc., New York, 1986. p. 250.
[147] N. Metropolis, A.W. Rosenbluth, W.N. Rosenbluth, A.H. Teller, E. Teller, Equation of state calculations by fast computing machines, J. Chem. Phys. 21 (1953) 1087–1092.
[148] A. Caille, A. Rapini, M.J. Zuckermann, A. Cros, S. Doniach, A simple model for phase transitions in monolayers and bilayers of lipid molecules, Can. J. Phys. 56 (1978) 348–357.
[149] R. Jerala, P.F.F. Almeida, R.L. Biltonen, Simulation of the gel–fluid transition in a membrane composed of lipids with two connected acyl chains: application of a dymer-move step, Biophys. J. 71 (1996) 609–615.
[150] R.M.F. Houtappel, Order–disorder in hexagonal lattices, Physica 16 (1950) 425–455.
[151] K. Binder, D. Stauffer, Monte Carlo studies of “random” systems, in: K. Binder (Ed.), Applications of the Monte Carlo method in statistical physics, Springer, Berlin, 1987 Chapter 8.
[152] J.N. Israelachvili, Refinement of the fluid–mosaic model of membrane structure, Biochim. Biophys. Acta 469 (1977) 221–225.
[153] M.M. Sperotto, O.G. Mouritsen, Monte Carlo simulation studies of lipid order parameter profiles near integral membrane proteins, Biophys. J. 59 (1991) 261–270.
[154] O.G. Mouritsen, M.M. Sperotto, Thermodynamics of lipid–protein interactions in lipid membranes: the hydrophobic matching condition, in: M. Jackson (Ed.), Thermodynamics of Cell Surface Receptors, CRC Press, Boca Raton, FL, 1992, pp. 127–181.
[155] H. Sun, I.P. Sugar, Acceleration of convergence to the thermodynamic equilibrium by introducing shuffling operations to the metropolis algorithm of Monte Carlo simulations, J. Phys. Chem. B 101 (1997) 3221–3227.

[156] E.J. Prenner, R.N.A.H. Lewis, L.H. Kondejewski, R.S. Hodges, R.N. McElhaney, Differential scanning calorimetric study of the effect of the antimicrobial peptide gramicidin S on the thermotropic phase behavior of phosphatidylcholine, phosphatidylethanolamine and phosphatidylglycerol lipid bilayer membranes, Biochim. Biophys. Acta 1417 (1999) 211–223.

[157] T. Heimburg, R.L. Biltonen, Thermotropic behavior of dimyristoylphosphatidylglycerol and its interaction with cytochrome *c*, Biochemistry 33 (1994) 9477–9488.

CHAPTER 8

Modeling of BLMs in Aspects of Phylogenetic Development of Vertebrates

Armen E. Zakharian and Naira M. Ayvazian*

Department of Biophysics, Faculty of Biology, Yerevan State University, A. Manukian Str.1, 375025, Yerevan, Armenia

Contents

> The main set of phospholipids, from which biological membrane are building, has formed very early during evolution of animal world...at the dawn of living cell' development, genetically fastened and passing through whole animal world.
>
> Kreps E.M. (1967)

Abstract

In 1967, the academic E.M. Kreps showed that the biochemical composition of lipid membranes in nervous tissues of vertebrates have undergone insignificant changes during the hundreds of millions of years of evolution. Until that time, numerous data on the lipid qualitative composition in vertebrate animals indicated that this component has been conserved in the course of the phylogenetic adaptation of animals. However, it can be assumed that the processes related to the conversion of lipids and lipid-containing structures and change in the membrane bilayer (BLM) state (namely, its microviscosity, thickness, permeability, etc.) underwent certain changes as the structural organization of vertebrates became more complicated.

* Corresponding author. Tel.: +374-1-278-180; Fax: +374-1-554-641; E-mail: taipan@ysu.am

ADVANCES IN PLANAR LIPID BILAYERS AND LIPOSOMES, VOLUME 2
ISSN 1554-4516 DOI: 10.1016/S1554-4516(05)02008-9

We isolated lipid fractions from the brain heart, liver, and muscle of vertebrates: crucian carp (*Carassius carassius*), marsh frog (*Rana ridibunda*), caucasian agama (*Stellio caucasicus*), and non-purebred white rats using the method of Keits. Model membranes were formed from the total lipid fraction on a Teflon aperture using the method of Muller. The electrical parameters of the BLMs were determined on an electrometric device equipped with a Keithley 301 differential feedback amplifier (USA) in a voltage-fixation mode. The potential of membrane rupture recorded in the experiments by a shielded camera was taken as the threshold value of the voltage applied; 0.1 M KCl, NaCl, LiCl, $CaCl_2$ served as the ionic media.

It was discovered that penetration of bilayers from lipids of nervous tissue decreases in a series of poikilotherm vertebrates and increases in mammals. Reverse conformity to a natural low was observed for bilayer membrane 'stability': electrical breakdown increases from fishes to amphibians to reptiles and decreases in membranes from lipids of mammals. In most cases, bilayer penetration is higher for K^+ ions than for Na^+ and Li^+. Interestingly, the electrical properties of BLMs from the heart in Ca^{+2} ions media demonstrated very high conductivity and we connect this fact with the specific function of heart muscle. It is noteworthy that the tendencies in the changes in the BLM electrical parameter correlate with the results of the studies on free-radical lipid peroxidation in the biological materials analyzed, as well as with data on changes of cholesterol level and the polyunsaturation index in the course of the evolution of vertebrates. Thus, the comparison of the parameters of the state of lipids and lipid-containing structures allowed us to assume that the electrical conductance and ionic permeability of the membranes are related to the degree of the phylogenetic organization of the studied organisms. Recent studies allow us to assume that modeling of artificial membranes from native lipids of different tissues of vertebrates can explain the problems of the evolution of function.

1. INTRODUCTION

The strategy of biochemical adaptation underlies all the processes determining the course of evolution development and can be subdivided into three main categories:

1. Quantitative adaptation, which is accommodative alterations of concentrations of macromolecular components.
2. Qualitative adaptation in answer to changes of environment with new types of macromolecules included in the acting system.
3. Modulate adaptation, which is the accommodative alterations of activity of the existing macromolecules.

For a long time, the role of lipids in adaptive strategy was considered only in the study of organisms with the property to form a specific tissue – fat [1,2]. The questions about the role of cell membrane lipids as a factor necessary for the process of the accommodation of an organism to its environment and about change in its composition and quantitative properties still remain open.

In the 1960s, the academic E.M. Kreps initiated comparative-biochemical investigations of cell membrane lipids [3,4]. He discovered a huge diversity of animal cells from prokaryotes to high eukaryotes based on a strictly limited collection of 'obligatory' phospholipids. Moreover, in an evolutionary aspect, he discovered a diversity of structure of phospholipid molecules, formatting biomembranes, abruptly narrowing as far as increasing the systematical condition of organisms. For vertebrates, the constant of the quantitative and qualitative composition of membrane lipids acquires a status of biological low [5–8]. In spite of such conservatism, it is believed that the biochemistry of lipids and their functional characteristics are strictly determined.

Since Mueller and Rudin [9] obtained a stable planar bilayer lipid membrane (BLM) from brain phospholipids in 1962, the electrical properties of membranes have been studied intensely, but mainly in terms of the membrane pathology related to genetic disorders [10], disturbance of cholesterol–phospholipid balance in atherosclerosis [11], cancer-related changes in the lipid composition of membranes [12], etc. Currently, there is no doubt that the functioning of membranes, including ionic permeability regulated by lipid complication, could also change the condition of the normal vitality of an organism [13,14].

However, it can be assumed that the processes related to the conversion of lipids and lipid-containing structures and change in the membrane bilayer state (namely, its microviscosity, thickness, permeability, etc.) underwent certain changes as the structural organization of vertebrates became more complicated. Also, BLMs formed from natural lipids isolated from tissues of vertebrate animals at different stages of evolutionary development represent an adequate model of the lipid moiety of biomembranes, which makes it possible to reconstruct some features characteristic of natural bilayer membranes [15,16].

In this work, we studied artificial model membranes formed from various lipids isolated from the tissues of vertebrate animals at different stages of evolutionary development. The selection of tissues was determined by their origin from different embryonic leaves: nervous tissue developed from ectoderm, liver from endoderm, and mesoderm is a beginning for all types of muscles.

The structure of this chapter is as follows: each part of the research discussions are completed by a short excursus to comparative physiology of vertebrates as a necessary item for illustration of the connection between the specificity of the properties of a model membrane and its possible reflection in functional changes taking place in the process of evolution of present tissue.

2. METHODS

Lipid fractions were purified from the brain, heart, liver, and striated muscles of the crucian carp (*Carassius carassius*), marsh frog (*Rana ridibunda*), Caucasian agama (*Stellio caucasicus*), and non-purebred male and female white rats using the method of Keits [17]. Model membranes were formed from the total lipid fraction on a Teflon aperture using the method of Mueller [18].

The electrical parameters of the BLMs were determined on a device equipped with a Keithley 301 differential feedback amplifier (United States) in a voltage-fixation mode, which allowed a membrane potential to be maintained on any level independently of ionic streams. The potential setting on the exit of the generator completely falls on the membrane, whose resistance is much higher than the resistance of the electrodes and the electrolyte and the effective resistance of the current's gauge. The electrometric device can measure a current through the membrane under a fixed value of the transmembrane difference of potentials. The resistance of the membrane is calculated by the following formula:

$$R_m = R_f U_m / U_f \text{ (Ohm)},$$

where U_f is the potential from source of direct current, U_m the difference of potentials on the membrane, R_m the membrane resistance, and R_f is the resistance of chain.

Conductivity of membrane:

$$g_m = 1/R_m \text{ (Ohm}^{-1}\text{)}$$

Specific electrical conductance (g) was expressed in Ohm^{-1}/mm^2 and 0.1 M KCl, NaCl, LiCl and $CaCl_2$ served as the ionic media.

The breaking potential of the membrane recorded in the experiments with a shielded camera was taken as the threshold value of the applied voltage. The potential of the membrane rupture is a criterion for evaluation of the natural defects of the model membranes. Under an electrical potential, the radius of these holes increased. There is a critical value of the radius (r_0), and when $r_\phi > r_0$ the membrane is destroyed.

3. COMPARATIVE ANALYSIS OF ELECTRICAL CONDUCTIVITY OF BLMs FROM DIFFERENT TISSUES OF VERTEBRATES

Study of the electrical conductivity of BLMs formed from different tissues of vertebrates showed that the resistance of model membranes from lipids of analogous tissues are not the same for vertebrates of different classes. However, the majority of bilayers obtained in 0.1 M KCl were characterized by an electrical resistance of approximately 10^9 Ohm, which is in good agreement with the known published data on BLMs consisting of natural lipids [19]. In our experiments, a high electrical resistance (10^{10}–10^{11} Ohm) was only characteristic of the membranes comprised of the lipids isolated from the nervous tissue. The difference in the permeability of black films formed from different lipids isolated from tissues of vertebrate animals is apparently due to the different chemical compositions of lipid membranes. In view of this, a relatively low electrical conductance of BLMs formed from brain lipids may be accounted for by the high cholesterol level (which is approximately an order of magnitude higher than in muscles and liver) and by its condensing effect on the lipid bilayer [20,21]. Such an effect of cholesterol on a membrane phase condition depends on its condensing action and is known by experiments with synthetic BLMs. Moreover, the effect of condensation (and also electrical resistance) strongly connects the structure of the lipid bilayer, which is determined by length, degree of unsaturation, and branching of the lipid's carbohydric chains. Cholesterol condenses with phospholipids only if these molecules in the monolayer are in neither a very stretched nor a very condensed condition in comparison to the whole trans-conformation of saturated lipids. Because the majority of native lipids are right in such intermediate structure, this effect of cholesterol also took place in our experiments.

Interestingly, we have observed a good correlation between changes of properties of BLMs from brains of different groups of vertebrates and the intensity of lipid peroxidation in the studied objects, which we have shown earlier using chemiluminescent analysis and TBA test (Fig. 1) [22,23]. In both cases, the decrease in the conductivity of BLMs formed from the nervous tissue in the poikilothermal animal series was noticeable (crucian carp, frog, lizard) and a slight increase of this parameter in rats (Table 1). If we compare our results with

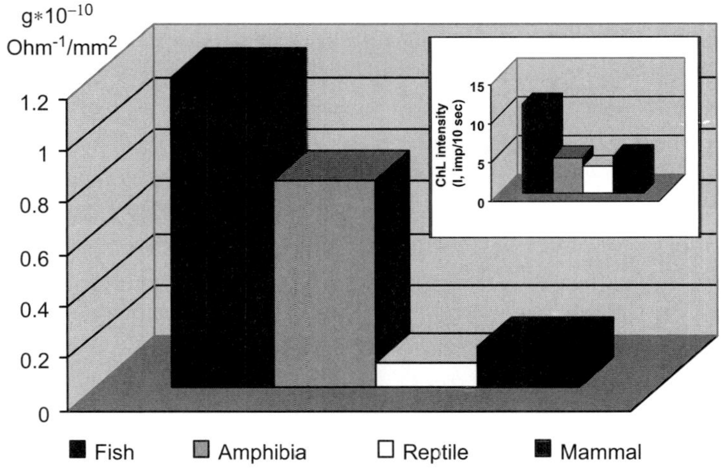

Fig. 1. Specific electrical conductance (g) of BLMs from brain lipids of vertebrates. *In the right corner*: spontaneous chemiluminescence (*ChL*) intensity of brain lipids of vertebrates in imp/10 s.

established data on the extent of cholesterol in the brain of this animal series (namely: fishes – 16.4%; frogs – 17.5%; lizards – 27%; rats – 16.7% from whole lipids), it is clear that during transition to the ground mode of life the amount of cholesterol in the brain increased unevenly (lizard), and we have seen some abrupt changes in the resistance of model membranes of reptiles in our experiments. However, in spite of the abrupt decrease of cholesterol concentration in rat's brain, the electrical resistance of its BLMs only diminishes slightly. Apparently it is bound with a very high level of saturated fatty acids in lipids of mammals' nervous tissue. This gave an opportunity to ensure a condensing condition of the microviscosity of the membranes of nervous tissue and to keep a low-level permeability for ions, notwithstanding a low level of cholesterol. This means that the level of electrical conductivity of the studied membranes is conditioned by the content of cholesterol on the one hand and of the fatty acid composition of membrane lipids on the other.

The data about research of model membranes from lipids of myocard summarized in Table 1 indicate that, for cold-blood animals, there are no changes of electrical conductivity in the course of evolution development. However, this parameter in membranes from heart lipids of mammals is a little higher, even more than for the analogous parameter in fishes. This is quite natural if we take into consideration that the content of saturated fatty acids in the heart of rats is 1.5 times greater than the corresponding parameters for fishes [7].

We were most intrigued by the fact that bilayers from the lipids of frog's liver has an electrical resistance almost exceeding the resistance of BLMs from the liver lipids of other studied groups of vertebrates. Obviously, this can be connected with an extremely high level of cholesterol in the liver of frogs (10% from whole lipids), but also the concentration is high in the liver of lizards where data on the electrical properties of BLMs shows no such coincidence. This is therefore a contradiction.

The electrical parameters (electrical resistance and specific conductance) of the bilayer membranes from muscle lipids tended to change in all classes of animals (the electrical resistance increased and the electrical conductance decreased) as the evolutionary organization of the animal became more complex.

Analyzing the results given, it should be taken into consideration that, unlike nervous tissue, which is conserved with respect to any environmental changes, the other studied tissues (especially the liver and muscle tissues) are fairly labile in terms of the changes of their fatty acid composition, which depends on some external factors such as food, seasonal and age-related changes, ecological characteristics of habitats, etc. So this is why it is not exceptional that some of the values given are not typical of the membrane permeability in each separate case and could be caused by indicated variations of the specific fatty acid composition of lipids.

Table 1. The electrical parameters of BLMs formed from the vertebrate-animal tissue lipids in the medium containing 0.1 M KCl

Tissue	Class							
	Fish		Amphibians		Reptiles		Mammals	
	R_m ($\times 10^9$ Ohm)	g ($\times 10^{-10}$ Ohm^{-1}/mm^2)	R_m ($\times 10^9$ Ohm)	g ($\times 10^{-10}$ Ohm^{-1}/mm^2)	R_m ($\times 10^9$ Ohm)	g ($\times 10^{-10}$ Ohm^{-1}/mm^2)	R_m ($\times 10^9$ Ohm)	g ($\times 10^{-10}$ Ohm^{-1}/mm^2)
Brain	17 ± 2	1.2	25 ± 2	0.8	200 ± 2	0.1	130 ± 0.7	0.16
Heart	1.1 ± 0.1	18	0.8 ± 0.2	25	0.7 ± 0.08	28	1.3 ± 0.1	16
Liver	0.6 ± 0.1	34	15 ± 2	1.4	1 ± 0.1	20	2 ± 0.08	10
Muscle	0.7 ± 0.05	28	0.8 ± 0.01	25	1.2 ± 0.05	16	2 ± 0.09	10

4. IONIC ELECTION OF BLMs FOR UNIVALENT IONS

According to contemporary notions about mechanisms of the homeostasing of the ionic content of cytoplasm, processes of active traffic in a stationary condition just recompense the losses determined by the leakage of ions that are processes of passive ionic traffic through [19]. Values of membrane permeability by canals are essentially different for various types of ions. As a rule, membrane permeability is to a marked degree higher for ions whose distribution is close to equilibrium. At the same time, for ions whose distribution is far from balanced, and therefore whose compensation for its escape needs a big expenditure of energy, the membranes are a considerable diffusion barrier.

One method for studying a membrane's ionic permeability is the investigation of the electrical properties of BLMs in different ionic solutions (Fig. 2). So we think that the parameters of planar layers from native lipids of different tissues of vertebrates forming in some important way for membrane action ionic media will give us the information that, to a certain degree, will let us judge the mechanism of passive traffic in native tissues with different functional specificity and their changes in the course of phylogenesis.

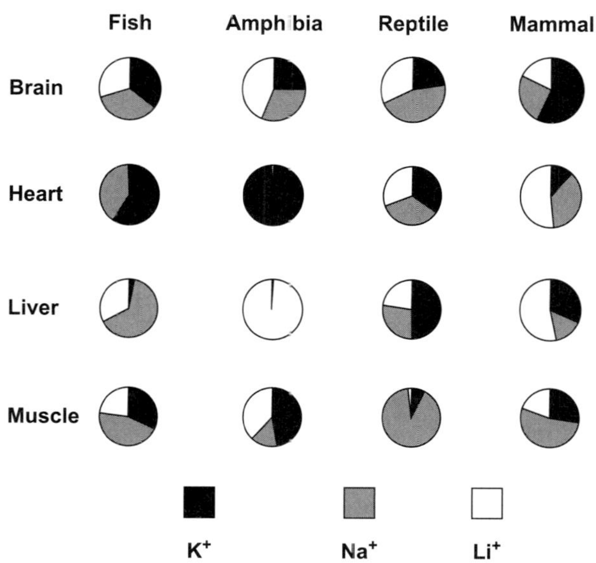

Fig. 2. Ionic penetration of BLMs from lipids of different tissues of vertebrates for univalent ions.

4.1. Ionic permeability of BLMs from brain lipids of vertebrates

Today it is clear that the mechanism of impulse conduction during the excitement and breaking in nervous cells is based on electrical currents in the membrane. Bearers of these currents in the excitement/breaking process are various types of ions. We have studied BLMs from brain lipids of vertebrates in K^+, Na^+ and Li^+ media and the results of these experiments are presented in Table 2. The electrical conductivity for univalent ions is extremely distinctive, especially for BLMs from brain of vertebrates:

$$K^+ > Na^+ > Li^+$$

Such a row of permeability of membranes for univalent ions is not dependent on the crystallization radius of ions nor the size of their hydrate covers:

$$K^+(0.133\ \text{nm}) > Na^+(0.095\ \text{nm}) > Li^+(0.060\ \text{nm})$$

Nevertheless, exactly such a series of permeability is typical for the majority of cell membranes under rest conditions and is similar to a so-called biological series of alkaline metal cations.

Interestingly, the difference in electrical conductivity of various ions for poikilotermal animals do not reach such big values. For example, for fishes, the difference is not noticeable at all. In the case of frogs and lizards, the reverse series of ionic permeability is observed and for the nervous tissue of lizards the electrical conductivity of membranes is very high for Na^+ ions. Possibly, such a scheme of ionic permeability underlies the phylogenetic change of the integrative function of the brain [24]. Therefore, for fishes and amphibians, the main integrative site of the brain is the middle brain (mesencefalon). For reptiles,

Table 2. Ionic permeability of BLMs from brain lipids of vertebrates

	Ion					
	K^+		Na^+		Li^+	
Class	R_m (×10^{10} Ohm)	g (×10^{-11} Ohm^{-1}/mm^2)	R_m (×10^{10} Ohm)	g (×10^{-11} Ohm^{-1}/mm^2)	R_m (×10^{10} Ohm)	g (×10^{-11} Ohm^{-1}/mm^2)
Fish	1.7±0.2	12	1.7±0.08	12	2±0.01	10
Amphibian	2.5±0.2	8	2±0.04	10	1.4±0.03	14
Reptile	20±2	1	10±0.08	2	14±0.01	1.4
Mammal	13±0.7	1.6	29±0.1	0.7	40±1	0.5

the more important site became the intermediate brain (diencefalon). In the brain of mammals appeared a new bark – a six-layer cover of the forebrain. Changes of physiology are based on changes of biochemistry.

4.2. Ionic permeability of BLMs from heart lipids of vertebrates

We have found very low values of electrical resistance for Li^+ and, especially, Na^+ from studying BLMs from the heart lipids of mammals (Table 3). In this case, the difference of conductivity is more than 10 times higher than in KCL media. For all that, in the case of poikylotermal vertebrates, the difference in the conductivity for K^+ and Na^+ is very small. BLMs from lipids of frog heart have very high permeability for Na^+.

Experimental results for the properties of BLMs from myocard lipids are most difficult to interprete because of the scarcity of data for the lipid biochemistry of this tissue. In spite of classification, by which myocard is a cross-striated muscle, it could not be allocated to one row with skeletal muscles because of its functional features. Our earlier researches of the free-radical oxidation of this tissue and studies of BLMs' properties are further proof of this. Interestingly, to compare our results with the known fact in the heart muscle of mammals is a so-called spike-current, which leads to depolarization represented by Na^+, but in the heart of poikylotermal animals, the main part of the current arises at the expense of Ca^{2+} ions [25]. Moreover, *in vitro* experiments with Purkinje fibers, typical only for the hearts of birds and mammals, it has been clear that, under substitution of Na^+ ions for Li^+, spike activity remains. As a result, all these changes lead to a physiological effect that the myocard of mammals conducts the wave of excitement much more speedily (5–6 m/s) than the poikylotermal heart (1 m/s).

4.3. Ionic permeability of BLMs from muscle lipids of vertebrates

Unlike the myocard, BLMs from lipids of cross-striated muscles, as in the case of the previous experiments, are the least variable tissue (Table 4). Almost all the studied classes of vertebrates are typical of the primary penetration for Na^+:

$$Na^+ > K^+ > Li^+$$

This is not surprising because in cross-striated muscles of all vertebrates the impulse, just as a nervous impulse, s triggered by an increase of the Na conductivity during the rise phase of the acting current, after which is the increase of K^+ conductivity. What attracts our attention is the very low permeability of lipid bilayers from amphibian muscles for Na^+ and the extreme instability of BLMs from the muscle lipids of frogs in KCl media (lifetimes of 2–3 min). At the same time,

Table 3. Ionic permeability of BLMs from heart lipids of vertebrates

Class	Ion					
	K^+		Na^+		Li^+	
	R_m ($\times 10^9$ Ohm)	g ($\times 10^{-9}$ Ohm^{-1}/mm^2)	R_m ($\times 10^9$ Ohm)	g ($\times 10^{-9}$ Ohm^{-1}/mm^2)	R_m ($\times 10^9$ Ohm)	g ($\times 10^{-9}$ Ohm^{-1}/mm^2)
Fish	1.1 ± 0.1	1.8	1.7 ± 0.2	1.2	0.008	0.025
Amphibian	0.8 ± 0.2	2.5	0.004 ± 0.001	0.005	0.008	0.025
Reptile	0.7 ± 0.08	2.8	0.7 ± 0.01	2.9	0.8 ± 0.01	2.5
Mammal	1.3 ± 0.1	1.6	38 ± 0.01	5	27 ± 2	7

Table 4. Ionic permeability of BLMs from muscle lipids of vertebrates

Class	Ion					
	K^+		Na^+		Li^+	
	R_m ($\times 10^9$ Ohm)	g ($\times 10^{-9}$ Ohm^{-1}/mm^2)	R_m ($\times 10^9$ Ohm)	g ($\times 10^{-9}$ Ohm^{-1}/mm^2)	R_m ($\times 10^9$ Ohm)	g ($\times 10^{-9}$ Ohm^{-1}/mm^2)
Fish	0.7±0.05	2.8	0.5±0.01	4	1±0.01	2
Amphibian	0.8±0.01	2.5	2.5±0.01	0.8	1±0.01	2
Reptile	1.2±0.05	1.6	0.1±0.001	20	4.3±0.8	0.46
Mammal	2±0.09	1	1±0.01	2	2.8±0.2	0.72

in Na^+ and Li^+ media, artificial membranes from the same lipids are stable enough (lifetimes of 20 min and more). The resistance of these planar bilayers is less than the resistance of BLMs in solutions of Na^+ and Li^+. However, membranes from muscle lipids of crucian carp are more stable for lower values of resistance.

4.4. Ionic permeability of BLMs from liver lipids of vertebrates

One more tissue for which abrupt differences of permeability for various ions takes place is liver tissue (Table 5). Interestingly, BLMs from fish liver lipids in this case demonstrate a very low conductivity of K^+ ions compared to Na^+ and Li^+, whose resistance is 10 times higher in K media than in other solutions. For amphibians and mammals, the ionic row of its transition through membrane activity value is distributed as follows:

$$Na^+ < K^+ < Li^+$$

Lipids from reptile liver are least penetrated for Li^+. Liver is a highly labile tissue, subject to a relatively quick change of lipid content with changing nourishment and ecological factors, unlike nervous tissue, which is most conservative in this sense. Some inadequate levels of resistance and conductivity of BLMs formed from different tissues of animals may be conditioned by the influence of changes in season or age and in the case of wild animals this alteration could not be identified every time. This study of the ionic penetration of planar membranes from liver lipids needs to take account of a number of additional factors for a more detailed analysis.

All our results reflected very well the literature data of heterogeneous lipids and of the fatty acid composition for both different organisms and separate tissues [8]. Of course, during discussion of our results we have not touched upon the question of a possible role of the protein component of the membrane.

5. IONIC PENETRATION OF BLMs FOR Ca^{2+}

It is hard to over-estimate the importance of Ca ions in the functioning of various types of biological membranes. For initiating such processes as muscle contraction, synaptic impulse transmission, etc., bivalent calcium ions play an essential role for the regulation of cell membranes' functional condition. Data for the binding of bivalent cations showed that Ca^{2+} can make clusters in lipid bilayers with negative charged phospholipids [13].

The study of BLM permeability for Ca^{2+} showed the specificity of its interaction with the membrane lipids depending on the functional characteristics of tissues. As is seen from Table 6, the electrical resistance of model membranes that formed

Table 5. Ionic permeability of BLMs from liver lipids of vertebrates

Class	Ion					
	K^+		Na^+		Li^+	
	R_m ($\times 10^9$ Ohm)	g ($\times 10^{-9}$ Ohm^{-1}/mm^2)	R_m ($\times 10^9$ Ohm)	g ($\times 10^{-9}$ Ohm^{-1}/mm^2)	R_m ($\times 10^9$ Ohm)	g ($\times 10^{-9}$ Ohm^{-1}/mm^2)
Fish	0.6 ± 0.1	3.4	0.025	80	0.045	40
Amphibian	15 ± 2	0.14	18 ± 0.3	0.11	0.7 ± 0.2	29
Reptile	1 ± 0.1	2	1.8 ± 0.08	1.1	2.2 ± 0.3	0.9
Mammal	2 ± 0.08	1	4 ± 0.7	0.5	1.2 ± 0.2	1.7

Table 6. The electrical parameters of BLMs formed from the vertebrate-animal tissue lipids in the medium containing 0.1 M $CaCl_2$

Tissue	Class							
	Fish		Amphibians		Reptiles		Mammals	
	R_m (Ohm)	g (Ohm^{-1}/mm^2)	R_m (Ohm)	g (Ohm^{-1}/mm^2)	R_m (Ohm)	g (Ohm^{-1}/mm^2)	R_m (Ohm)	g (Ohm^{-1}/mm^2)
Brain	$5 \pm 0.3 \times 10^6$	0.4×10^{-6}	$1.7 \pm 0.08 \times 10^{10}$	1.2×10^{-10}	$0.7 \pm 0.09 \times 10^9$	2.9×10^{-9}	$1 \pm 0.04 \times 10^{10}$	2×10^{-10}
Heart	–	–	$1.25 \pm 0.07 \times 10^7$	1.6×10^{-7}	$1 \pm 0.1 \times 10^7$	2×10^{-7}	$0.19 \pm 0.01 \times 10^7$	1×10^{-6}
Liver	$4.4 \pm 0.4 \times 10^9$	0.45×10^{-9}	$0.95 \pm 0.08 \times 10^6$	2.1×10^{-6}	$2.5 \pm 0.01 \times 10^9$	0.8×10^{-9}	$4.5 \pm 0.7 \times 10^9$	0.4×10^{-9}
Muscle	$2 \pm 0.4 \times 10^9$	1×10^{-9}	$2.5 \pm 0.01 \times 10^8$	0.8×10^{-8}	$1 \pm 0.07 \times 10^9$	2×10^{-9}	$5 \pm 0.8 \times 10^9$	0.4×10^{-9}

from brain lipids of amphibians and mammals in a $CaCl_2$-containing media was much greater compared to other tissues. A similar tendency with respect to the low conductance of Ca^{2+} was observed in the lipids from the liver and muscle of amphibians.

The electrical resistance of the studied BLMs have mostly high values. A possible mechanism underlying this phenomenon may involve the direct binding of Ca^{2+} to phosphate in polar lipids, which leads to a partial neutralization of their negative charge and restricted mobility of the polar group on the whole. Thus, Ca^{2+} inducing the transition of lamella phase to hexagonal is a reason for the increase in temperature of the phasic transition [26].

It is known that increasing Ca^{2+} concentration in media leads to a decrease in the erythrocyte membrane penetration for univalent ions and a limit of the values of the potentials difference on the membrane displacing to the higher levels [19]. Such an effect of Ca^{2+}, both in the case of erythrocytes' $\Delta\mu$ changing and the alteration of *g* values in our results, reminds one of the influence of these ions on the value of the threshold potentials difference evoking the excitement of nervous and muscle filaments. However, the regulation of these processes in organisms is realized by means of enzymatic systems of active ion transport (namely Ca-ATPase) and the intensity of its functioning will inevitably depend on the accumulation of Ca^{2+} in the environment.

The fact that model membranes from brain lipids of fishes have an extremely low degree of resistance aroused our curiosity. A comparative analysis of data by electrical conductivity of other bilayers indicates a gradual decrease of BLM resistance in the course of increasing evolution development as well as an increase of BLM penetration for Ca^{2+}. However, values of electrical properties for rats go to a higher level than fish BLM, but not higher than the heart lipids of mammals, where the membranes have even more penetration than that of reptiles.

It should be noted that a comparative analysis of BLM permeability for both univalent ions and this case demonstrates abrupt differences in the tissue of frogs against the analogous tissue of other studied animals. Thus, if the liver membrane of an amphibian has a very high resistance for K^+ and Na^+, for Ca^{2+} it is quite the reverse because bilayers have an extremely low electrical resistance, approximately 10^6 Ohm. Why are liver lipids of frogs so clearly different from lipids of the same tissue from other representatives of vertebrates? We are inclined to think that the reason for this phenomenon is a season of purification of lipids from frogs, which takes place in spring. At this time, the liver of frogs is in a condition of high hormonal "phone" because of the massive shedding of sex hormones and a high level of sterins, which are known as BLM stabilizators in KCl media and a guarantee of its high resistance. We have noticed an analogous tendency in works about seasonal changes of lipid peroxidation and chemiluminescence in native membranes of rats' hepathocytes [27].

The data obtained by some authors [28] showed that the threshold Ca^{2+} concentration that caused contraction in the striated muscles of homoiothermal animals is an order of magnitude lower than that required for the muscles of poikilothermal animals (10^{-6} and 10^{-7} M, respectively). These data are in good agreement with our results: less penetration of BLMs from muscle lipids of rats for Ca^{2+} than in the case of poikylotermal animals because rats' membranes need more accumulation of Ca-ions for realizing muscle contraction.

The BLMs formed from heart lipids, unlike those containing lipids from other tissues, display a high conductance of Ca^{2+} in the electrical field (approximately 10^{-7}). On the face of it, the very low resistance of membranes from heart lipids in our experiments runs counter to the nature of the conductivity of BLMs from skeletal muscle lipids. However, the most important factor of the specific conductance of myocard is the automatic compensation of loading increase, that is, if the ventricle of the heart is strained by high volume, the power of its contraction increases. Thus, the heart muscle does not relax like a common cross-striated muscle, but has a phase of decompensation [25] and the stream of Ca^{2+} through the membrane hardly weakens. Hystologically, the heart system of vertebrates adapted to the fast conductivity of excitement and surely the qualitative content of lipids closely connects with its function specificity, which is extremely high Ca penetration of membranes from heart muscle lipids. We believe the key role for this belongs to cardiolipin, whose percentage in heart lipids is extremely high [7]. The availability of four fatty acid tails with one phosphate head and of negative charge forms the specificity of membranes with a predominance of these phospholipids. Obviously, with such a configuration of lipids, the membrane should be more capable of electroporation and binding of positive bivalent ions. Possibly the same reason underlies the low stability of these membranes under high voltage, as shown below.

6. COMPARATIVE INVESTIGATION OF BREAKING POTENTIAL OF BLMs FROM LIPIDS OF VERTEBRATES

All living cells are characterized by the availability of a stationary difference of the electrical potential caused by ionic asymmetry on both sides of the membrane and its transmembrane diffusion. The absolute value of the membrane potential may be very high (more than 200 mV in some freshwater algae, bacteria, and energized mitochondria). In the course of the excitement of nervous and muscle cells, the quick depolarization and polarization of membranes by influence of temper occurs. Its own or an external electrical potential can be a reason for the instability or even the destruction of the lipid bilayer.

In the last stage of our investigation we have tried to establish a connection between the variation of the membrane lipid composition comparing different groups of vertebrates with a change in the stability of BLMs formed from lipids of corresponding tissues by definition of the break-potential value for studying model membranes.

Results of research into the electrical abruption of membranes from different tissues of vertebrates in media of 0.1 M univalent ions K^+, Na^+, Li^+, and Ca^{2+} are shown in Tables 7 and 8. It is supposed that the process of electrical rupture of BLM underlies the process of electroporation, in the course of which huge defects like inverted pores appear, which have even penetrated such big molecules as proteins [29].

BLMs from brain lipids are mostly stable, with the breaking potential having its highest value in the brain membranes of reptiles, which surely depend on cholesterol content. Interestingly, BLMs from lipids of organisms connected by their physiological function with a water environment (fishes, amphibian) are more stable in K^+ media than Na^+, Li^+, and Ca^{2+}. However, for ground vertebrates this correlation changes; values of electrical rupture for BLMs by the brain lipids of lizards and mammals are higher in solutions of Na^+ and Li^+ than of K^+ and Ca^{2+}:

Fish $K^+ > Na^+ > Li^+ > Ca^{2+}$
Amphibian $K^+ > Ca^{2+} > Na^+ > Li^+$
Reptile $Li^+ > Na^+ > K^+ > Ca^{2+}$
Mammal $Na^+ > Li^+ > K^+ > Ca^{2+}$

The same peculiarity of break-potential is also observed for BLMs from the heart lipids of different vertebrates. For model membranes by lipids of liver for all studied groups of animals, there is more stability in solutions with K^+ compared with other ions. Lipid bilayers from muscle of fishes and amphibians are unstable and have a short lifetime, so generally we could not record values of electrical rupture for these membranes.

In our experiment, right before the rupture of BLM, a fall of electrical resistance of the membrane was noted. This does not contradict the literature data because compression of the membrane in a powerful electrical field led to relief of the ion penetration in the lipid bilayer [30]. Such changes of penetration are explained by the appearance of transmembrane defects, not because of the availability in the bilayer of qualitatively new lipid admixtures, but only by the properties and structural condition of the BLM lipid itself. Thus, we may suggest that BLMs of highly unsaturated fatty acids of fishes and reptiles have a more friable structure than model membranes from lipids of more highly organized vertebrates. In a high electrical field, these lipids form a pore much more easily than lipids of membranes in more condensing conditions. On the other hand, during electroporation of BLMs with rigid structure, Na^+ and Li^+ with a less crystalline radius penetrate through small pores easily without any compression of the membrane. At the same time, for penetration through the membrane of big ions

Table 7. The values of electrical rupture (mV) of the BLMs formed from the vertebrate-animal tissue lipids in media with univalent ions

Tissue	Class											
	Fish			Amphibian			Reptile			Mammal		
	K^+	Na^+	Li^+	K^+	Na^+	Li^+	K^+	Na^+	Li^+	K^+	Na^+	Li^+
Brain	325	216	210	347	290	247	360	366	386	255	350	310
Heart	189	174	152	200	157	123	168	180	243	200	285	198
Liver	170	146	–	225	314	136	216	183	152	219	205	200
Muscle	160	–	–	–	–	–	186	–	144	180	128	127

Table 8. The values of electrical rupture (mV) of the BLMs formed from the vertebrate-animal tissue lipids in media with Ca^{2+}

	Class			
Tissue	Fish	Amphibian	Reptile	Mammal
Brain	200	304	232	242
Heart	–	140	152	187
Liver	122	400	650	190
Muscle	178	–	135	148

like K^+ and Ca^{2+}, the formation of huge pores is necessary, which also leads to faster electrical rupture of bilayers.

We should note that an abrupt increase of voltage in these experiments led to different values of the break-potential compared with a gradual increase. Moreover, the difference between fast and gradual potential in separate cases could be even as much as 100 mV. It has been pointed out that the mechanism of the rupture process in this time is not similar to the explained one [13]. During fast increase of potential, BLM is regarded as a volume electrical layer, able to compress under an external electrical field. By equilibrium in the case of the influence of the electrical field, the elastic power is balanced by electrical potential:

$$-\varepsilon V^2/2l^2 = E \ln (l/L),$$

where V is the potential, ε the dielectric constant in the center of the membrane, l the primary thickness of film, L the thickness of film after compression, and E is the Hung module.

By sharply increasing the potential to an extreme value, rupture of the membrane occurs in accordance with the condition:

$$-\varepsilon V^2/2l^2 \approx 0.18.$$

According to Berestovsky [31], in the course of an uneven increase of potential on the membrane, the bilayer became thicker because of the volume compression of its carbohydric part. Then compression spread evenly on the surface in 0.1 ms. Also, determination of the breaking potential of BLM and the speed of increase of potential differences has been discovered [32].

7. CONCLUSION

There has been 200 million years of vertebrate development from fishes to mammals and a great diversity of species and forms, and all this life is based on a universal bimolecular structure of membranes, with both rigid and labile formation responsible for all vital processes taking place in our complicated organism.

So, what can we say when analyzing the results of our investigation about the condition of lipids in biomembranes?

At first, all our experimental data suggested that in the course of the development of vertebrates the structure of lipid bilayers became more rigid and its barrier function for ion permeability increased, especially for K^+. This process goes through the substitution of unsaturated fatty acids by a saturated one in a percentage or, alternatively, through the gradual effect of free-radical oxidation processes.

The conformity to a natural low that our studies revealed is mostly expressed in the nervous tissue of the studied animals because it is this tissue that is most conservative in the sense of the variation in the membrane's lipid composition and it hardly reacts to changes in external factors such as the lipid composition of food and conditions. The importance of cholesterol as a temperature-adaptation factor is evident [33], but the evolutionary mechanism of the transition to homoiothermy decreased this key role of cholesterol. Thus, the penetration of K^+ in the course of these changes increase, slightly, but, at the same time, for Ca^{2+}, whose accumulation is a first step for the reaction of fast impulse transference by a nervous route and synaptic transmissions, the membranes of mammals are a more valid diffusion barrier than the membranes of reptiles.

We conclude from the present investigation that the skeletal muscles have undergone fewer changes in the course of the evolution of vertebrates. As far as tissues of liver and heart are concerned, the studied BLMs reflect well enough the heterogeneous nature of the lipid content, which underlies the functional specificity of these tissues and its variations in organisms at different stages of evolutionary development.

Thus, comparison of the parameters of the state of lipids and lipid-containing structures allowed us to assume that the electrical conductance and ionic permeability of the membranes are related to the degree of phylogenetic organization of the organisms studied (and some of their tissues). It is noteworthy that the tendencies in the changes in the BLM electrical parameter correlate with the results of the studies on free-radical lipid peroxidation in the biological materials analyzed [34].

The presented data are only a first step to a wide range of investigations of the phylogenetic aspects of development based on a biophysical method to explore this problem. Also, one of the most important conclusions of this work is information on the method of artificial membrane modeling from native lipids for studying the functional specificity of tissues in their phylogenetic development.

REFERENCES

[1] E. Folk, Textbook of Environmental Physiology, 2nd edition, University of Iowa, Lea Febiger, Philadelphia, 1974, 213 pp.
[2] E.M. Kreps, Lipidy kletochnykh membran (Lipids of Cell Membranes), Nauka, Leningrad, 1981, p. 339.
[3] E.M. Kreps, Fosfolipidi kletochnych membran nervnoi sistemi w razwitii zhivotnogo mira (Cell Membrane Phospholipids of Nervous System in Animal World Development), Bach Readings XXII, Nauka, Leningrad, 1967, p. 75.
[4] E.M. Kreps, Evolucionnaja Fiziologia (Evolutional Physiology), Nauka, Leningrad, 1979, p. 603.
[5] A.N. Shulyndin, Evolution diversity of phospholipid structures, Usp. Sovr. Biol. 95 (1) (1983) 16–31.
[6] N.F. Avrova, Biochemical mechanisms of vertebrate adaptation to changing conditions of environment: role of lipids, J. Evol. Biochem. Physiol. 35 (3) (1999) 170–180.
[7] S.A. Zabelenskii, M.A. Chebotareva, A.I. Krivchenko, Comparative studying of mitochondrial phospholipids of vertebrates living in different conditions of provision of oxygen: new glance on fatty acid role, J. Evol. Biochem. Physiol. 34 (2) (1998) 163–172.
[8] S.A. Zabelenskii, M.A. Chebotareva, A.I. Krivchenko, New glance on a role of phospholipid's fatty acids: possibility of electrical charge transfer in membrane monolayer, J. Evol. Biochem. Physiol. 36 (3) (2000) 202–209.
[9] P. Mueller, D.O. Rudin, H.Ti. Tien, W.C. Wescott, Reconstruction of cell membranes structure in vitro and its transformation into an excitable system, Nature 194 (1962) 979–980.
[10] L. Bolis, A. Lif, Membrany i bolezn' (Membranes and Disease), Meditsyna, Moscow, 1980.
[11] D. Selkoe, Aging, amyloid, and Alzheimer's disease: a perspective in honor of Carl Cotman, Neurol. Res. 28 (11) (2003) 1705–1715.
[12] A. Chanturiya, M. Witteker, J. Zimmerberg, Fusion of secretory granules with planar lipid bilayer membranes, Biophys. J. 66 (2) (1994) 284.
[13] V.F. Antonov, Lipidy i ionnaya pronitsaemost' membran (Lipids and Ion Permeability of Membranes), Nauka, Moscow, 1982, p. 150.
[14] H. Tien, A. Ottova, Planar Lipid Bilayers (BLMs) and Their Applications, Elsevier, Amsterdam, 2003, p. 1043.
[15] H. Tien, A. Ottova, The lipid bilayers concept and its experimental realization: from soap bubbles, the kitchen sink to Black lipid membranes, J. Memb. Sci 189 (2001) 83–117.
[16] J.S. Ye, A. Ottova, H.T. Tien, F.S. Sheu, Nanostructured platinum-lipid bilayer composite as biosensor, Bioelectrochemistry 59 (1-2) (2003) 65–72.
[17] M. Kates, Techniques of Lipidology, Elsevier, Amsterdam, 1972.
[18] P. Mueller, D. Rudin, Action potentials induced in bimolecular lipid membranes, Nature 217 (1968) 713–719.
[19] A.A. Lev, Ionnaya izbiratel'nost' kletochnykh membran (Ion Selectivity of Cell Membranes), Nauka, Moscow, 1975, p. 323.
[20] T.N. Medvedeva, Sravnitel'naya biokhimiya obmena veshchestv u zhivotnykh (Comparative Biochemistry of Metabolism in Animals), 1986, Kuibyshev, p. 603.
[21] S. Koronkiewicz, S. Kalinowski, Influence of cholesterol on electroporation of bilayer lipid membranes: chronopotentiometric studies, Biochim. Biophys. Acta 1661 (2) (2004) 196–203.
[22] A.E. Zakarian, N.M. Ayvazian, K.G. Karageuzyan, Comparative analysis of free-radical lipid oxidation processes proceed's intensity of poikilotherm vertebrates, Reports (Doklady) of Russian Academy of Sciences 374 (1) (2000) 111–114.

[23] A.E. Zakharian, N.M. Ayvazian, Ionic permeability of model bilayer membranes from lipids of vertebrate tissues for univalent ions, Biophysika 47 (6) (2002) 1068–1072.
[24] K. Schmidt-Nielsen, Animal Physiology: Adaptation and Environment, Cambridge University Press, New York, 1982, p. 799.
[25] L. Prosser, Comparative Animal Physiology, W.B. Saunders Company, Philadelphia, 1978, p. 653.
[26] J. Katsaras, X-ray diffraction studies of oriented lipid bilayers, Biochem. Cell Biol. 75 (5–6) (1995) 209–218.
[27] G.H. Pogossian, A.E. Zakharian, Season dependence of 17 β-estradiol influence on rats liver cell's nuclear membrane chemiluminescence, Reports of Vth Meeting Arm. Physiol. Soc. 1994, p. 11.
[28] A.D. Slonim, Evolyutsiya termoregulyatsii (The Evolution of Thermoregulation), Nauka, Moscow, 1986, p. 75.
[29] M. Kozlov, Intermediates of membrane fusion and fission: structure, energy and the role for proteins, Advanced Course "Lipid Signalling and Membrane Traffic", S. Maria Imbaro, Italy, 2003, June 20–25, pp. 42–45.
[30] E. Neuman, A. Sower, C. Jordan, Electroporation and Electrofusion in Cell Biology, Plenum Press, New York, 1989.
[31] G.N. Berestovsky, M.Z. Gyulkhandanyan, V.G. Ivkov, V.D. Razin, Voltage-induced reflectivity relaxation of bilayer lipid membranes: on changes of bilayer thickness, J. Membr. Biol. 43 (1978) 107–126.
[32] Y.A. Shipunov, G.Y. Drachev, Dependance of BLM's electrical rapture from voltage increasing speed, Biofizika 25 (1980) 921–923.
[33] V.A. Tyurin, Y.Y. Tyurina, T.V. Sokolova, A. Arduini, Participation of palmitinic acid in synaptosomal membrane preparation in course of oxidative stress: role of palmitoil-carnitin in adaptation mechanism, J. Evol. Biochem. Physiol. 34 (1) (1998) 3–10.
[34] N.M. Ayvazian, A.E. Zakarian, K.G. Karageuzyan, Intensity of lipid's free-radical oxidation processes and superoxiddismutase's activity in nerve tissue of vertebrates, Neurochemistry (4) (2002) 278–284.

CHAPTER 9

Electrochemical Study of the Bilayer Lipid Membrane

Erkang Wang* and Xiaojun Han

State key Laboratory of Electroanalytical Chemistry, Changchun Institute of Applied Chemistry, Chinese Academy of Sciences, Changchun, Jilin 130022, China
E-mail: xjhan@ciac.jl.cn

Contents

Abstract

We summarized the recent work of our group with electrochemistry as the main study method. This chapter was divided into seven sections.

1. *The ion-channel behavior of gramicidin in bilayer lipid membrane (BLM)*. Gramicidin was incorporated in a BLM. The behavior of the ion channel was studied by cyclic voltammetry. At a very low concentration of gramicidin in a BLM, the channel behavior was controlled by the voltage applied across the membrane. When the voltage is higher than 75 mV, the channel is closing, while lower than 75 mV, the channel is opening, but when the concentration of the gramicidin in the BLMs is high, the channel behavior is changed to voltage-independent.

2. In this section the method of forming a BLM was described, and its character was evaluated by electrochemical methods.

3. *The permeation of a BLM regulated by ion*. Interaction of lanthanide ions with supported BLMs (s-BLMs) was investigated by cyclic voltammetry (CV) and a.c. impedance spectroscopy. Lanthanide can affect the conformation of s-BLM and make it form some pores through which $Fe(CN)_6^{3-/4-}$ can reach the electrode surface. The interaction ability of three lanthanides with s-BLM follows the sequence: $Eu^{3+} > Tb^{3+} > La^{3+}$.

* Corresponding author. Tel.: +86-431-5689711; Fax: +86-431-5689711;
E-mail: ekwang@ciac.jl.cn

ADVANCES IN PLANAR LIPID BILAYERS AND LIPOSOMES, VOLUME 2
ISSN 1554-4516 DOI: 10.1016/S1554-4516(05)02009-0

4. *The interaction of peptide with a s-BLM*. We described two kind of peptides interacted with a s-BLM in this section. These peptides include microperoxidase and nisin.

The interaction of microperoxidase-11 (MP11) with cationic lipid vesicles of didodecyldimethylammonium bromide induces an α-helical conformation from random coils conformations in solution and then this change makes heme macrocycle more distorted. The binding of MP11 in solution to DDAB vesicles and the ordered structure formation are driven mostly by electrostatic interaction between negatively charged residues in the undecapeptide and positively charged lipid headgroups on the membrane surface. Upon binding to DDAB, its half-peak potential was also changed.

Nisin is a positively charged antibacterial peptide which binds to the negatively charged membranes of Gram-positive bacteria. The initial interaction of the peptide with a model membrane of negatively charged dipalmitoylphosphatidylglycerol (DPPG) model lipid membranes was studied by voltammetry and a.c. impedance spectroscopy. Adsorption of nisin into BLM destroyed the insulating capability of BLM, causing membrane resistance to decrease and membrane capacitance to increase. Experimental results suggested that the selective interaction of positively charged nisin with negatively charged BLM can induce some pores.

5. *The interaction of drug with s-BLM*. Amphotericin B (AmB) is a popular drug frequently applied in the treatment of systemic fungal infections. In the presence of ruthenium (II) as the marker ion, the behavior of AmB to form ion channels in sterol-free and cholesterol- or ergosterol-containing supported phosphatidylcholine bilayer model membranes were studied by CV, a.c. impedance spectroscopy, and UV/visible absorbance spectroscopy. In a fixed cholesterol or ergosterol content (5 mol%) in glassy carbon electrode-supported model membranes, experimental results show that, no matter what form of AmB, monomeric or aggregated, it could form ion channels in supported ergosterol-containing phosphatidylcholine bilayer model membranes. However, AmB could not form ion channels in its monomeric form in sterol-free and cholesterol-containing supported model membranes. When AmB is present as an aggregated state, it can form ion channels in cholesterol-containing supported model membranes. So the state of AmB played an important role in forming ion channels in sterol-free and cholesterol-containing supported phosphatidylcholine bilayer model membranes.

6. *The biosensor based on BLM*. BLM was constructed on the surface of a glassy carbon electrode. Horseradish peroxidase was embedded into the s-BLM to develop a kind of mediator-free biosensor for H_2O_2. The biosensor exhibited fine electrochemical response, stability, and reproducibility due to the presence of the s-BLM. As a model of biological membrane, s-BLM could supply a biological environment for enzyme maintaining its activity.

A novel type of potassium sensor based on the capacitance change of valinomycin-incorporated bilayer supported on a gold electrode has been developed and characterized. Investigating the capacitance change allows a simple and specific technique for the measurement of potassium ion in solution. Especially, the homemade capacitance meter is used to monitor the bilayer membrane formation and detect K^+.

7. *Constructing a hybrid bilayer membrane based on carbon substrate*. Carbon material is inert and has a wide potential window in electrochemistry, so modification of carbon materials is of interest to material science and electrochemistry. Primary alkylamine was chemically modified onto the surface of carbon electrode by electrochemical scans, and thus a monolayer was formed on the electrode. Because alkane chains section is towards the outside, a hydrophobic surface is constructed. Then a self-assembled phospholipid monolayer was formed onto the hydrophobic surface. Thus, a new kind of hybrid bilayer membrane (HBM) system was constructed. The formed HBM was

characterized by electrochemical and ATR–FTIR methods. From ATR–FTIR results, the lipid order parameter (*S*) of 0.73 was obtained. This kind of hybrid membrane has the advantages of lipid/alkanethiol HBM, such as stability, electrochemical detectability, etc.

The electrochemical study of a bilayer lipid membrane (BLM), especially CV study, originated from a series of innovative work by Tien *et al.* [1–3] in the early 1980s. The unmodified BLM is good insulation; its resistance is more than 10^8 ohm. People were enlightened by the structure of protein embedding in biomembrane, and they used something such as polypeptide, protein, dye, iodine, organic metal, semiconductor particles, etc. to modify the BLM, which made BLM possess a certain function and enhance its current response.

The group of Tien [4–6] firstly studied the electron transfer of BLM using an electrochemistry method. The good CV response emerged when BLM was modified by an electron conductor and an electron donor and a receptor were introduced to the two sides of BLM. They thought BLM modified by an electron conductor displays the behavior of a double dipole electrode. One of the sides of BLM is cathode and the other side is anode. The redox reaction in both sides results in electron transfer across BLM.

Ion transfer is another problem and has two types: one is the ionophore-facilitated ion transfer, the other is non-ionophore-facilitated ion transfer. Ionophore includes nature circle bacteriophage (valinomycin [7], etc.) and some synthetic ionophores (crownether [8], etc.). Non-ionophores include several kinds of special lipophilic ions, such as TPB^- and dyes. Tien and Bender [9] used CV to study BLM. They found I_2 modified BLM was easily penetrated by I^-, but it had no response for F^-, Cl^-, and Br^-. This phenomenon is also occurred for other halogens (Br_2, etc.) [10]. Ding and Wang [11] studied the process of lipophilic TPB^- penetrating BLM. They investigated the effect of pH and cholesterol to this process. Certainly, there are still a few investigations on ion transfer in BLM using the CV method. A lot of work needs to be done in this field.

A supported lipid membrane means modification of a lipid monolayer or bilayer on the substrate. If the substrate is an electrode, the supported lipid membrane can be studied by electrochemistry. The most important advantage lies in its stability. It is more stable than BLM, which gives it its practicality. Furthermore, it is suitable for *in situ* spectra study.

Thiol/lipid self-assembles on a golden electrode to form a supported membrane system. Because of the block of thiol, electrochemical active species cannot transfer an electron with an electrode. Thus, there is no cyclic voltametric response. This is not only a shortcoming but also merit. People can use the high resistance of the thiol/lipid bilayer to carry out voltage and capacitance measurement.

Since 1989 Tien *et al.* [12] succeeded in fabricating supported bilayer membrane on new cleaved metal, the direct electron transfer of protein [13] has been realized by the good biocompatibility of lipid. Salamon *et al.* [14] investigated the effect of anion and lipid concentration on electron transfer. Furthermore, they analyzed the tropism and active center of the protein in the lipid membrane.

1. THE ION-CHANNEL BEHAVIOR OF GRAMICIDIN IN BLM

Gramicidin is a linear polypeptide antibiotic that forms specific channels in membranes for the transport of monovalent cations. Although its structure is very simple [15], the mechanism of gramicidin insertion into lipid bilayers and ion permeation has not been understood thoroughly. The structure of the channel within the BLMs is formyl-NH to formyl-NH (N-to-N) terminal dimers of b-helices as suggested by Urry [16], whereas gramicidin dissolved in organic solvents exists as a mixture of parallel and antiparallel interwined dimers and disorder monomers [17], depending upon the solvent type and gramicidin concentration. Therefore, gramicidin undergoes a conformational change at the transition from a dissolved molecule to one dispersed in water [18] and finally to a membrane-spanning channel. Different experimental techniques have been used to elucidate the permeability characteristics of gramicidin channel. Single [19] and many-channel [20] current–voltage measurements have been particularly popular.

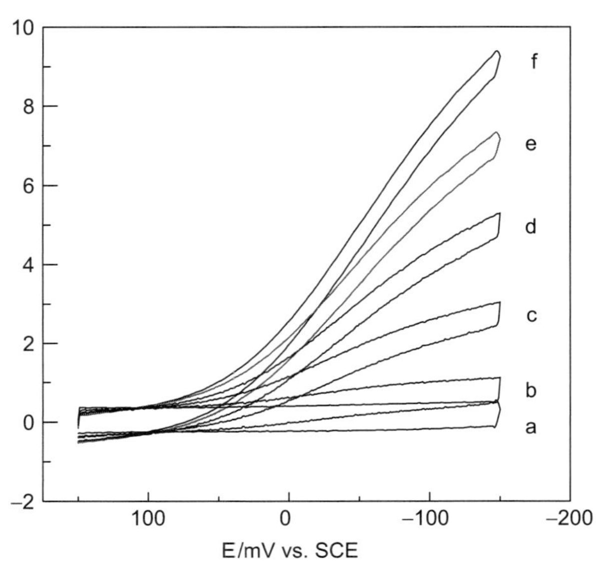

Fig. 1. Cyclic voltammograms of 10^{-4} M H^+ transport across BLMs at different times. (a) 0 min, (b) 15 min, (c) 25 min, (d) 30 min, (e) 35 min and (f) 40 min, scan rate 100 mV/s.

The following content describes the electrochemical behavior of the channel incorporated into the BLMs by CV [21].

The gramicidin channel in a membrane can selectively transport monovalent cations and the sequence of conducting is $H^+ > NH_4^+ > Cs^+ > K^+ > Na^+ > Li^+$ [22]. So $LiClO_4$ was selected as the bathing solution. The BLM-forming solution was a mixture of PC and cholesterol (3:1) dissolved in 2:1 decane/ethanol solution at 5 mg PC per ml solution. GD was first dissolved in ethanol and then added to the BLM-forming solution at a weight ratio of 1000:1 (named solution A) and 10:1 (named solution B), respectively. When solution A was used to form the BLMs, it can be seen from curve a in Fig. 1 that there is no ion transfer across the membrane because the membrane conductance reflects the permeability to ions in the bathing solution. It also indicates that Li^+ cannot transport through the membrane under this condition. Although the concentration of gramicidin in membrane-forming solution is known, we do not know the precise concentration of GD in the BLMs; we can only say it is very small or small enough. Adding 5 ml 0.1 M H^+ into the inner compartment to give a final concentration in the aqueous solution of 10^{-4} M, there is no current at first, but after 15 min, the positive current increased dramatically only in the range of -150–75 mV while in the range of 150–75 mV, the current remained zero. Thus, H^+ can transport easily across the membrane and the channel is controlled by the voltage applied across the membrane. Because we add H^+ to the inner compartment, the current is considered positive if it is passing the membrane in the direction from the inner compartment to the outer compartment. As can be seen from Fig. 1, the voltage 75 mV is the switch for

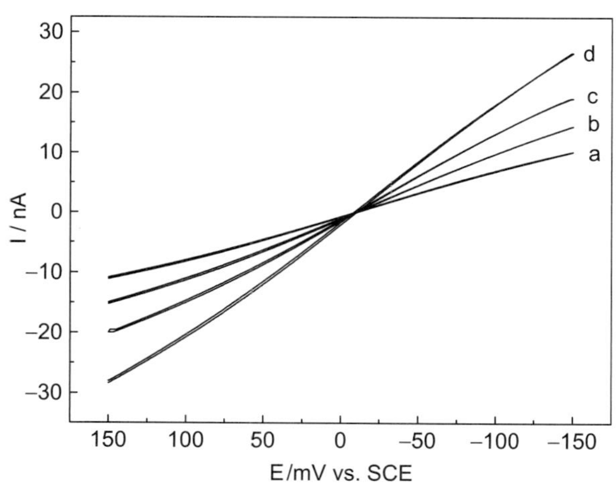

Fig. 2. Cyclic voltammograms of 10^{-2} M Li^+ transport across BLMs at different times. (a) 20 min, (b) 30 min, (c) 40 min and (d) 50 min, scan rate 100 mV/s.

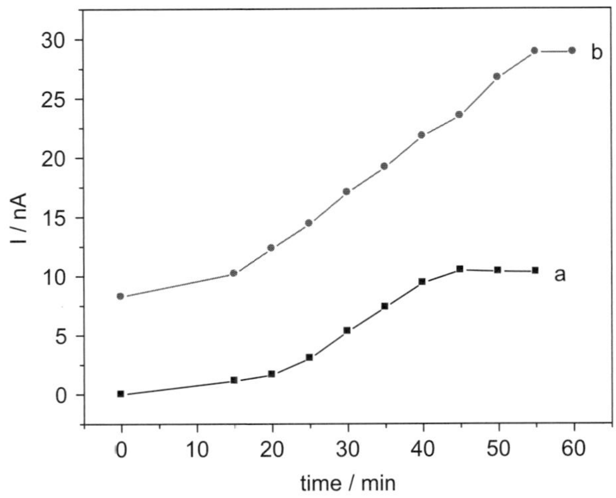

Fig. 3. Currents of (a) H^+ and (b) Li^+ across the membrane as a function of time with the voltage of -150 mV applied across the membrane.

the channel. When the voltage is higher than 75 mV, the switch is turning off and when the voltage is lower than 75 mV, the switch is turning on.

However, when the concentration of GD contained in the BLMs is high using solution B, the characteristic of the channel is different. As can be seen from Fig. 2, even without adding H^+, the current through the membrane increased extensively in the whole range of the voltage as if the membrane is 'leaking'. This indicates that Li^+ can transport across the membrane when the concentration of GD in the BLMs is high. The channel behavior is not only concentration-dependent but also time-dependent. Figure 3 shows the membrane current as a function of time with the voltage of 150 mV applied across the membrane. The membrane conductance increased with time and reached steady state after 40–50 min.

2. CONFIRMATION OF BLM ON THE ELECTRODE BY ELECTROCHEMICAL METHODS

The method for forming BLM on a glassy carbon electrode (GCE) was described as follows [23]. Dimyristoylphosphatidylcholine (DMPC) was dissolved in chloroform to give a final concentration of 2 mg ml^{-1}. Prior to supported BLM (s-BLM) formation, a GCE was polished with 1.0, 0.3, and 0.05 μm alumina slurry, respectively, then sonicated for 1 min in de-ionized water and acetone successively. Then the GC electrode was immersed in the phosphate buffer (pH 7.0) and the potential was held at 1500 mV for 3 min in order to polarize the electrode.

After the electrode was polarized, it was dried under purified nitrogen. Subsequently, a 5 μl aliquot of the lipid solution was dropped onto the surface of the electrode by a microsyringe and the electrode was immediately transferred in the phosphate buffer in which the supported lipid layer was formed spontaneously.

Figure 4(a) shows the cyclic voltammetric response of a bare GC electrode in 5 mmol l^{-1} $K_3[Fe(CN)_6]$ solution. After the GC electrode coated with lipid solution had been immersed in the phosphate buffer for about half an hour, it was put in 5 mmol l^{-1} $K_3[Fe(CN)_6]$ solution and the cyclic voltammetric response was observed (Fig. 4(b)). Comparing Fig. 4(b) with Fig. 4(a), it was found that a decrease in the amperometric response of the electrode and an increase in the peak-to-peak separation between the cathodic and anodic waves of $K_3[Fe(CN)_6]$ occurred. Thus, a conclusion can be made that $[Fe(CN)_6]^{3-}$ could be precluded from reaching the surface of the GC electrode to a certain degree, which implied that the supported lipid membranes had been successfully formed on the surface of the electrode.

Impedance spectroscopy is an effective method for probing the feature of a surface-modified electrode [24]. The complex impedance can be presented as the sum of the real, Z_{re}, and imaginary, Z_{im}, components that originate mainly from the resistance and capacitance of the cell, respectively. Figure 5 illustrates the results of impedance spectroscopy on (a) a bare electrode and (b) a modified electrode with supported lipid membranes in the presence of equimolar 10 mmol l^{-1}

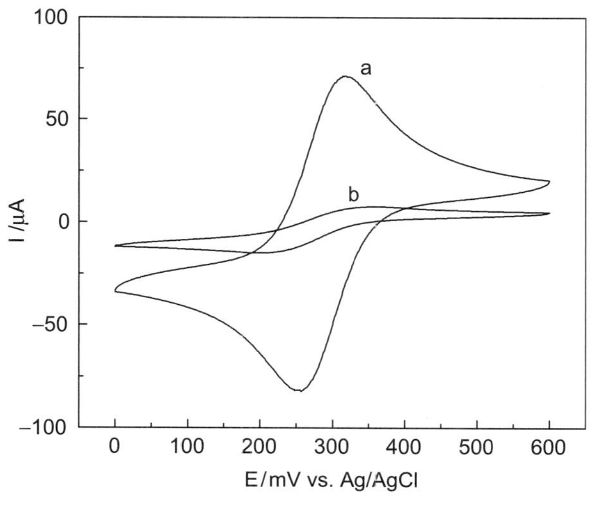

Fig. 4. Cyclic voltammograms of GC electrode in 5 mmol l^{-1} $K_3[Fe(CN)_6]$ solution containing 1 mol l^{-1} KCl: (a) bare GC electrode; (b) GC electrode coated with lipid membranes. All data were recorded in 5 mmol l^{-1} phosphate buffer (pH 7.0). Scan rate: 100 mV s^{-1}.

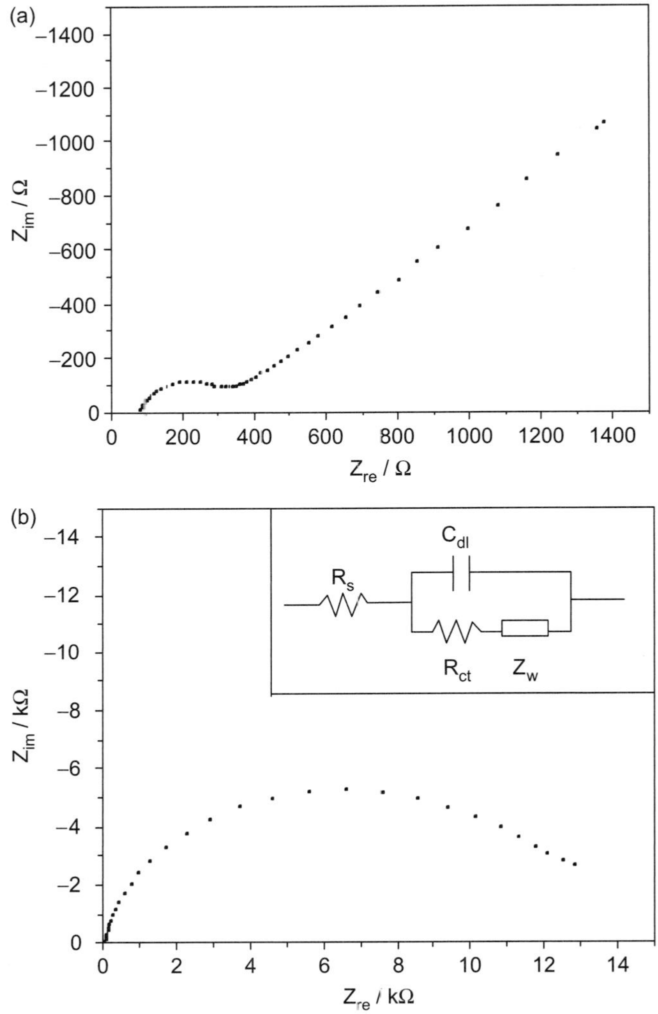

Fig. 5. Complex plane impedance plots in 10 mmol l^{-1} $K_3[Fe(CN)_6]/K_4[Fe(CN)_6]$ (1:1) mixture containing 0.1 mol l^{-1} KCl + 5 mmol l^{-1} phosphate buffer (pH 7.0) at (a) bare GC electrode; (b) modified electrode with supported lipid membranes. Inset: Modified Randle's equivalent circuit used to model impedance data in the presence of redox couples.

$[Fe(CN)_6]^{4-/3-}$ + 0.1 mol l^{-1} KCl, which are measured at the formal potential of $[Fe(CN)_6]^{4-/3-}$. It can be seen from Fig. 5(a) that the bare GC electrode exhibits an almost straight line that is characteristic of a diffusional limiting step of the electrochemical process. With respect to the modified electrode, a significant difference in the impedance spectra was observed (Fig. 5(b)).

In order to give more detailed information about the impedance property of the membranes, a modified Randle's equivalent circuit (inset of Fig. 5(b)) was chosen to fit the measured results. The total impedance is determined by several parameters: (1) electrolyte resistance, R_s; (2) the double-layer capacitance, C_{dl}; (3) charge transfer resistance R_{ct}; and (4) Warburg element Z_w.

To determine the thickness of the lipid membranes, the capacitance was chosen to show this feature. The double-layer capacitance (C_{dl}) consists of the unmodified GC electrode (C_{GC}) and a capacitance originated from the DMPC membranes (C_m) on the surface of the electrode. Their relationship can be expressed by equation (1) [25,26]:

$$\frac{1}{C_{dl}} = \frac{1}{C_{GC}} + \frac{1}{C_m} \tag{1}$$

From equation (1), we can get the value of C_m as 0.37 μF cm^{-2}. With the capacitance value of the lipid membranes and an estimate of its dielectric constant κ, $\kappa=2.05$ [27], the thickness d of the lipid membranes was estimated according to following equation (3):

$$C_m = \varepsilon_0 \kappa / d \tag{2}$$

where ε_0 is the permittivity of free space ($\varepsilon_0=8.85\times10^{-14}$ F cm^{-1}). From equation (2), we can get that the thickness of the DMPC membranes is about 5 nm. Because the thickness of a DMPC monolayer should be approximately 2.4 nm [28], it could be concluded that the DMPC membranes formed on the surface of the GC electrode were bilayer membranes. In the following content, this method was used to form BLM on a GCE.

3. THE PERMEATION OF BLM REGULATED BY ION

The biomembrane is an essential element of all living organisms. All biomembranes are composed of a lipid bilayer intercalated with other constituents, such as proteins, carbohydrates, and their complexes of lipid. Simplified models of biomembranes have been the subject of intense study. Many kinds of model membranes have been used in research, including lipid vesicles or liposomes, BLMs, Langmuir–Blodgett (LB) layers, and so on.

Because the s-BLM has been proved to be very useful and easy to work in the field of membrane research and solved the shortcomings of the conventional BLM, it will be an ideal model for investigating the interaction of ion and lipid membranes.

The interaction of metals with biomembranes has received much attention in view of the effects of metals in physiologically important membrane processes such as membrane fusion and transport. Lanthanide (Ln^{3+}) can promote the growth of crops and increase their production. However, Ln^{3+} can also affect animal tissues and has some toxic functions at certain concentrations. The mechanism by which Ln^{3+} exerts biological effects remains unclear. Lanthanide ions can react with glutathione and regulate the permeability of the glutathione assembly monolayer [29]. Lanthanide ions can also affect the conformation of lipid [30], the phase-transition properties of phospholipid bilayers [31], and induce the formation of stripes domain structure in phospholipid LB monolayer [32].

Interaction between lanthanide and phospholipid has been studied by nuclear magnetic resonance (NMR) [30], atomic force microscopy (AFM) [32], differential scanning calorimetry (DSC) [31], laser-excited europium (III) luminescence [33], etc. Here, we use the ion-channel sensor principle proposed by Umezawa and co-workers in 1987 [34] to investigate the interaction of lanthanide ions with s-BLM [35]. The principle is to mimic biomembranes containing ion-channel proteins, which can selectively bind ligands to open their usually closed ion channels and allow a large amount of ions or molecules to pass biomembranes. Similarly, analyte (usually called stimulus) binding to receptor membranes can control the access of electroactive species (often referred to as markers). Through the electrochemical response of a marker, the interaction of lanthanide with s-BLM can be studied. The BLM can be considered as the receptor membrane, lanthanide as the stimulus and $Fe(CN)_6^{3-/4-}$ as the markers.

When the GC electrode modified with a BLM was inserted in a solution containing 1 mM $Fe(CN)_6^{3-/4-}$, the cyclic voltammogram was recorded as shown in Fig. 6(a). Compared with a bare GC electrode for the electrochemical behavior in a $Fe(CN)_6^{3-/4-}$ solution (Fig. 6(d)), almost no redox peaks appeared, which indicated the electronic communication from $Fe(CN)_6^{3-/4-}$ to the electrode was inhibited. This also demonstrated BLMs formed on the surface of a GCE. After adding 1 μM Eu^{3+}, the cyclic voltammetric response of $Fe(CN)_6^{3-/4-}$ was recorded as shown in Fig. 6(b). A pair of distinct current peaks emerged due to the marker ions reaching the surface of the GC electrode, which resulted from two possible reasons: one was that the pores of the s-BLM were formed by lipid reacting with the Eu^{3+}, the other was desorption of lipid from the surface of the GC electrode. In order to clarify the actual reason, the electrode was transferred from a solution with 1 μM Eu^{3+} into a solution without Eu^{3+}. The cyclic voltammetric response of $Fe(CN)_6^{3-/4-}$ was recorded as shown in Fig. 6(c). It was almost the same as in Fig. 6(a), which implied that the s-BLM was still on the surface of the GC electrode. Thus, undoubtedly Eu^{3+} could react with DMPC bilayer membrane as demonstrated by the electrochemistry method. Lanthanide ions bound to the polar head group of lipid [30,33,36], resulting in the formation of more ordered (and hence more compact) surface domain structures and formation of some pores which permitted $Fe(CN)_6^{3-/4-}$ transmembrane mass transfer. Certainly, there could be other factors as well. The binding of cationic Eu^{3+} to a lipid polar head group rendered the lipid surface (partially) positively charged, which might then facilitate the transport of anionic $Fe(CN)_6^{3-/4-}$ to the surface of the electrode. On the other

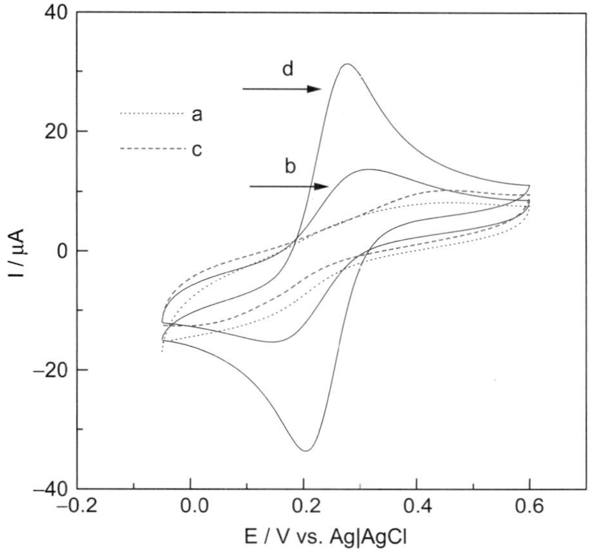

Fig. 6. Cyclic voltammograms of 1 mM $Fe(CN)_6^{3-/4-}$ at s-BLM-coated GC electrode in several solutions: (a) 0.1 M KCl; (b) 0.1 M KCl and 1 μM Eu^{3+}; (c) 0.1 M KCl. (d) represents the cyclic voltammogram of 1 mM $Fe(CN)_6^{3-/4-}$ at bare GC electrode with 0.1 M KCl as supporting electrolyte. Scan rates are all 50 mV s^{-1}.

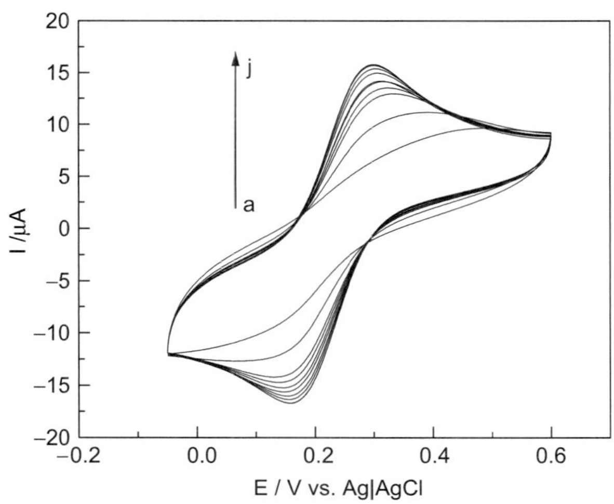

Fig. 7. Cyclic voltammograms of 1 mM $Fe(CN)_6^{3-/4-}$ at s-BLM-coated GC electrode in solutions containing (a) 0.2 μM Eu^{3+}, (b) 0.4 μM Eu^{3+}, (c) 0.6 μM Eu^{3+}, (d) 0.8 μM Eu^{3+}, (e) 1.0 μM Eu^{3+}, (f) 1.2 μM Eu^{3+}, (g) 1.4 μM Eu^{3+}, (h) 1.6 μM Eu^{3+}, (i)1.8 μM Eu^{3+}, (j) 2.0 μM Eu^{3+}. Scan rate, 50 mV s^{-1}.

hand, the lipid charge state might affect the electron-tunneling kinetics of anionic $Fe(CN)_6^{3-/4-}$ due to the electrostatic interactions.

The cyclic voltammograms of the GCE modified by BLM in $Fe(CN)_6^{3-/4-}$ solution containing different concentration of Eu^{3+} were recorded in Fig. 7. Each peak current was obtained after an interaction between Eu^{3+} and lipid reaching equilibrium. With increasing concentration of Eu^{3+}, the redox peaks were increasingly obvious. Simultaneously, peak separation (Fig. 8(a)) decreased

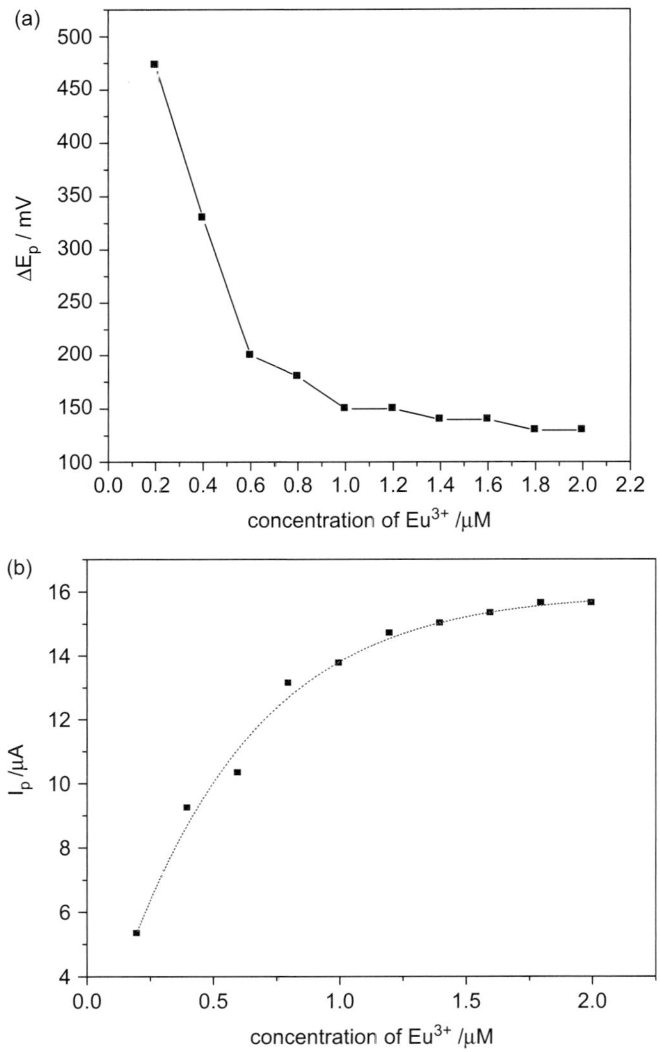

Fig. 8. (a) The peak separations of redox, (b) cathodic peak currents obtained in Fig. 7 as a function of concentration of Eu^{3+}.

and peak currents (Fig. 8(b)) of the redox increased. From Fig. 8(a), we can see that, from 0.2 μM (473 mV) to 1.0 μM (150 mV), the peak separation decreased rapidly, and from 1.0 μM (150 mV) to 2.0 μM (130 mV) it decreased slowly. The decrease of peak separation meant the electro-transfer velocity increased. The peak currents of $Fe(CN)_6^{3-/4-}$ increased from 0.2 to 1.2 μM and reached a plateau above 1.2 μM as seen from Fig. 8(b). Clearly, with increasing the concentration of Eu^{3+}, the number of pores of s-BLM increased, so the amount of $Fe(CN)_6^{3-/4-}$ which reached the surface of the electrode was increasing, resulting in the increase in the peak current and the decrease in the peak separation. However, because there were limited lipid molecules on the surface of the GC electrode, the interaction site between the Eu^{3+} and lipid was limited. When the interaction sites were all occupied, the peak current reached a plateau.

From the experimental results, the other lanthanide ions such as Tb^{3+} and La^{3+} also had the ability to react with s-BLM. Figure 9 showed cyclic voltammograms of La^{3+}, Tb^{3+}, and Eu^{3+} at the same concentration, 0.8 μM. In the same condition, the lanthanide possessing a higher peak current has a stronger interaction ability with s-BLM. From their peak currents, we can deduce that the interaction of Eu^{3+} with s-BLM was stronger than that of the other two lanthanides (Tb^{3+} and La^{3+}), and that Tb^{3+} was stronger than La^{3+} in the interaction with s-BLM. This might result from the binding constants of lanthanide ions with phosphatidylcholine. Lanthanide ions react with phosphatidylcholine with a 1:2 stoichiometry [37].

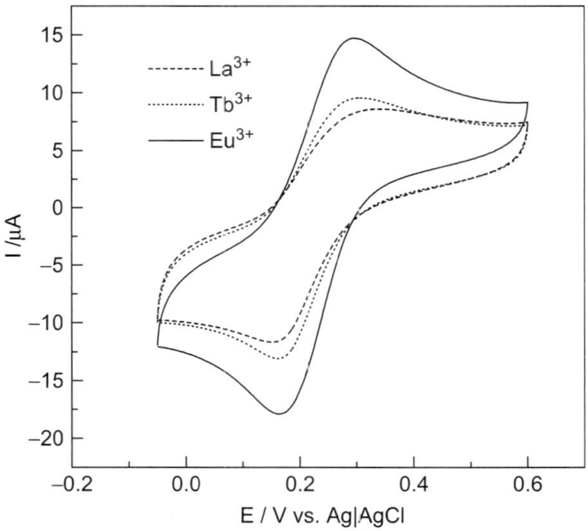

Fig. 9. Cyclic voltammograms of 1 mM $Fe(CN)_6^{3-/4-}$ at s-BLM-coated electrode in solutions containing 0.8 μM different lanthanide ions: dash, La^{3+}; dot, Tb^{3+}; solid, Eu^{3+}.

Table 1. The binding constants of lanthanide ions with lipid membrane

Lanthanide ions	Binding constants, Kα, (l/mol)	References
La^{3+}	120	
Tb^{3+}	565	[39]
Eu^{3+}	284–9400	[40]

Table 1 shows the binding constants of lanthanide with a lipid membrane [38]. From it, we could see that the binding constants of Eu^{3+}, Tb^{3+}, and La^{3+} were 284–9400, 565, 120 l/mol, respectively. Our results are basically consistent with the sequence of the binding constants' sequence.

The other ions, such as perchlorate ion [41,42], ferricyanide anion [43], Ca^{2+} [44,45], PF_6^- [46], heteropolyanion [47], etc., have a similar interaction with BLM. Polyamidoamine (PAMAM) dendrimers also can react with BLM [48].

4. THE INTERACTION OF PEPTIDE WITH BLM

As is known, membranes in living organisms are about half phospholipids and half protein and many enzymes function in nature bound to membranes. Membrane interfaces have a potent ability to induce secondary structure in a wide range of membrane-active peptides, such as hormones, toxins, and antimicrobial peptides. The bilayer of phospholipid is an integral structural unit of biomembranes. A variety of bilayer structure formed by synthetic lipids have been used to mimic membrane properties, and they furnish a unique opportunity to investigate the relationship between structure and function. The vesicle system is a self-closed bilayer aggregate and constitutes one of the most explored models of membranes.

The peroxidases are ferric heme enzymes, which catalyze the oxidation of a variety of substrates by hydrogen peroxide. One of the heme model species for the peroxidases can be obtained from the enzymatic cleavage of cytochrome C, microperoxidase-11, which still exhibits peroxidase activity [49]. The heme undecapeptide is a good model to study the electrochemical behavior of heme iron in hemoproteins [50]. Its primary structure where numbers refer to positions in the native cytochrome C (cyt. *c*) is shown in Fig. 10. The heme group is covalently bound to the undecapeptide *via* two thioether bonds between two cysteines and the vinyl groups of the heme; His-18 is coordinated to the heme iron on the 'proximal' site, as in native cytochrome.

MP11 affords the opportunity to mimic the behavior of the parent macromolecule 'bottom–up', and to study the metalloporphyrin when relatively unshielded but still water-soluble over an appreciable pH and concentration range.

The redox potentials of cytochromes are determined by several factors, such as the type of axial ligands, hydrogen bonding by axial histidines [51], and the effect

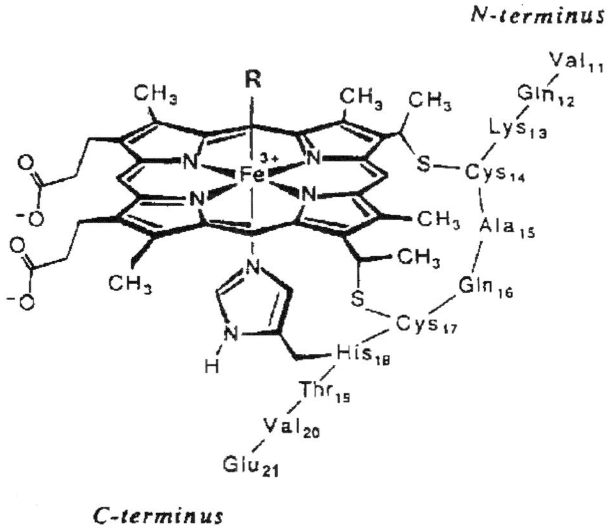

Fig. 10. The structure of microperoxidase-11. The numbering of the amino acid residues refers to the original sequence of cytochrome *c*.

of the pH of the medium. Especially, the non-planar structure of the heme influences relevant chemical and photophysical properties (e.g., axial ligand affinity, redox potentials, etc.) [52,53].

In the following text [54], DDAB-induced MP11 conformations were investigated by CV, UV/vis spectrometry and circular dichroism (CD).

Cyclic voltammograms, as a function of scan rate for MP11–DDAB film on a GC electrode in PBS, are shown in Fig. 11(a). It was also found that neither the bare GC electrode nor the DDAB-modified GC electrode gave any peaks. The electrochemical response for MP11–DDAB film on a GC electrode shows pairs of quasi-reversible ($\Delta E\text{p} = 100$ mV) oxidation/reduction wave. The average half-peak potential for MP11–DDAB on GC electrode is -143 ± 5 mV (*versus* Ag/AgCl/Sat. KCl). The half-peak potential is more positive than the value when MP11 was immobilized on the electrode or free in solution [50,55,56]. A positive shift of half-peak potential for MP11 in the presence of surfactant was also reported in the literature [57]. CV results of a positive shift of half-peak potential and quasi-reversible characteristic of electrochemistry for MP11 on GC electrodes indicate that conformation change from random coils to α-helix for MP11 hinders the electron transfer between MP11 and the electrode. It may prove that the exposure of the heme of MP11 diminishes and the distance between the electrode and the active site increases. Further more, peak current i_p is proportional to the scan rate ν (shown in Fig. 11(b)), as predicted by thin-layer electrochemistry theory [58].

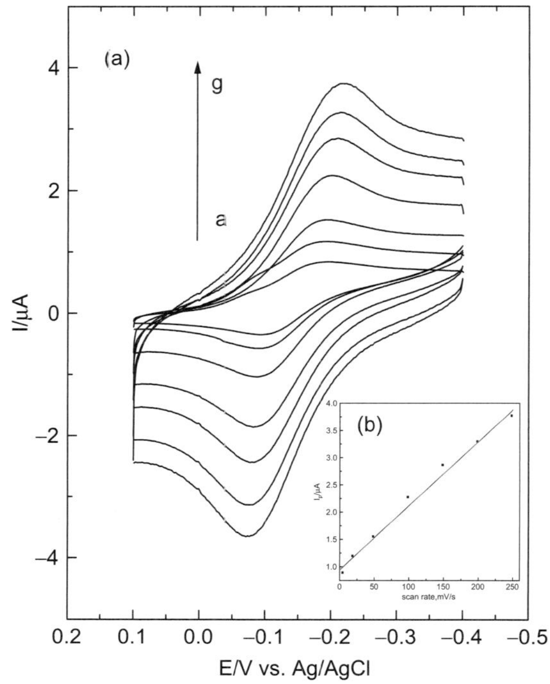

Fig. 11. (a) Cyclic voltammograms of MP11–DDAB film modified GC electrode in pH 6.9 phosphate buffer 10 mmol l^{-1}, with 0.1 mol l^{-1}KCl at 25 °C. The scan rates are 10, 20, 50, 100, 150, 200, 250 mV s^{-1} from a to g. (b) Influence of scan rate on cathodic peak current for MP11–DDAB film on GC electrode, scan rates: 10, 20, 50, 100, 150, 200, 250 mV s^{-1}.

The catalytic reduction of hydrogen peroxide was performed on the GC electrode with MP11–DDAB film (shown in Fig. 12) as soon as hydrogen peroxide was added to the buffered solution. Cyclic voltammograms for the electrode with and without H_2O_2 show that MP11 in DDAB film keeps its catalytic activity.

Heme absorption is a very useful conformational probe for the study of heme proteins as well as positions of the Soret absorption band, providing information about the environment of heme [59] on the binding of MP11 to DDAB vesicles. The 400 nm band (shown in Fig. 13, solid line), characteristic of the native random coils MP11 in solution [60], is shifted to 410 nm (shown in Fig. 13, dotted line). This red shift is consistent with a more non-planar porphyrin macrocycle [61–63]. This change indicates that a heme status for the binding DDAB vesicles is distinctly different from that in the native protein. The same experiment was done with DMPC vesicles, a kind of neutral charge lipid, but no shift was found at 400 nm, only a slight change in peak height (data not shown). This indicates a heme state for MP11 in a DMPC vesicle is similar to that in the native one. Structural and

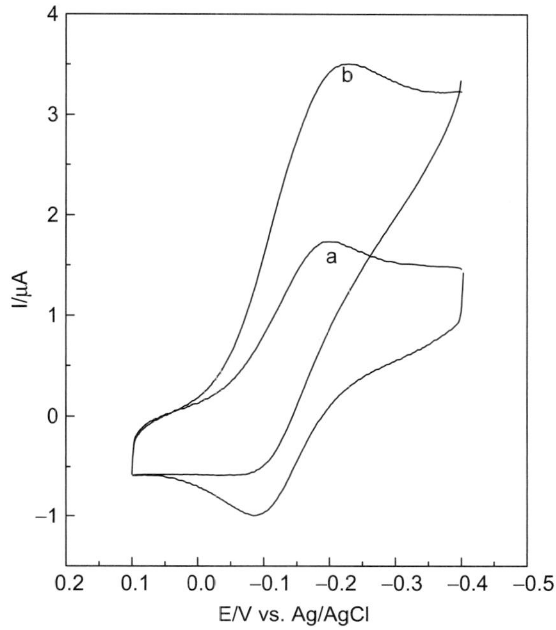

Fig. 12. Cyclic voltammograms of H_2O_2 in pH 6.9 phosphate buffer 10 mmol l^{-1}, with 0.1 mol l^{-1} KCl 25 °C, at a GC electrode modified with MP11–DDAB film: (a) no H_2O_2 ; (b) 1 mmol l^{-1} H_2O_2; scan rate: 50 mV s^{-1}.

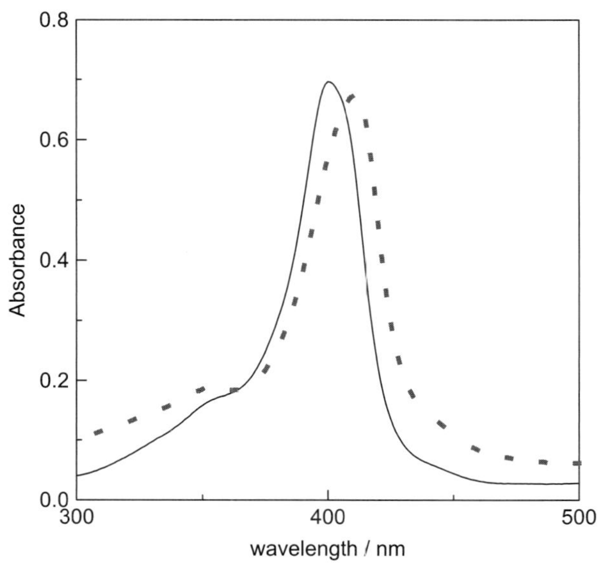

Fig. 13. UV–vis absorption spectra of MP11 in pH 6.9 phosphate buffer (solid line) and when bound to DDAB vesicles (dotted line), MP11 4.3 μM, DDAB 44 μM. Spectra were obtained at room temperature (25 °C), cell path lengths 1 cm.

conformational change may happen when cyt. *c* is binding to opposite-charge lipid vesicles; meanwhile, the Soret absorbance in UV/vis of cyt. *c* was also changed. This change was driven by the electrostatic interaction between positively charged residues in the protein and the negatively charged lipid headgroups on the membrane surface [64].

The pI of MP11 is 4.7 [65]. When it is in PBS at pH 6.8, MP11 is negatively charged, while DDAB is positively charged. The electrostatic interaction between MP11 and DDAB vesicles may alter the conformation of MP11. At the same time, the heme status is also changed by the interaction. The Soret band red shift also indicates a more stable complex between MP11 and DDAB vesicles than random coils MP11 in solution buffer and it may be consistent with a more non-planar porphyrin macrocycle [61].

CD measurements provided an excellent means of monitoring the interaction between protein and other molecules. CD was used to monitor the effects of the interaction with lipid vesicles on the structural and conformational properties of MP11. The spectra in the far ultraviolet band provide information about the conformation of the peptide backbone [66]. The far UV/CD spectrum of MP11 in solution shows the presence of only random coils conformations (shown in Fig. 14, solid line) [60,65,67]. Upon binding to DDAB vesicles, a dramatic change

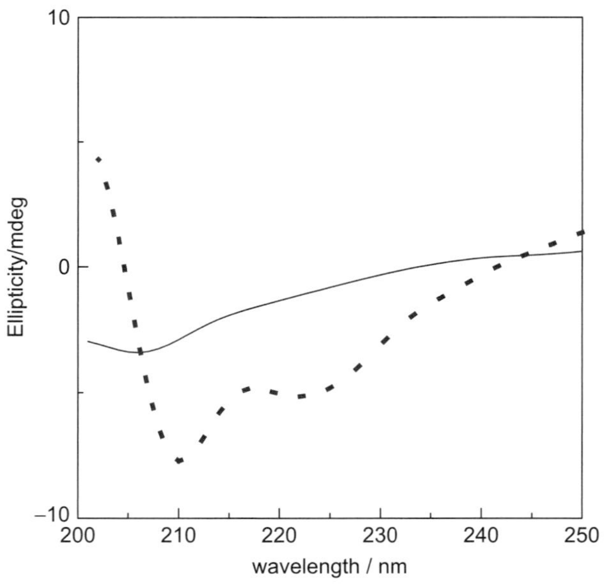

Fig. 14. Far–UV/CD spectra of MP11 in pH 6.9 phosphate buffer (solid line) and when bound to DDAB vesicles (dotted line), MP11 14 μM, DDAB 100 μM. Spectra were obtained at room temperature (25 °C), cell path lengths 1 cm; four scans were averaged per spectrum.

was observed in the far UV/CD band at 209 and 222 nm (shown in Fig. 14, dotted line). It shows the typical features characteristic of proteins containing mainly α-helical structure [68,69], in which the 222 nm dichroic band is predominantly associated with α-helical n–π^* amide transitions, and the negative minimum at around 209 nm are the dichroic bands corresponding to the π–π^* amide transitions [68,69]. The spectral change around the minimum at 209 nm may be due to the presence of optically active heme transition other than that associated with the amide transition of the polypeptide chain [64]. The fraction of α-helix in MP11 binding to DDAB was 62.5%. The data were calculated from the formula: $\%\alpha = 100\ (\Delta\varepsilon_{222} - \Delta\varepsilon_{R})/(\Delta\varepsilon_{H} - \Delta\varepsilon_{R})$ [66,70].

The CD spectra in the Soret region can provide further insight into the integrity of the heme. From the CD spectra of MP11 in PBS (shown in Fig. 15, solid line), two proximate bands of opposite sign appear in the Soret region. When bound upon DDAB vesicles, the two symmetric bands increase dramatically on the CD spectrum (shown in Fig. 15, dotted line). MP11 does not contain aromatic amino acid residues, so its considerable optical activity cannot be due to coupling of heme π–π^* transitions with its aromatic residues [71]. Free unbound heme is not optically active because of the plane of symmetry in its ensemble-average structure. However, when bound to a chiral macromolecule, small but systematic

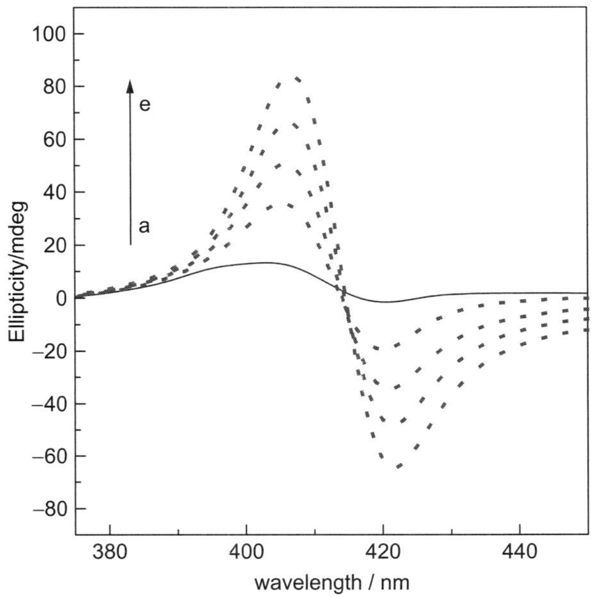

Fig. 15. CD spectra of the Soret region of MP11 in pH 6.9 phosphate buffer (solid line) and when bound to DDAB vesicles (dotted line), MP11 14 μM; DDAB: 50, 100, 200, 400 μM from b to e.

distortions from planarity could generate significant rotational strength in the heme transition [71]. The MP11 retains the thioether linkages and the histidy coordinated to the heme iron. These three points of attachment may suffice to induce a systematic deviation from planarity in the heme in the solution of MP11 [71]. When binding to DDAB vesicles, the extent of non-planar distortion for heme in MP11 increases. This change leads to an enhanced CD signal in the Soret region for MP11. The electrostatic interaction and hydrophobic environment provided by DDAB lipid vesicles may induce the native folding of the MP11, because of folding of MP11 itself, the peptide backbone exerts the force on the heme macrocycle ruffling and distortion inside MP11.

MP11 becomes an ordered and compact structure when binding to DDAB by electrostatic interaction. α-Helix formation in the peptide backbone distorted and ruffled the porphyrin macrocycle. When DMPC was introduced to MP11 solution, no obvious change in UV/vis and CD (data not shown). That indicates that only hydrophobic interaction between DMPC and MP11 cannot influence the MP11 conformation. The interaction between DDAB and MP11 may consist of three main steps (shown in Fig. 16).

1. *Unfolded partitioning*. The primary cause of favorable partitioning of MP11 from water into a membrane is non-polar interactions due to the expulsion of non-polar compounds (Val11, Ala15, and Val20 of non-polar amino acid) and electrostatic attraction between basic negatively charged amino acid residues and positively charged DDAB headgroups. From the above UV/vis experimental result, it can be concluded that electrostatic interaction plays an important role in the partitioning.

2. *Partly folded to α-helix*. Upon binding to the membrane, the MP11 can change its conformation to α-helix. According to the data obtained by the experiment, $\%\alpha = 62.5$, it can be seen that MP11 is partly folded when binding to the DDAB vesicles. In addition, there can be an electrostatic effect arising from the differences of the dielectric constants between the water and membrane related to the cost of partitioning H-bonded peptide bonds [72].

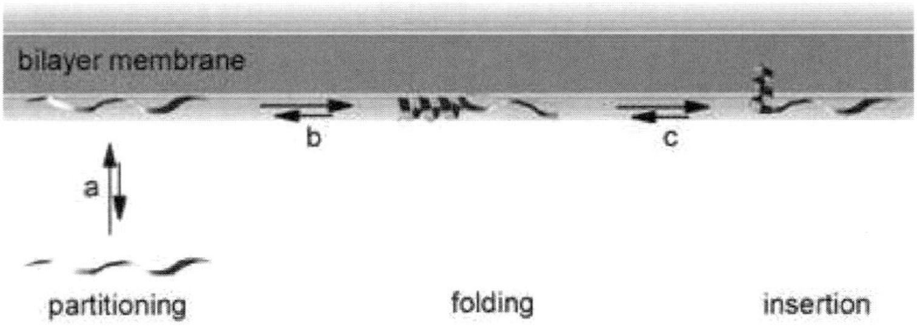

Fig. 16. Scheme of steps about the interaction between DDAB lipid vesicles and MP11: (a) unfolded partitioning; (b) partly folded to α-helix; (c) partial insertion.

3. *Partial insertion*. Actually, for membrane-active peptides, insertion is the last step in folding when binding to a membrane [73]. Furthermore, the thermodynamic cost of transferring charged or highly polar uncharged compounds into the oil-like hydrocarbon interior of a bilayer membrane is very high. Increasing ionic strength caused by dissociation, the same result is obtained for cyt. *c* [74]. However, absorbance spectrum in the Soret band of MP11 is not fully recovered even with a very high ionic strength. Moreover, N-terminus amino acid of the peptide backbone of MP11 is non-polar (Val11), which gives a possibility for MP11 insertion into the lipid membrane. However, the extent of insertion into lipid vesicles for MP11 is unknown and needs to be further investigated.

The lipid membranes of cells are the sites of biological activity for many natural and synthetic cytolytic peptides. One particular class of peptides includes positively charged amphipathic molecules, which bind to negatively charged lipid membranes in a non-specific way. This class includes a number of toxins and antibacterial peptides, which exercise their lytic action *via* the formation of non-selective trans-membrane pores. Melittin [75–77] and magainin [78], for example, have been studied extensively and have become models for the design of a number of synthetic analogs, amphipathic in nature and positively charged, with enhanced cytolytic or bactericidal activity [79].

Of all the antimicrobial peptides known, only a very few of them are actually allowed to be used either as a preservative in the food industry or as an antibiotic in health care. The 34-residue-long peptide nisin is one of these few, and has been used as a food preservative for a long time. There are two naturally occurring forms of nisin, nisin A and nisin Z, differing in only one residue (His or Asn at position 27) and with very similar bactericidal activities. The peptide shares similar characteristics with other antimicrobial peptides. It is overall positively charged (+4) and its structure possesses amphipathic properties. However, some structural properties make nisin rather special. Nisin is post-translationally modified such that serine and threonine residues are dehydrated to become dehydroalanine and dehydrobutyrine. Subsequently, five of the dehydrated residues are coupled to upstream cysteines, thus forming the thioether bonds that produce the characteristic lanthionine rings (Fig. 17). The thioether bonds give nisin two rigid ring systems, one N-terminally and one C-terminally located. A hinge region (residues 20–22) that is often found in antimicrobial peptides separates the ring systems. Nisin kills its target by pore formation in the target membrane. It inhibits the growth of a wide range of Gram-positive microorganisms and also inhibits the germination and/or outgrowth of spores of *Bacillus* and *Clostridium* species. It has been widely used as a preservative in the food industry for a number of years [80,81]. Nisin activity is not well understood. Its microbial action is due to an interaction with the phospholipid membrane [82–84], but it does not seem to require a specific membrane receptor. The phospholipid composition

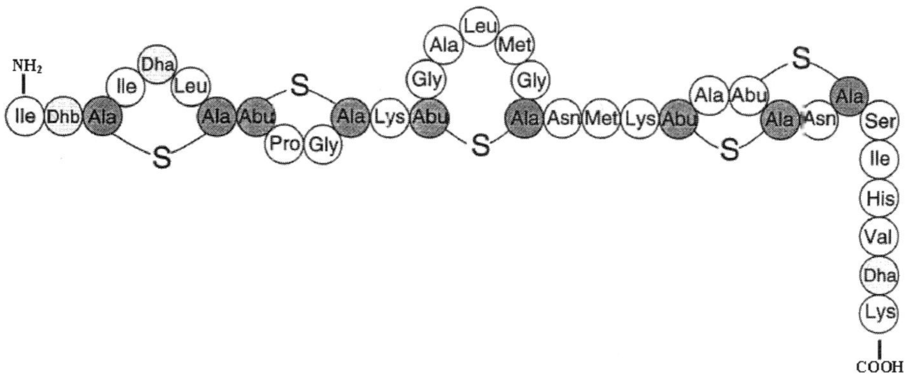

Fig. 17. The primary structure of nisin Z. Dark gray: lanthionine residues; light gray: dehydrated residues in lanthionins. Dha, dehydroalanine; Dhb, dehydrobutyrine, Ala-*S*-Ala, lanthionine; Abu-*S*-Ala, -methyllanthionine.

of the membrane influences nisin activity [85–87]. The molecular details of the pore-forming mechanism of nisin remain unclear.

In the following work [88], a GCE supporting BLM as a biomembrane model was used to study the interaction between nisin and the model membrane. The current study was undertaken in an attempt to better understand the molecular mechanism of the interaction of nisin with cell membranes. Nisin interacted strongly selectively with negatively charged lipid membrane and it made pores in the BLM. Pore formation made by nisin in the BLM let $Fe(CN)_6^{3-/4-}$ cross the biomembrane easily.

In order to investigate the interaction between nisin and s-BLM, $Fe(CN)_6^{3-/4-}$, was chosen as the marker ions. The amperometric response of the marker ions on s-BLM at different concentrations of nisin is shown in Fig. 18. At first, there was no obvious amperometric response of the marker ions on s-BLM when the concentration of nisin was below 150 μM (Fig. 18, curve a), except for the characteristic of a cyclic voltammogram of only a capacitor's charging and discharging. However, when the concentration of nisin was above 150 μM, an amperometric response of $Fe(CN)_6^{3-/4-}$ complex was found. Increasing the concentration of nisin, the enhanced amperometric response of $Fe(CN)_6^{3-/4-}$ was found gradually (Fig. 18, curves b–f). With the concentration of nisin reaching 0.75 mM, a dramatic enhanced amperometric response of $Fe(CN)_6^{3-/4-}$ was found. Above this value, no obvious further enhanced amperometric response of the marker ions was found. Increasing the concentration of nisin continuously to 1 mM, the characteristic of amperometric response of $Fe(CN)_6^{3-/4-}$ on bare GCE appeared (Fig. 18, curve g) and a pair of well-defined reversible waves of $Fe(CN)_6^{3-/4-}$ couple was obtained again. It indicated that nisin corrupted totally the lipid membrane that was built on the GCE surface previously. The behavior of nisin acted as some kind

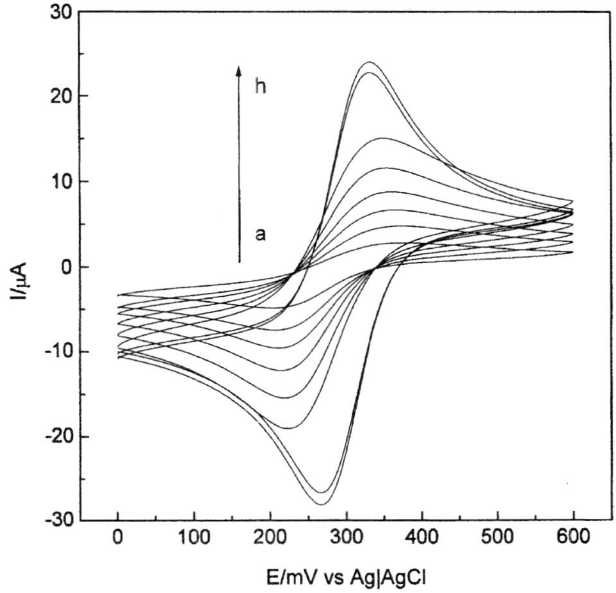

Fig. 18. Cyclic voltammeric responses of 1 mM $Fe(CN)_6^{3-/4-}$ at the GCE coated with BLM with 0.1 M KCl, 10 mM phosphate buffer at pH 6.0 in the presence of different concentration of nisin (a) 0, (b) 150, (c) 250, (d) 300, (e) 500, (f) 700, (g) 750, (h) 1000 μM. Scan rate, 100 mV s^{-1}.

of surfactant that removed the lipid molecules from the GCE that resulted in the marker ions flushing to the GCE without any hindrance. Diao [89] analyzed the pore formation induced by Ca^{2+} on the s-BLM. Compared to Ca^{2+}, nisin has a larger molecule volume and more charge. Nisin could contact with much more lipid molecules and give the lipid membrane stronger disturbance to make the surface of the lipid membrane more inhomogeneous and remove finally the lipid molecule from the GCE.

Figure 19 shows nisin concentration-dependent changes against the reduction peak current of the marker ions. It contains three different domains: low, middling, and high concentration domains (stages A, B, and C). At low concentration domains, the current value for the reduction peaks shows almost a small flat roof with only a capacitor's charging and discharging of s-BLM. This was because that small amount of nisin molecules were immersed in the 'sea' of a large amount of lipid molecules and they could not affect the lipid membrane greatly. That is to say, nisin could not induce pores in the lipid membrane so the marker ions had no chance to cross the lipid membrane. At middling concentration domains, the current value for the reduction peaks showed almost a linear relationship with the concentration of nisin in the range 150–600 μM. The peak current initially increases with

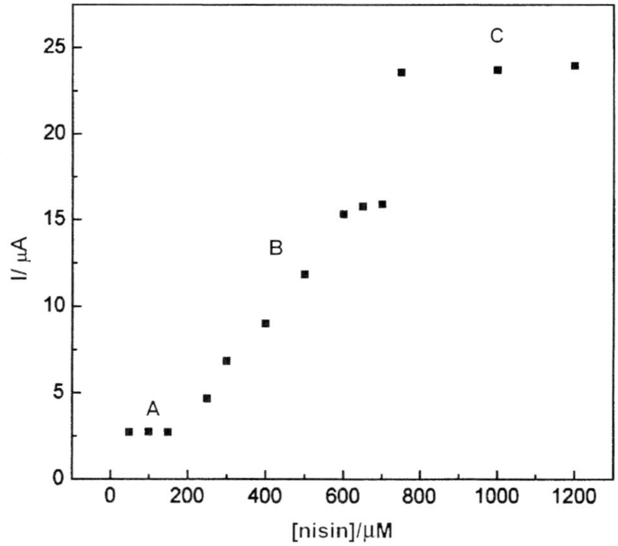

Fig. 19. The plot of the peak current *I* *versus* [nisin]/μM in the presence of 1 mM $Fe(CN)_6^{3-/4-}$ in 0.1 M KCl, 10 mM phosphate buffer at pH 6.0, Scan rate, 100 mV s^{-1}.

the concentration of nisin, then increases linearly and finally reaches a plateau. These results may be explained as follows: nisin molecules accumulated and aggregated to form hydrophilic pores of molecular dimensions for marker ions in the hydrophobic membrane surrounding and it led to formation of the pores in the s-BLM. At the same time, the extent of mismatched bilayer regions between nisin-associated and nisin-free lipids was intensified by the electrostatic attraction between the nisin and acidic headgroups of lipids. However, at high concentration domains, the current value for the reduction peaks increased sharply to a plateau whose shape was almost the same as that obtained on the bare GCE. This indicated that the lipid membrane on GCE collapsed and was removed from the substrate by a relatively high concentration of nisin except for the remaining small amount of separated lipid molecules on the surface of GCE.

5. THE INTERACTION OF DRUG WITH s-BLM

Amphotericin B (AmB) is one of the main polyene antibiotics widely used to treat deep-seated fungal infections [90]. Unfortunately, the classical formulation of AmB, Fungizone, has negative side effects (e.g., nephrotoxicity) that seriously impair its efficacy. The mechanism of the biological action of AmB is most probably directly related to the ability of the drug to form hydrophilic pores in the hydrophobic membrane core, where it increases the permeability of the cells to ions and small molecules [91–93]. It was proposed in the 1970s that the interaction between membrane sterols and AmB is responsible for the selectivity of the drug. It has therefore been assumed that the selective toxicity of AmB for fungi results from its capacity to bind more strongly to ergosterol, the principal fungal sterol, than to cholesterol, the principal sterol of mammalian cells. However, the detailed molecular mechanisms of the interaction of AmB with the membrane, as well as the formation of a transmembrane pore structure, are still imperfectly understood. The evidence for channel formation without sterols challenges the idea that the most widely accepted and oldest AmB channel models have included sterols as 'staves' in a barrel-type structure [94–96].

In the following work of our group [97], synthetic chromatographically pure lipids were used to show that the channels formed are actually due only to AmB and not to sterol or other lipid contaminants. A s-BLM on a GCE and SUVs (small unilamellar vesicles) as biomembrane models were used to study the interaction between AmB and the model membrane. AmB could form an ion channel in s-BLM and the states of AmB controlled the ion channel activities. The following experiment was undertaken in an attempt to understand better the molecular mechanism of the formation of AmB ion channels in lipid bilayers in the absence of sterol.

The formation of ion-permeable channels by AmB was monitored by measuring the responses of ruthenium (II) complex cations as the marker ions on the GCE. Six typical concentrations of AmB were used: 10^{-7}, 2×10^{-7}, 1×10^{-6}, 5×10^{-6}, 2×10^{-5}, and 5×10^{-5} M. From the absorbance spectroscopy of AmB, it can be concluded that they are in the form of monomers at the first two concentrations and in the form of aggregates at the last four concentrations. In each case, the supported bilayer model membrane systems are as follows: GCE-supported pure DMPC (S-DMPC), GCE-supported DMPC plus cholesterol 5% (S-CHOL), and GCE-supported DMPC plus ergosterol 5% (S-ERGO). After the BLM was formed on the surface of GCE, it was immersed in buffer solution containing 0.5 mM *tris*(2,2-bipyridine)ruthenium (II). Figure 20 shows the cyclic voltammograms in solutions with different concentrations of AmB upon S-DMPC, S-CHOL, and S-ERGO. At concentrations of 10^{-7} and 2×10^{-7} M for S-DMPC and S-CHOL, no obvious characteristic of redox peaks of ruthenium (II) complex appeared on the cyclic voltammograms (Fig. 20, A and B, curves a and b). The suppression of CV peaks of the ruthenium (II) complex appeared to be due to the closed channels of the lipid membrane. However, for S-ERGO, the enhanced amperometric response of ruthenium (II) complex was found gradually (Fig. 20, A and B, curve c). AmB molecules can open some kind of channel for the ruthenium (II) complex to cross the lipid membrane to reach the surface of the electrode. When the concentration of AmB was increased to 1×10^{-6} M, the amperometric response of ruthenium (II) complex was also found in S-CHOL (Fig. 20 C, curve b). It meant that AmB

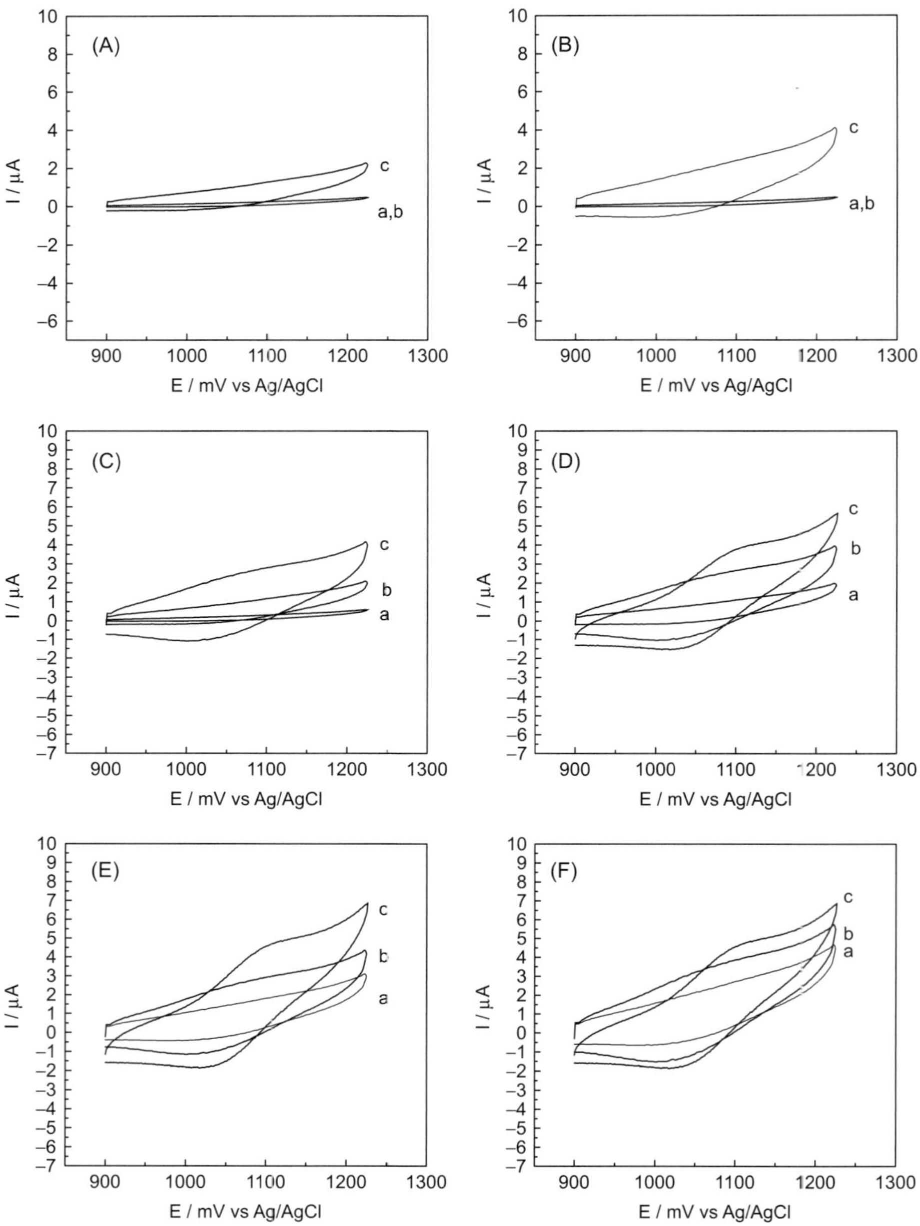

Fig. 20. Cyclic voltammetric responses of 0.5 mM ruthenium (II) complex cation containing 0.1 M KCl, 10 mM phosphate buffer, pH 7.18 at S-DMPC (curve a), S-CHOL (curve b), and S-ERGO (curve c) in different concentrations of AmB: (A) 10^{-7} M; (B) 2×10^{-7} M; (C) 1×10^{-6} M; (D) 5×10^{-6} M; (E) 2×10^{-5} M; (F) 5×10^{-5} M. The scan rate was 50 mV s^{-1}.

could form ion channels in S-CHOL at this concentration. As for S-DMPC, no cyclic voltammetric response of the ruthenium (II) complex yet appeared (Fig. 20 C, curve a), while for S-ERGO, the enhanced amperometric response of the ruthenium (II) complex was again obtained (Fig. 20 C, curve c). At a concentration of 5×10^{-6} M, the suppression of CV peaks of the ruthenium (II) complex became dramatically higher (Fig. 20 D, curve a *versus* Fig. 20 A–C, curve a), indicating that AmB began to form ion channels in S-CHOL. At the same time, the enhanced amperometric response of the ruthenium (II) complex was again obtained in S-CHOL and S-ERGO (Fig. 20 D, curves b and c). However, when the concentration of AmB reached 2×10^{-5} M, a maximal amperometric response of the ruthenium (II) complex was found in S-ERGO (Fig. 20 E, curve c). When the concentration of AmB was increased to 5×10^{-5} M, increased current responses of ruthenium (II) complex were also found in S-DMPC and S-CHOL From the result, we can see that AmB can form ion channels in S-ERGO at the lowest concentration among the three different supported model membranes. However, AmB can form ion channels in S-ERGO at the highest concentration. In other words, AmB can form ion channels most easily in S-ERGO and with the most difficulty in S-DMPC among the three membrane systems investigated. However, the potential for AmB to form ion channels depends on the different constituents of the supported model membranes and the nature of the component (cholesterol or ergosterol). The binding of AmB to biological and model membranes has been studied over a long time. Its mechanism of action is thought to involve interaction with sterols, leading to pore formation and increased permeability and, ultimately, to membrane disruption and cell lysis. The AmB pore was proposed to consist of antibiotic molecules intercalated with a sterol molecule [98]. The higher affinity for fungal ergosterol over mammalian cholesterol has been invoked as the explanation for the antibiotic's selectivity for the fungal membrane, because the equilibrium between monomers and aggregates appears to play a key role in drug activity. In more recent years, work has aimed at understanding the membrane effects of the antibiotic in terms of its aggregation state and of the differences in its affinities for cholesterol and ergosterol as a function of the aggregation state. Bolard *et al.* [99] found that the binding of the antibiotic to membranes in monomeric or in aggregated form depended on the lipid composition, and toxic effects were ascribed to the aggregated state. At low AmB concentration, a rigid molecule of AmB interacts with the lipid alkyl chains by van der Waals force [100] and acts like other similar structure-modifying agents, such as cholesterol or membrane-spanning polar carotenoids. Hydrophobic interactions of alkyl chains with AmB are most probably responsible for the formation of certain fractions of the lipid. Furthermore, the AmB molecules were immersed in the sea of lipid and were separated by lipid molecules. They could not contact each other while they remained in their monomeric form. Thus, AmB could act only as a helper for stabilizing the s-BLM, and the redox probe could not cross the membrane to give an electrochemistry response at the surface

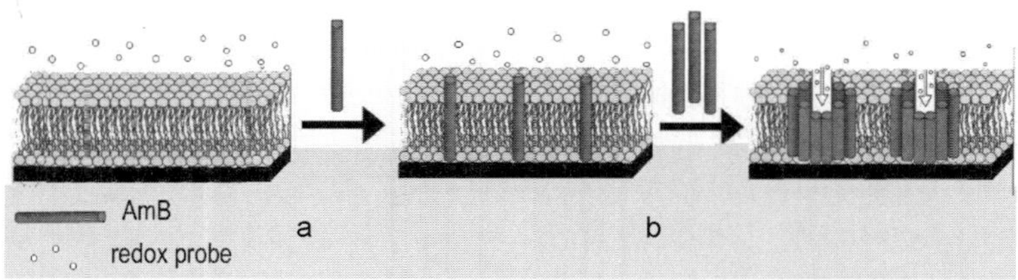

Fig. 21. The process of AmB forming ion channels in the GCE-supported sterol-free supported phosphatidylcholine bilayer model membrane. (a) At low concentration, AmB molecules were isolated by lipid molecules and could not form ion channels for the redox probe to pass. (b) When performing at a relatively highly aggregated state, AmB molecules aggregated each other to form ion channels.

of the GCE. However, this is not the case when ergosterol was present in the supported bilayer model membrane. It has been postulated that preferential destruction of fungal cells by AmB is due partly to its higher affinity to the ergosterol-rich fungal membranes than to the cholesterol-containing membranes typically found in mammalian cells [101]. In addition, AmB has a higher binding affinity constant for ergosterol than for cholesterol [102], so AmB could interact with ergosterol contained in S-ERGO and lead to formation of ion channels even at such a low concentration.

One of the processes where AmB formed ion channels in the sterol-free GCE-supported BLM (defect-free) was predicted (Fig. 21). This, involved the initial accumulation of AmB at the membrane surface or inside the membrane, and then it was assumed that several molecules associated with the membrane to form a pore. Whether this aggregation of molecules began before insertion or in the membrane after insertion is unknown. As yet, it is also unknown how many monomers were required to form a pore. The AmB molecules were thought to align around a central channel, with the hydrophobic faces toward the lipid bilayer and the hydrophilic faces toward the pore center.

6. THE BIOSENSOR BASED ON BLM

The development of the electrochemical biosensor has grown at a rapid pace since the 1980s. In the biomimetic approach, a lipid bilayer is used [103]. The functions of biomembranes are mediated by specific modifiers, which assume their active conformations only in the lipid bilayer environment. Further, the presence of the lipid bilayer greatly reduces the interference and effectively excludes hydrophilic electroactive compounds from reaching the detecting surface, which may cause undesired reactions. Thus, from the specificity, selectivity, and design points of view, BLM is a natural environment for embedding a host of materials of interest for the biosensor development. However, the fragile structure of BLM system precludes its extensive use in biosensors. As has been demonstrated by Tien *et al.* [104–108], the solid s-BLM system not only possesses the advantage of a conventional BLM structure but additionally gains new important properties such as long-time stability and very dynamic structure. So s-BLM system is a better choice to develop a new class of electrochemical biosensors. Providing a natural environment with s-BLM for protein immobilization and the maintenance of protein functions is a suitable option for the design of biosensors.

From the work of Siontorou *et al.* [109], it is known that DNA could be incorporated in s-BLM based on GC electrode. However, little investigation of protein in s-BLM based on a GC electrode has occurred. In the following text, a facile method to immobilize protein in s-BLM was developed in order to investigate the behavior of the protein in s-BLM based on a GC electrode with electrochemical methods. This facile method to immobilize protein in s-BLM is especially suitable to develop the electrochemical biosensors based on a GC electrode.

The electrode modified with BLM was immersed in 2 mg ml^{-1} HRP solution for 10 h to make HRP incorporated into s-BLM, and the current response of the enzyme electrode to H_2O_2 was tested. A current–time plot of the enzyme electrode

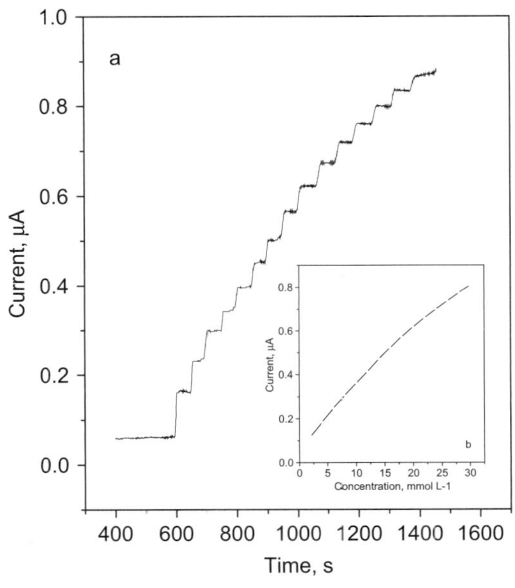

Fig. 22. (a) Typical current–time response of the sensor with increasing concentration of H_2O_2 in 2×10^{-3} mol l^{-1} steps in 5 mmol l^{-1} phosphate buffer (pH 7.0). Potential -100 mV. (b) Calibration curve of the biosensor.

on successive step changes of H_2O_2 concentration is demonstrated in Fig. 22(a). According to previous work [110], −100 mV was selected as the working potential. When an aliquot of H_2O_2 was added to the buffer solution, the reductive current rise steeply to reach a stable value. The response time (reaching 90% of the maximum response) was 10 s, which indicated a fast diffusional process and high activity of the HRP in this enzyme-membrane system.

The calibration curve of the biosensor is shown in Fig. 22(b). The linear range of H_2O_2 concentration spans between 0.5 and 18 mmol l^{-1} with a correlation coefficient of 0.998, and the sensor has a detection limit of 1.7×10^{-4} mol l^{-1} at a signal-to-noise ratio of 3. The detection limit of the biosensor is a little higher compared with other HRP-H_2O_2 biosensors [111,112].

The effect of a substance that might interfere with the response of the enzyme was studied at −100 mV. The interference was measured by a current obtained for each interfering substance at a concentration of 2 mmol l^{-1} compared with the current obtained for equimolar H_2O_2, and this ratio was used as a criterion for the selectivity of the sensor. In the experiment, glucose, sucrose, lactic acid, and ethanol did not cause any observable interference; only ascorbic acid interfered slightly (ratio 0.08).

The reproducibility of the sensor was examined at a H_2O_2 concentration of 2×10^{-3} mol l^{-1} with the same sensor, and the relative standard deviation is 6.3% ($n = 9$). The storage stability of the enzyme electrode was tested over a 20-day period. When the electrode was stored in the refrigerator at 4 °C, no apparent decrease in the response to 2×10^{-3} mol l^{-1} H_2O_2 was found over this period.

The good electrochemical response, stability, and reproducibility of the enzyme electrode demonstrate the high activity of the HRP, which can be attributed to the s-BLM system that supplies a biological environment on the surface of the GC electrode to maintain the active configuration of the enzyme. So the high activity of HRP is retained. This also demonstrates that the s-BLM is an ideal choice to immobilize the enzyme for constructing the third-generation biosensor based on a GC electrode.

Valinomycin is suitable for a potassium ionophore in a number of natural and artificial membranes including the BLM system [113–115]. In the valinomycin-incorporated BLM, permeability for K^+ was 400-fold greater than that for Na^+, which is valuable to develop electrochemical biosensors for potassium detection [116]. The concept of capacitance measurement has recently been developed [117–120]. Compared with other electrochemical methods (conductance, potential, etc.), the capacitance measurement is made for some special purposes: (i) membrane thickness measurements, (ii) investigation of membrane formation processes, (iii) suitability for studying the insulating material, such as the antibody–antigen, biotin–avidin, enzyme–substrate interaction, and (iv) suitability for detection of small molecules at very low detection limits [121]. For electrolytic capacitors, the capacitance depends on the thickness and dielectric behavior of a dielectric layer on the surface of a metal plate. This technique is especially well suited for the study of mimetic lipid membrane because the deposition of a lipid bilayer at the electrode surface introduces a large change in the double-layer structure and hence a large change in the electrode capacitance [117]. For example, the protein adsorption on an electrode may cause the change of the electrode capacitance [118]. The capacitance can be calculated from equation (3)

$$C = A\varepsilon_0\varepsilon_r/d \tag{3}$$

ε_0 represents the permittivity of free space, ε_r the relative dielectric permittivity of the material that separates the electrode from the mobile charges, *A* the surface area, and *d* the distance of the closest approach of the mobile charges to the electrode surface [119]. It is believed that a charge change occurs when K^+ binds to valinomycin. A capacitive transducer should be able to detect this change directly. The ways to determine capacitance are mostly based on impedance and potential step measurements. In this work [122], a homemade capacitance meter was developed. The K^+ sensor based on a supported phospholipid membrane on the alkanethiol SAM-modified gold electrode has been investigated using this new technique.

A supported DODAB (*N*,*N*-dimethyl-*N*,*N*-dioctadecylammonium bromide) membrane may be deposited at the surface of the electrode by the LB method [123], the painted method [124,125], or fusing lipid vesicles onto the alkylated substrate [126,127]. The painted method was chosen. The process could be monitored by capacitance change, which was widely used as an indicator of

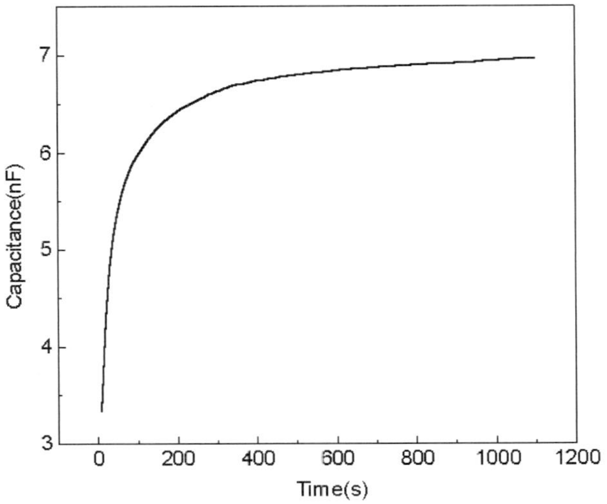

Fig. 23. The process of DODAB painted on ME (methyl sulfide) modified gold electrode in 0.1 M *Tris* solution.

membrane formation [128]. Figure 23 shows the capacitance change during the formation of the supported bilayer. From the curve, two distinct stages could be observed. A significant capacitance increase occurred at the first stage. After 10 min, the capacitance attained a constant value, indicating a suitable film has been formed. This was in agreement with the literature [129]: the membrane was initially thick when the drop of forming solution was placed on the ME (methyl sulfide)-modified gold electrode surface. For this reason, its capacitance was low. Then the solvent and the excess DODAB passed into the aqueous solution, which led to the thinning of the membrane with time. So its capacitance increased slowly until it reached a steady state.

From the above analysis, it can be seen that this kind of solid supported membrane has low capacitance and high resistance. It provides a suitable condition for capacitance measurement. There are many methods for detecting potassium ion for its leading role in membrane transport [116,130,131]. However, there is a need for fast, easy, and reliable methods for the detection of potassium ion. The following work was based on the new measurement of capacitance using a homemade capacitance meter. Upon the addition of K^+ to the stirred buffer solution, the steady-state capacitance response of a valinomycin-incorporated gold electrode was measured. Figure 24 shows a typical recording of a trace of capacitance change as a function of different concentrations of K^+ injected to a

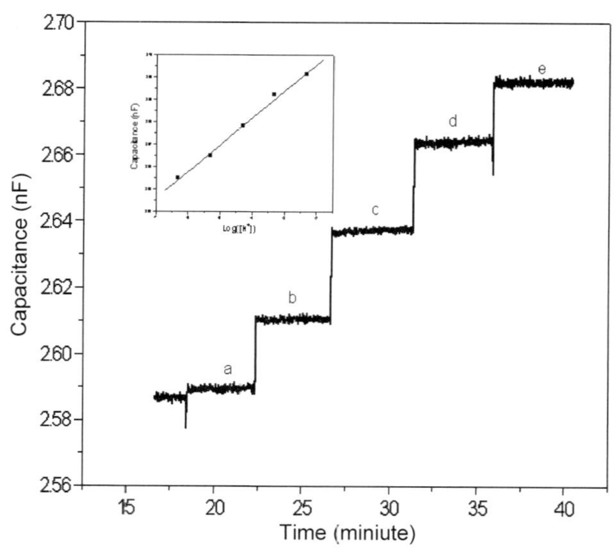

Fig. 24. Capacitance change of the Au/ME/DODA+valinomycin electrode *versus* K^+ concentration: (a) 5.0×10^{-7} M, (b) 5.0×10^{-6} M, (c) 5.0×10^{-5} M, (d) 5.0×10^{-4} M, (e) 5.0×10^{-3} M. The measurements were performed in 0.01 M *Tris*–HCl (pH 8.28) buffer solution.

stirred buffer solution of 0.01 M *Tris*–HCl. A minimum concentration of 5.0×10^{-8} M was required to induce the significant signal change compared to the background capacitance. As can be seen, the electrode modified with a valinomycin-incorporated membrane shows a very sensitive response to K^+. However, the response is very low for potassium ion in the absence of valinomycin in the membrane (data not shown here). This was consistent with the report by Galla and co-workers [131]. In addition, it is obvious that the steady-state capacitance was reached in a short time between each step. In the concentration range 5.0×10^{-7}–5.0×10^{-3} M, the steady-state capacitance increases linearly with the increase of the logarithm of the potassium ion concentration (the inset of Fig. 24). A number of methods have been successfully applied to detect potassium ion. Most of them are based on the measurements of conductivity and Nernst potential [116,132]. Usually, the detection limit using these methods is 10^{-6} M. As reported previously, the detection limit using capacitance measurements could be down to femtomolar concentration for some ions, which is more sensitive than that of other methods [121]. In this system, the detection limit obtained was 5.0×10^{-8} M, which is lower than other electrochemical methods. It is known that valinomycin is a neutral carrier, which is made up by α-amino acids and α-hydroxyl acids in alternating sequences. It acts as a carrier for alkali ions. The structure of valinomycin is ring-shaped and the carrier itself is uncharged. While valinomycin interacts with alkali ions, a valinomycinalkali ion complex would be formed. It means that the carrier–ion complex is the only charged component in the membrane [131]. The binding of metal ions to the initially neutral layer increases the relative permittivity of the layer (ε_r), which leads to the increase of the capacitance according to equation (1). So, the total capacitance of the membranes should increase with the addition of potassium ion.

7. CONSTRUCTING A HYBRID BILAYER MEMBRANE BASED ON A CARBON SUBSTRATE

As is known, one kind of s-BLM is one in which both parts of the bilayer are phospholipids, and the other kind of s-BLM is a hybrid bilayer membrane (HBM). The HBM system consisting of both natural (phospholipids) and synthetic (alkanethiol) components is easily formed by self-assembly supported on a conductive metal surface, and is stable for very long periods of time.

Carbon material is inert and has a wide potential window in electrochemistry, so modification of carbon materials is of interest to material science and electrochemistry. Deinhammer [133] reported a method for the modification of GCEs with primary amines. The modifying layer is shown to be a compact packing monolayer of amino groups [133–135]. This method was used to construct a hydrophobic surface of alkane amine on carbon electrodes, then a self-assembled

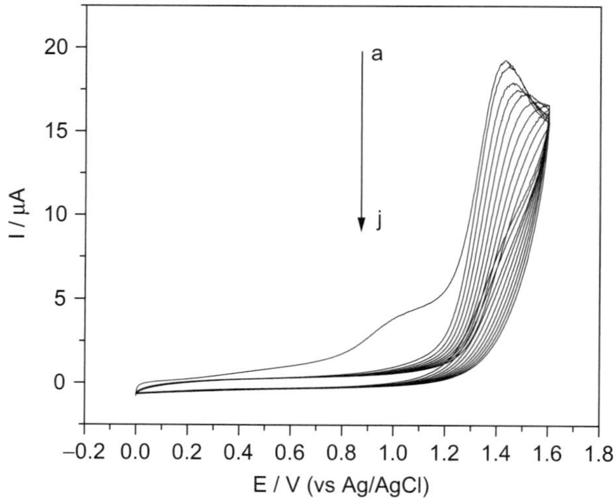

Fig. 25. Cyclic voltammograms of a freshly polished GC electrode upon scanning in 5 mM dodecylamine +0.1 M $LiClO_4$ ethanol solution. Scan rate 10 mV s^{-1}.

phospholipid monolayer onto the hydrophobic surface. Thus, a new kind of HBM system was constructed, and impedance spectroscopy were used to characterize it [136].

Figure 25 shows the typical cyclic voltammetric curves obtained on a GC electrode continuously scanning 10 circles in an ethanolic solution containing 5 mM dodecylamine and 0.1 M $LiClO_4$. A well-defined and irreversible oxidation peak occurs at 1.42 V (curve a). The peak currents decrease (b–j) when the potential is repeatedly scanned. The peak potential moves positively and finally disappears. This chemical process is attributed to the oxidation of the amino group of dodecylamine to the amino cation radical, subsequently forming carbon–nitrogen bonds on the GC electrode surface [133,134]. Further evidence for the explanation mentioned above derives from the XPS data and from the stability test of the dodecylamine monolayer.

From the XPS data (Fig. 26), the position of the peak maximum (399.8 eV) is consistent with the formation of a carbon–nitrogen bond between the amine cation radical and an aromatic moiety of the GC electrode surface. As a comparison, the N (1s) binding energies for aliphatic primary amines are ~398.0 eV [137]. Furthermore, the XPS experiment is done after sonicating the modified electrode in phosphate buffer for 15 min, so the carbon–nitrogen bond between the amine cation radical and GC surface indeed existed and was very stable.

The electrochemistry experiments also demonstrated that the alkylamine monolayer was formed on the surface of a GC electrode. Figure 27(a) demonstrated the electrochemistry behavior of the fresh GC electrode in 0.1 mM

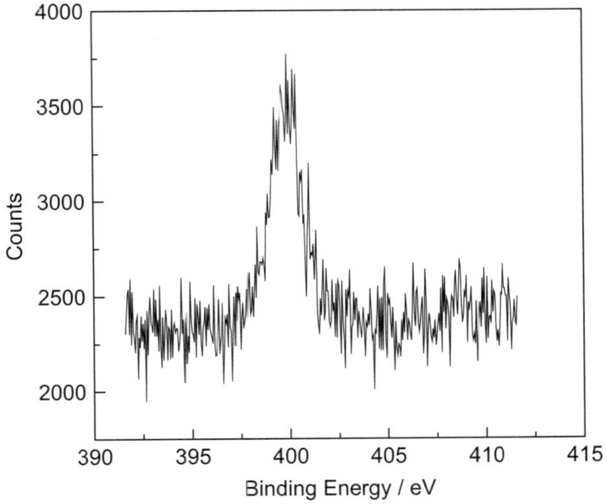

Fig. 26. X-ray photoelectron spectra in the *N* (1s) region for alkylamine monolayer modified GC.

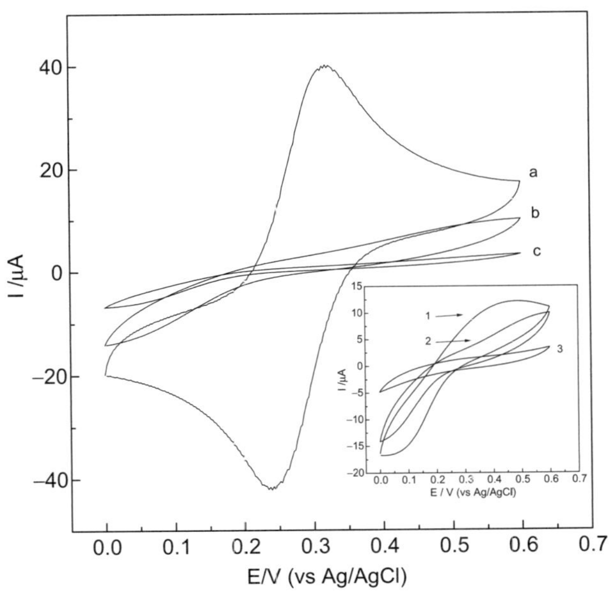

Fig. 27. Cyclic voltammograms of 1 mM $Fe(CN)_6^{3-/4-}$ at (a) bare GC electrode; (b) dodecylamine monolayer modified GC electrode; (c) DMPC/dodecylamine HBM modified GC electrode with 0.1 M KCl as supporting electrolyte. Inset: Cyclic voltammograms of 1 mM $Fe(CN)_6^{3-/4-}$ at octylamine (1), dodecylamine (2), octadecylamine (3) monolayer modified GC electrode. Scan rate 50 mV s^{-1}.

$Fe(CN)_6^{3-/4-}$ solution, which showed a quasi-reversible redox couple. After modified chemically with dodecylamine, the electrode was scanned in the same solution (Fig. 27(b)). Comparing Fig. 27(b) with Fig. 27(a), the peak currents decreased remarkably, which indicated the dodecylamine monolayer was formed on the GCE. The GCE modified with different alkylamine was studied in 0.1 mM $Fe(CN)_6^{3-/4-}$ solution (inset of Fig. 27). Curves 1, 2, and 3 represent octylamine, dodecylamine, and octadecylamine-modified electrode, respectively. With the length of alkylamine increasing, the peak currents decrease, because the longer chains make the amine layer thicker, which leads to a slower electron transfer of ferricyanide. This kind of behavior is similar to a gold electrode modified with alkyl thiol. After the phospholipid layer modified on the alkylamine layer (Fig. 27(c)), the peak current decreased remarkably. From electrochemical impedance results, the lipid monolayer (d_{DMPC}) was obtained as 2.3 ± 0.2 nm, which coincides with the thickness of the DMPC monolayer (2.4 nm [28]).

HBM based on carbon electrode has the advantages of lipid/alkanethiol HBM, such as good stability, electrochemical detectability, etc. In addition, because of the variety of carbon electrode, HBM-based carbon electrode can be applied extensively.

REFERENCES

[1] H.T. Tien, Cyclic voltammetry of bilayer lipid membranes, J. Phys. Chem. 88 (1984) 3172–3174.
[2] H.T. Tien, 715 – Cyclic voltammetry of electron-conducting bilayer lipid membranes, Bioelectrochem. Bioenerg. 13 (1984) 299–316.
[3] J.W. Chen, H.T. Tien, A comparative study between a two-compartment cell and the standard electrode system, Bioelectrochem. Bioenerg. 19 (1988) 491–497.
[4] H.T. Tien, Redox reactions in bilayer lipid membrane as determined by cyclic voltammetry, Bioelectrochem. Bioenerg. 12 (1984) 529–533.
[5] J. Kotowski, H.T. Tien, Sucrose influence on lecithin and polypyrrole lecithin bilayer membranes, Bioelectrochem. Bioenerg. 22 (1989) 69–74.
[6] P. Butko, Z. Salamon, H.T. Tien, Adsorption of gentamicin onto a bilayer lipid membrane, Bioelectrochem. Bioenerg. 23 (1990) 153–160.
[7] H. Minami, N. Sato, M. Sugawara, Y. Umezawa, Comparative-study on the potentiometric responses between a valinomycin-based bilayer lipid-membrane and a solvent polymeric membrane, Anal. Sci. 7 (1991) 853–862.
[8] J. Takashi, K. Masataka, K. Tomiyasu, H. Hajime, Selective Na^+ transport through phospholipid bilayer membrane by a synthetic calix[4]arene carrier, Langmuir 12 (1996) 2684–2689.
[9] C.J. Bender, H.T. Tien, Dynamic polarization of lipid bilayers as a special case in the electrochemistry of liquid/liquid interfaces, Anal. Chim. Acta 198 (1987) 259–269.
[10] C.J. Bender, H.T. Tien, Electrical properties of bilayer lipid membranes containing iodine and iodide, investigated by cyclic voltammetry, Anal. Chim. Acta 201 (1987) 51–58.
[11] L. Ding, E. Wang, Electrochemical studies of lipophilic ion transport through BLM. The influence of sterols on its transport, Electroanalysis 8 (1996) 922–926.

[12] H.T. Tien, Z. Salamon, Formation of self-assembled lipid bilayers on solid substrates, Bioelectrochem. Bioenerg. 22 (1989) 211–218.

[13] Z. Salamon, G. Tollin, Cyclic voltammetric behavior of [2Fe–2S] ferredoxins at a lipid bilayer modified electrode, Bioelectrochem. Bioenerg. 27 (1992) 381–391.

[14] Z. Salamon, G. Tollin, Interfacial electrochemistry of cytochrome *c* at a lipid bilayer modified electrode: effect of incorporation of negative charges into the bilayer on cyclic voltammetric parameters, Bioelectrochem. Bioenerg. 26 (1991) 321–334.

[15] R. Sarges, B. Witkop, Gramicidin A. IV. Primary sequence of valine and isoleucine gramicidin A, J. Am. Chem. Soc. 86 (1964) 1862–1863.

[16] D.W. Urry, Gramicidin A transmembrane channel: a proposed (L,D) helix, Proc. Natl Acad. Sci. USA 68 (1971) 672–676.

[17] D.A. Langes, X-ray crystal structures for two different antiparallel double-helical pores, Science 241 (1988) 188–191.

[18] G. Kemp, C. Wenner, Solution, interfacial, and membrane properties of gramicidin A, Arch. Biochem. Biophys. 176 (1976) 547–555.

[19] O.S. Andersen, Ion movement through gramicidin A channels. Single-channel measurements at very high potentials, Biophys. J. 41 (1983) 119–133.

[20] O.S. Andersen, M. Fuchs, Potential energy barriers to ion transport within lipid bilayers: studies with tetraphenylborate, Biophys. J. 15 (1975) 795–830.

[21] W. Jing, Z. Wu, E. Wang, Electrochemical study of gramicidin D forming ion-permeable channels in the bilayer lipid membranes, Electrochim. Acta 44 (1998) 99–102.

[22] O.S. Andersen, Gramicidin channels, Annu. Rev. Physiol. 46 (1984) 531–548.

[23] Z. Wu, B. Wang, Z. Cheng, X. Yang, S. Dong, E. Wang, A facile approach to immobilize protein for biosensor: self-assembled supported bilayer lipid membranes on glassy carbon electrode, Biosens. Bioelectron. 16 (2001) 47–52.

[24] X. Ren, P.G. Pickup, An impedance study of electron transport and electron transfer in composite polypyrrole + polystyrenesulphonate films, J. Electroanal. Chem. 420 (1997) 251–257.

[25] A.L. Plant, Supported hybrid bilayer membranes as rugged cell membrane mimics, Langmuir 15 (1999) 5128–5135.

[26] F. Patolsky, M. Zayats, E. Katz, I. Willner, Precipitation of an insoluble product on enzyme monolayer electrodes for biosensor applications: characterization by faradaic impedance spectroscopy, cyclic voltammetry, and microgravimetric quartz crystal microbalance analyses, Anal. Chem. 71 (1999) 3171–3180.

[27] R. Fettiplace, D.M. Andrews, D.A. Haydon, The thickness, composition and structure of some lipid bilayers and natural membranes, J. Membr. Biol. 5 (1971) 277–296.

[28] J.K.H. Horber, C.A. Lang, T.W. Hansch, W.M. Heckl, H. Mohwald, Scanning tunneling microscopy of lipid films and embedded biomolecules, Chem. Phys. Lett. 145 (1988) 151–158.

[29] K. Takebara, M. Aibara, N. Ueda, An ion-gate response of a glutathione monolayer assembly highly sensitive to lanthanide ions, Electroanalysis 6 (1994) 1083–1086.

[30] H. Hauser, M.C. Phillips, B.A. Levine, R.J.P. Williams, Conformation of the lecithin polar group in charged vesicles, Nature 261 (1976) 390–394.

[31] X.M. Li, Y.F. Zhang, J.Z. Ni, J.W. Chen, F. Hwang, Effect of lanthanide ions on the phase-behavior of dipalmitoylphosphatidylcholine multilamellar lipisomes, J. Inorg. Biochem. 53 (1994) 139–149.

[32] C. Yuan, X. Li, D. Ding, B. Li, H. Zhang, Z. Lu, J. Liu, J. Ni, Lanthanide ion induced formation of stripes domain structure in phospholipid Langmuir–Blodgett monolayer film observed by atomic force microscopy, Surf. Sci. 366 (1996) L729–L734.

[33] T.R. Herrmann, A.R. Jayaweera, A.E. Shamoo, Interaction of europium(III) with phospholipid vesicles as monitored by laser-excited europium(III) luminescence, Biochemistry 25 (1986) 5834–5838.

[34] M. Sugawara, K. Kojima, H. Sazawa, Y. Umezawa, Ion-channel sensors, Anal. Chem. 59 (1987) 2842–2846.

[35] X. Han, Y. Tong, W. Huang, E. Wang, Study of the interaction between lanthanide ions and a supported bilayer lipid membrane by cyclic voltammetry and ac impedance, J. Electroanal. Chem. 523 (2002) 136–141.
[36] C.B. Yuan, D.Q. Zhao, F.K. Pei, J.Z. Ni, B. Zhao, F. Huang, The effect of metal ions to polar head conformation of sphingolipid bilayer, Chin. Sci. Bull. 39 (1994) 1482–1484.
[37] H. Hauser, C.C. Hinckley, J. Krebs, B.A. Levine, M.C. Phillips, R.J.P. Williams, The interaction of ions with phosphatidylcholine bilayers, Biochim. Biophys. Acta 468 (1977) 364–377.
[38] J. Ni, Bioinorganic chemistry of rare earth elements, Science Publishing Corporation, China, 1995. p. 238.
[39] H. Grasdalen, L.E.G. Eriksson, J. Westman, A. Ehrenberg, Surface potential effects on metal ion binding to phosphatidylcholine membranes. ^{31}P NMR study of lanthanide and calcium ion binding to egg-yolk lecithin vesicles, Biochim. Biophys. Acta 469 (1977) 151–162.
[40] Y.K. Levine, A.G. Lee, N.J.M. Birdsall, J.C. Metcalfe, J.D. Robinson, The interaction of paramagnetic ions and spin labels with lecithin bilayers, Biochim. Biophys. Acta 291 (1973) 592.
[41] Z. Wu, J. Tang, Z. Cheng, X. Yang, E. Wang, Ion channel behavior of supported bilayer lipid membranes on a glassy carbon electrode, Anal. Chem. 72 (2000) 6030–6033.
[42] X. Han, G. Xu, S. Dong, E. Wang, Studies of perchlorate triggered ion-gate behavior of sBLM by electrochemiluminescence and its application to a sensor for perchlorate, Electroanalysis 14 (2002) 1185–1190.
[43] X. Han, E. Wang, Ion-channel sensing of ferricyanide anion based on a supported bilayer lipid membrane, Anal. Sci. 17 (2001) 1171–1174.
[44] X. Han, Y. Song, Z. Wu, B. Zhang, E. Wang, Induction effect of Ca^{2+} on ion-channel behavior of supported phospholipid membranes, Chem. J. Chinese U. 22 (2001) 1298–1300.
[45] Y. Tong, Z. Wu, X. Han, E. Wang, J. Bioenerg. Biomembr. 34 (2002) 185–191.
[46] Y. Tong, X. Han, E. Wang, Study of the ion channel behavior of didodecyldimethylammonium bromide formed bilayer lipid membrane stimulated by PF^{6-}, Chin. J. Chem. 21 (2003) 62–65.
[47] J. Wang, L. Wang, S. Liu, X. Han, W. Huang, E. Wang, Interaction of $K_7Fe^{3+}P_2W_{17}O_{62}H_2$ with supported bilayer lipid membranes on platinum electrode, Biophys. Chem. 106 (2003) 31–38.
[48] W. Huang, X. Han, E. Wang, Defect formation induced by PAMAM dendrimers on Pt-supported bilayer lipid membranes investigated by electrochemistry, J. Electrochem. Soc. 150 (2003) E218–E221.
[49] O.Q. Munro, H.M. Marques, Heme-peptide models for hemoproteins. 1. Solution chemistry of *N*-acetylmicroperoxidase-8, Inorg. Chem. 35 (1996) 3752–3767.
[50] R. Santucci, H. Reinhard, M. Brunori, Direct electrochemistry of the undecapeptide from cytochrome *c* (microperoxidase) at a glassy carbon electrode, J. Am. Chem. Soc. 110 (1988) 8536–8537.
[51] R. Quinn, J.M. Smith, J.N. Burstyn, J.S. Valentine, Influence of hydrogen bonding on the properties of iron porphyrin imidazole complexes. An internally hydrogen bonded imidazole ligand, J. Am. Chem. Soc. 106 (1984) 4136–4144.
[52] K.M. Barkigia, L. Chantranupony, K.M. Smith, J. Fajer, Structural and theoretical models of photosynthetic chromophores. Implications for redox, light-absorption properties and vectorial electron flow, J. Am. Chem. Soc. 110 (1988) 7566–7567.
[53] R.G. Alden, M.R. Ondrias, J.A. Shelnutt, Influences of pi–pi complex formation, dimerization, and binding to hemoglobin on the planarity of nickel(II) porphyrins, J. Am. Chem. Soc. 112 (1990) 691–697.
[54] W. Huang, Z. Zhang, X. Han, J. Tang, S. Dong, E. Wang, Electrochemistry and spectroscopy study on the interaction of microperoxidase-11 with lipid membrane, Biophys. Chem. 94 (2001) 165–173.

[55] L. Jiang, A. Glidle, C.J. McNeil, J.M. Cooper, Characterization of electron transfer reactions of microperoxidase assembled at short-chain thiol-monolayers on gold, Biosens. Bioelectron. 12 (1997) 1143–1155.
[56] T. Ruzgas, A. Gaigalas, L. Gorton, Diffusionless electron transfer of microperoxidase-11 on gold electrodes, J. Electroanal. Chem. 469 (1999) 123–133.
[57] D.K. Das, O.K. Medhi, Effect of surfactant and pH on the redox potential of microperoxidase 11 in aqueous micellar solutions, J. Chem. Soc. Dalton Trans. 10 (1998) 1693–1698.
[58] R.W. Murray, Electroanalytical Chemistry, Marcel Dekker, New York, 1984.
[59] G. Irace, E. Bismuto, F. Savy, G. Colonna, Unfolding pathway of myoglobin: molecular properties of intermediate forms, Arch. Biochem. Biophys. 244 (1986) 459–469.
[60] A.M. Jehanli, D.A. Stotter, M.T. Wilson, Magnetic studies on the ferri-haem undecapeptide of cytochrome *c*, Eur. J. Biochem. 71 (1976) 613–616.
[61] J.A. Shelnutt, C.J. Medforth, M.D. Berber, K.M. Barkigia, K.M. Smith, Relationships between structural parameters and Raman frequencies for some planar and nonplanar nickel(II) porphyrins, J. Am. Chem. Soc. 113 (1991) 4077–4087.
[62] L.D. Sparks, C.J. Medforth, M.S. Park, J.R. Chamberlain, M.R. Ondrias, M.O. Sengo, K.M. Smith, J.A. Shelnutt, Metal dependence of the nonplanar distortion of octaalkyltetraphenylporphyrins, J. Am. Chem. Soc. 115 (1993) 581–592.
[63] C.J. Medforth, M.O. Senge, K.M. Smith, L.D. Sparks, J.A. Shelnutt, Nonplanar distortion modes for highly substituted porphyrins, J. Am. Chem. Soc. 114 (1992) 9859–9869.
[64] T.J.T. Pinheiro, G.A. ElÖve, A. Watts, H. Roder, Structural and kinetic description of cytochrome *c* unfolding induced by the interaction with lipid vesicles, Biochemistry 36 (1997) 13122–13132.
[65] M.T. Wilson, R.J. Ranson, P. Masiakowski, E. Czarnecka, M. Brunori, A kinetic study of the pH-dependent properties of the ferric undecapeptide of cytochrome *c* (microperoxidase), Eur. J. Biochem. 77 (1977) 193–199.
[66] J.M. Scholtz, Q. Hong, E.J. York, J.M. Stewart, R.L. Baldwin, Parameters of helix-coil transition theory for alanine-based peptides of varying chain lengths in water, Biopolymers 31 (1991) 1463–1470.
[67] R. Santucci, A. Picciau, G. Antonini, L. Campanella, A complex of microperoxidase with a synthetic peptide: structural and functional characterization, Biochim. Biophys. Acta 1250 (1995) 183–188.
[68] N. Greenfield, G.D. Fasman, Computed circular dichroism spectra for the evaluation of protein conformation, Biochemistry 8 (1969) 4108–4116.
[69] N.J. Greenfield, Estimation of globular protein secondary structure from circular dichroism, Biochemistry 20 (1981) 33–37.
[70] B. Gopal, K.G. Papavinasasundaram, G. Dodson, M.J. Colston, S.A. Major, A.N. Lane, Spectroscopic and thermodynamic characterization of the transcription antitermination factor NusE and its interaction with NusB from myccbacterium tuberculosis, Biochemistry 40 (2001) 920–928.
[71] G. Blauer, N. Sreerama, R.W. Woody, Optical activity of hemoproteins in the Soret region. Circular dichroism of the heme undecapeptide of cytochrome *c* in aqueous solution, Biochemistry 32 (1993) 6674–6679.
[72] M. Dathe, T. Wieprecht, Structural features of helical antimicrobial peptides: their potential to modulate activity on model membranes and biological cells, Biochim. Biophys. Acta 1462 (1999) 71–87.
[73] N. Ben-Tal, A. Ben-Shaul, A. Nicholls, B. Honig, Free-energy determinants of alpha-helix insertion into lipid bilayers, Biophys. J. 70 (1996) 1803–1812.
[74] M. Rytomaa, P.K. Kinnunen, Evidence for 2 distinct acidic phospholipid-binding sites in cytochrome-*c*, J. Biol. Chem. 269 (1994) 1770–1774.

[75] J. Dufourcq, J.F. Faucon, G. Fourche, J.L. Dasseux, M.L. Maire, T.G. Krzywicki, Morphological changes of phosphatidylcholine bilayers induced by melittin: vesicularization, fusion, discoidal particles, Biochim. Biophys. Acta 859 (1989) 33–48.
[76] C.E. Dempsey, A. Watts, A deuterium and phosphorus-31 nuclear magnetic resonance study of the interaction of melittin with dimyristoylphosphatidylcholine bilayers and the effects of contaminating phospholipase A2, Biochemistry 26 (1987) 5803–5811.
[77] G. Beschiaschvili, J. Seelig, Melittin binding to mixed phosphatidylglycerol/phosphatidylcholine membranes, Biochemistry 29 (1990) 52–58.
[78] K. Matsuzaki, Magainins as paradigm for the mode of action of pore forming polypeptides, Biochim. Biophys. Acta 1376 (1998) 391–400.
[79] D.J. Hirsh, J. Hammer, W.L. Maloy, J. Blazyk, J. Schaefer, Secondary structure and location of a magainin analogue in synthetic phospholipid bilayers, Biochemistry 35 (1996) 12733–12741.
[80] K. Rayman, A. Hurst, in: E.J. Vandamme (Ed.), Biotechnology of Industrial Antibiotics, Marcel Dekker, New York, 1984, p. 607.
[81] J. Delves-Broughton, P. Blackburn, R.J. Evans, J. Hu-genholtz, Applications of the bacteriocin, nisin, Antonie Van Leeuwenhoek 69 (1996) 193–202.
[82] S. Hennings, R. Metz, W.P. Hammes, Studies on the mode of action of nisin, Int. J. Food Microbiol. 3 (1986) 121–134.
[83] F.H. Gao, T. Abee, W.N. Konings, Mechanism of action of the peptide antibiotic nisin in liposomes and cytochrome-*c* oxidase-containing proteoliposomes, Appl. Environ. Microbiol. 57 (1991) 2164–2170.
[84] M.J. Garcia-Garcera, M.G. Elferink, A.J. Driessen, W.N. Konings, Invitro pore-forming activity of the lantibiotic nisin-role of protonmotive force and lipid- composition, Eur. J. Biochem. 212 (1993) 417–422.
[85] A.J. Driessen, H.W. van den Hooven, W. Kuipers, M. van de Kamp, H.G. Sahl, R.N. Konings, W.N. Konings, Mechanistic studies of lantibiotic-induced permeabilization of phospholipid vesicles, Biochemistry 34 (1995) 1606–1614.
[86] T. Abee, F.H. Gao, W.N. Konings, in: G. Jung, H.G. Sahl (Eds.), Nisin and Novel Lantibiotics, ESCOM Science, Leiden, 1991, pp. 373–385.
[87] M. Kordel, F. Schueller, H.G. Sahl, Interaction of the pore forming-peptide antibiotics Pep 5, nisin and subtilin with non-energized liposomes, FEBS Lett. 244 (1989) 99–102.
[88] W. Huang, Z. Zhang, X. Han, J. Wang, J. Tang, S. Dong, E. Wang, Concentration-dependent behavior of nisin interaction with supported bilayer lipid membrane, Biophys. Chem. 99 (2002) 271–279.
[89] P. Diao, D. Jiang, X. Cui, D. Gu, R. Tong, B. Zhong, Cyclic voltammetry and a.c. impedance studies of Ca^{2+}-induced ion channels on Pt-BLM, Bioelectrochem. Bioenerg. 45 (1998) 173–179.
[90] J. Brajtburg, W.G. Powderly, G.S. Kobayashi, G. Medoff, Amphotericin B: delivery systems, Antimicrob. Agents Chemother. 34 (1990) 381–384.
[91] B.D. Kruijff, W.J. Gerritsen, A. Oerlemans, R.A. Demel, L.L.M.V. Deenen, Polyene antibiotic-sterol interactions in membranes of Acholeplasma laidlawii cells and lecithin liposomes. II. Temperature dependence of the polyene antibiotic-sterol complex formation, Biochim. Biophys. Acta 339 (1974) 44–56.
[92] M.B. Marin, M.M. Bello, I.O. Blake, A microscopic electrostatic model for the amphotericin B channel, Biochim. Biophys. Acta 1061 (1991) 65–77.
[93] G. Fujii, J.E. Chang, T. Coley, B. Steere, The formation of amphotericin B ion channels in lipid bilayers, Biochemistry 36 (1997) 4959–4968.
[94] A.V. Croquin, J. Bolard, M. Chabbert, C. Gary-Bobo, Differences in the interaction of the polyene antibiotic amphotericin B with cholesterol- or ergosterol-containing phospholipid vesicles. A circular dichroism and permeability study, Biochemistry 22 (1983) 2939–2944.

[95] S.C. Hartsel, W.R. Perkins, G.J. McGarvey, D.S. Cafiso, A selective cholesterol-dependent induction of proton/hydroxide currents in phospholipid vesicles by amphotericin B, Biochemistry 27 (1988) 2656–2660.
[96] T. Ruckwardt, A. Scott, J. Scott, P. Milulecky, S.C. Hartsel, Lipid and stress dependence of amphotericin B ion selective channels in sterol-free membranes, Biochim. Biophys. Acta 1372 (1998) 283–288.
[97] W. Huang, Z. Zhang, X. Han, J. Tang, Z. Peng, S. Dong, E. Wang, Electrochemistry and spectroscopy study on the interaction of microperoxidase-11 with lipid membrane, Biophys. Chem. 94 (2001) 165–173.
[98] A. Finkelstein, R. Holz, in: G. Eisenman (Ed.), Membranes, Lipid Bilayers and Antibiotics, Marcel Dekker, New York, 1973, Vol. 2, p. 377.
[99] J. Bolard, P. Legrand, F. Heitz, B. Cybulska, One-sided action of amphotericin B on cholesterol-containing membranes is determined by its self-association in the medium, Biochemistry 30 (1991) 5707–5715.
[100] Y. Aracava, I.C.P. Smith, S. Schreier, Effects of amphotericin B on membrane permeability-kinetics of spin probe reduction, Biophys. Chem. 14 (1981) 325–332.
[101] D.B. Archer, E.F. Gale, Antagonism by sterols of the action of amphotericin and filipin on the release of potassium ions from Candida albicans and Mycoplasma, J. Gen. Microbiol. 90 (1975) 187–190.
[102] J.D. Readio, R. Bittman, Equilibrium binding of amphotericin B and its methyl ester and borate complex to sterols, Biochim. Biophys. Acta 685 (1982) 219–224.
[103] A. Ottova, V. Tvarozek, J. Racek, J. Sabo, W. Ziegler, T. Hianik, H.T. Tien, Self-assembled BLMs: biomembrane models and biosensor applications, Supramol. Sci. 4 (1997) 101–112.
[104] X. Lu, A.L. Ottova, H.T. Tien, Biophysical aspects of agar-gel supported bilayer lipid membranes: a new method for forming and studying planar bilayer lipid membranes, Bioelectrochem. Bioenerg. 39 (1996) 285–289.
[105] M. Rehak, M. Snejdarkova, M. Otto, Self-assembled lipid bilayers as a potassium sensor, Electroanalysis 5 (1993) 691–694.
[106] M. Otto, M. Snejdarkova, M. Rehak, Hydrogen-peroxide oxygen biosensor based on supported phospholipid-bilayer, Anal. Lett. 25 (1992) 653–662.
[107] D.P. Nikolelis, U.J. Krull, Bilayer-lipid membranes for electrochemical sensing, Electroanalysis 5 (1993) 539–545.
[108] D.P. Nikolelis, C.G. Siontorou, U.J. Krull, P.L. Katrivanos, Ammonium ion minisensors from self-assembled bilayer lipid membranes using gramicidin as an ionophore. Modulation of ammonium selectivity by platelet-activating factor, Anal. Chem. 68 (1996) 1735–1741.
[109] C.G. Siontorou, A.M.O. Brett, D.P. Nikolelis, Evaluation of a glassy carbon electrode modified by a bilayer lipid membrane with incorporated DNA, Talanta 43 (1996) 1137–1144.
[110] Q. Deng, S. Dong, Mediatorless hydrogen peroxide electrode based on horseradish peroxidase entrapped in poly(*o*-phenylenediamine), J. Electroanal. Chem. 377 (1994) 191–195.
[111] J. Hlavay, G. Guilbault, A fiberoptic biosensor for determination of hydrogen-peroxide-effect of the membrane type on the immobilization of POD enzyme, Acta Chim. Hung. 130 (1993) 83–93.
[112] D. Emerson, S.F. Peteu, R.M. Worden, A catalase microbiosensor for detecting hydrogen peroxide, Biotechnol. Tech. 10 (1996) 673–678.
[113] R. Benz, D. Cros, Influence of sterols on ion transport through lipid bilayer membranes, Biochem. Biophys. Acta 506 (1978) 265–280.
[114] N.R. Clement, J.M. Gold, Modulation by small hydrophobic molecules of valinomycin-mediated potassium transport across phospholipid vesicle membranes, Biochemistry 20 (1981) 1539–1543.

[115] U.J. Krull, R.N. Koilpillai, C.O. Rae, E.T. Vandenberg, Electrostatic and structural alterations of lipid monolayers and bilayers by interaction with valinomycin and phloretin, Thin Solid Films 195 (1991) 279–288.

[116] L. Ding, J. Li, S. Dong, E. Wang, Supported phospholipid membranes: comparison among different deposition methods for a phospholipid monolayer, J. Electroanal. Chem. 416 (1996) 105–112.

[117] C. Berggren, G. Johansson, Capacitance measurements of antibody-antigen interactions in a flow system, Anal. Chem. 69 (1997) 3651–3657.

[118] P. Bataillard, F. Gardies, N. Jaffrezic-Renault, C. Martelet, B. Colin, B. Mandrand, Direct detection of immunospecies by capacitance measurements, Anal. Chem. 60 (1988) 2374–2379.

[119] V. Billard, C. Martelet, P. Binder, J. Therasse, Toxin detection using capacitance measurements on immunospecies grafted onto a semiconductor substrate, Anal. Chim. Acta 249 (1991) 367–372.

[120] A. Gebbert, M. Alvarez-Icaza, W. Stcklein, R.D. Schmid, Real-time monitoring of immunochemical interactions with a tantalum capacitance flow-through cell, Anal. Chem. 64 (1992) 997–1003.

[121] I. Bontidean, C. Berggren, G. Johansson, E. Csoregi, B. Mattiasson, J.R. Lloyd, K.J. Jakeman, N. Brown, Detection of heavy metal ions at femtomolar levels using protein-based biosensors, Anal. Chem. 70 (1998) 4162–4169.

[122] Z. Cheng, L. Luo, Z. Wu, E. Wang, X. Yang, A new kind of potassium sensor based on capacitance measurement of mimic membrane, Electroanalysis 13 (2001) 68–71.

[123] B.W. Gregory, R.A. Dluhy, L.A. Bottomley, Structural characterization and nanometer-scale domain formation in a model phospholipid bilayer as determined by infrared spectroscopy and scanning tunneling microscopy, J. Phys. Chem. 98 (1994) 1010–1021.

[124] E.L. Florin, H.E. Gaub, Painted supported lipid membranes, Biophys. J. 64 (1993) 375–383.

[125] K. Seifert, K. Fendler, E. Bamberg, Charge transport by ion translocating membrane proteins on solid supported membranes, Biophys. J. 64 (1993) 384–391.

[126] A.L. Plant, Self-assembled phospholipid/alkanethiol biomimetic bilayers on gold, Langmuir 9 (1993) 2764–2767.

[127] A.L. Plant, M. Gueguetchkeri, W. Yap, Supported phospholipid/alkanethiol biomimetic membranes: insulating properties, Biophys. J. 67 (1994) 1126–1133.

[128] M. Zviman, H.T. Tien, Formation of a bilayer lipid-membrane on rigid supports – an approach to BLM-based biosensors Biosens. Bioelectron. 6 (1991) 37–42.

[129] S. Kalinowski, Z. Figaszewski, A new system for bilayer lipid membrane capacitance measurements: method, apparatus and applications, Biochim. Biophys. Acta 1112 (1992) 57–66.

[130] P. Biihlmann, H. Aoki, K.P. Xiao, S. Amemiya, K. Tohda, Y. Umezawa, Chemical sensing with chemically modified electrodes that mimic gating at biomembranes incorporating ion-channel receptors, Electroanalysis 10 (1998) 1149–1158.

[131] C. Steinem, A. Janshoff, K. von dem Bruch, K. Reihs, J. Goossens, H.J. Galla, Valinomycin-mediated transport of alkali cations through solid supported membranes, Bioelectrochem. Bioenerg. 45 (1998) 17–26.

[132] C. Miller, P. Cuendet, M. Gratzel, K^+ sensitive bilayer supporting electrodes, J. Electroanal. Chem. 278 (1990) 175–192.

[133] R.S. Deinhammer, M. Ho, J.W. Anderegg, M.D. Porter, Electrochemical oxidation of amine-containing compounds: a route to the surface modification of glassy carbon electrodes, Langmuir 10 (1994) 1306–1313.

[134] J. Liu, S. Dong, Grafting of diaminoalkane on glassy carbon surface and its functionalization, Electrochem. Commun. 2 (2000) 707–712.

[135] B. Borbier, J. Pinson, G. Desarmot, M. Sanchez, Electrochemical bonding of amines to carbon fibers surface towards improved carbon–epoxy composites, J. Electrochem. Soc. 137 (1990) 1757–1763.

[136] X. Han, L. Wang, B. Qi, X. Yang, E. Wang, A new strategy for constructing hybrid bilayer membrane based on carbon substrate, Anal. Chem. 75 (2003) 6566–6570.

[137] N. Nordberg, R.C. Albridge, T. Bergmark, U. Ericson, J. Hedman, C. Nordling, K. Siegbahn, B.J. Lindberg, Molecular spectroscopy by means of electron spectroscopy for chemical analysis. Charge distribution in nitrogen compounds, Ark. Kemi. 28 (1967) 257–278.

CHAPTER 10

Mechanisms of Membrane Permeabilization by Apoptosis-Regulatory Proteins of the BCL-2 Family

Oihana Terrones, Aitor Etxebarria, and Gorka Basañez*

Unidad de Biofísica (Centro Mixto Consejo Superior de Investigaciones Cientificas—Universidad del Pais Vasco/Euskal Herriko Unibertsitatea), Universidad del Pais Vasco/Euskal Herriko Unibertsitatea (UPV/EHU), P.O. Box 644, 48080 Bilbao, Spain

Contents

Abstract

Proteins of the BCL-2 family are crucial components of the intracellular apoptotic machinery which primarily act by regulating the cytoplasmic release of a variety of death-promoting proteins normally confined to the mitochondrial intermembrane space. Here, we review possible molecular mechanisms by which BCL-2 family members may increase outer mitochondrial membrane permeability in order to modulate this intermembrane protein release. The focus is on the information gathered from studies using model membrane systems. Two major mechanisms of membrane permeabilization by BCL-2 family proteins are discussed: (1) formation of purely proteinaceous channels, and (2) generation of lipid-containing pores.

1. INTRODUCTION

Apoptosis is a genetically encoded and evolutionarily conserved form of cell death that is essential for normal development and maintenance of tissue homeostasis in multicellular organisms [1]. Inappropriate apoptosis can be the cause or

* Corresponding author. Tel.: +34-946013355; Fax: +34-946013360; E-mail: gbzbaasg@lg.ehu.es

ADVANCES IN PLANAR LIPID BILAYERS AND LIPOSOMES, VOLUME 2
ISSN 1554-4516 DOI: 10.1016/S1554-4516(05)02010-7

contributing factor to many human diseases including cancer, autoimmunity, AIDS, myocardial infarction, and acute and chronic neurological disorders.

A pivotal event in many mammalian apoptotic pathways is the release of a number of apoptogenic macromolecules (cytochrome *c*, SMAC/DIABLO, OMI/HTRA2, AIF, and others) from the mitochondrial intermembrane space into the cytosol [2]. This crucial process for progression of the apoptotic cascade is tightly regulated by BCL-2 family proteins. Members of the BCL-2 family can be classified into three main sub-groups based on the presence of up to four evolutionarily conserved regions named BCL-2 homology (BH) domains which determine structure and function [3]. Members of the first sub-group, exemplified by BCL-2 and its close homologue BCL-X_L contain all four BH domains, block the release of mitochondrial intermembrane apoptogenic proteins into the cytoplasm, and inhibit cell death (BCL-2 type proteins). Members of the second sub-group, exemplified by BAX contain BH1–BH3 domains, induce the cytoplasmic release of mitochondrial pro-death factors, and promote cell death (BAX-type proteins). Members of the third sub-group, including BID and BIM, share only the BH3 domain, and also promote efflux of mitochondrial cytotoxic molecules, as well as cell death (BH3-only proteins).

A large body of evidence suggests that BCL-2 family proteins modulate the release of mitochondrial intermembrane apoptogenic factors through changes in the permeability of the outer mitochondrial membrane (OMM) [4]. Elucidating how exactly proteins of the BCL-2 family influence OMM permeability is one of the most intense areas of research in the apoptosis field. Here, we summarize current knowledge on the molecular mechanisms by which BCL-2 family members affect membrane permeability. The primary focus is on the information gathered from studies using model membrane systems (i.e., liposomes and planar phospholipid bilayers). Two distinct mechanisms of membrane permeabilization by BCL-2 members are discussed: (1) opening of a proteinaceous channel, defined as a transmembrane aqueous conduit lined exclusively by a fixed set of amino acids, and (2) formation of a lipidic pore, in which the edge of the transmembrane aqueous pathway is composed, at least partially, by lipid molecules (Fig. 1). We have left aside most aspects related to modulation by BCL-2 family members of channels endogenous to mitochondrial and endoplasmic reticulum membranes, which are described in detail in other excellent reviews [4,5].

2. BCL-2 FAMILY MEMBERS AS PROTEINACEOUS CHANNELS

2.1. BCL-2 type proteins

The first evidence suggesting that BCL-2 family members possess the intrinsic capacity to increase membrane permeability arose when the three-dimensional

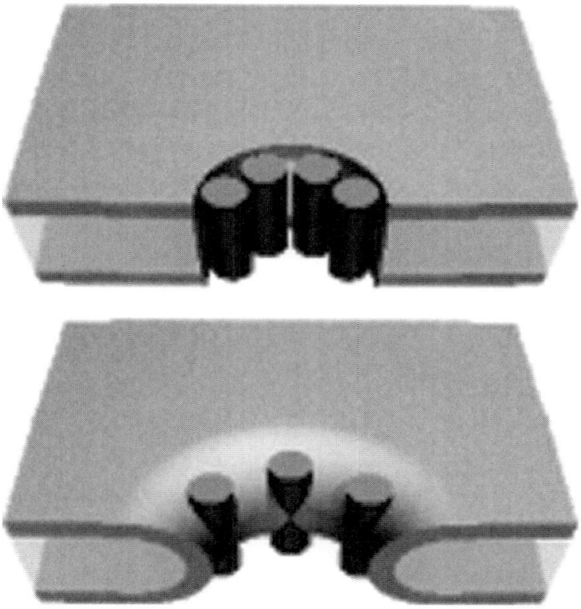

Fig. 1. Pictorial representation of a purely proteinaceous channel (top) and a lipidic pore (bottom). Grey layers represent headgroup regions of bilayer lipids. Protein helices are represented by black cylinders. Note that unlike in a proteinaceous channel (i) the surface of the lipidic pore is lined by lipid headgroups, (ii) membrane monolayers are bent at the edge of the lipidic pore, and (iii) the two leaflets of the bilayer become continuous at the lipidic pore. See text for more details (Obtained from Ref. [37] with permission of Biophys. J.).

structure of BCL-X_L was solved [3]. BCL-X_L consists of a globular bundle of α-helices connected by flexible loops, with two largely hydrophobic α-helices forming a helical hairpin at the core of the structure, which is protected from aqueous solution by surrounding amphipathic helices. This structure bears a striking similarity to those of the pore-forming domains of bacterial toxins such as colicins and the diphtheria toxin (DT), known to form ion channels in target membranes. Consistent with this structural homology, subsequent studies demonstrated that BCL-X_L forms ion-conducting channels of up to a few hundreds picosiemens in planar lipid bilayer membranes [6], and that it causes a permeability pathway of relatively small size in the membrane of pure lipid vesicles [7]. Further studies showed that BCL-2 also (i) possesses a three-dimensional structure similar to pore-forming domains of bacterial toxins [3], (ii) forms ion-conducting channels in planar phospholipid bilayer membranes [8], and (iii) induces leakage of small water-soluble fluorescent dyes entrapped within liposomes [9].

In colicins, membrane association leads to unfolding of the globular helical bundle and exposure of the hydrophobic helical hairpin for insertion into the lipid bilayer in a transmembrane orientation, while the rest of the molecule folds up to rest on the membrane surface [10]. Solid-state NMR experiments performed in lipid bilayer samples under conditions appropriate for BCL-X_L channel formation revealed that the molecule inserts its central pair of hydrophobic α-helices into the membrane spanning the lipid bilayer, while the remaining amphipatic helices of the protein form a helical network on the membrane surface [11]. Similarly, a chemical labeling study showed that induction of apoptosis provokes a conformational change in BCL-2 that results in insertion of its hydrophobic helical hairpin into the lipid bilayer in a transmembrane orientation [12].

The central hydrophobic hairpin of DT constitutes the minimal region required to recapitulate the channel properties of the entire molecule [13]. Interestingly, deletion of the two central hydrophobic α-helices of BCL-2 not only abrogated its capacity to permeabilize planar lipid bilayer membranes and liposomes, but also abolished the anti-apoptotic function of this molecule [14]. DT channel properties can also be altered by reversal of the charge of the loop connecting the two central hydrophobic helices, due to the prevention of the hydrophobic helical hairpin from traversing the hydrophobic matrix of the membrane [15]. A mutant of BCL-X_L possessing an analogous charge reversal at the corresponding loop region displayed a reduced capacity to preserve cell viability and produced channels with shorter lifetimes relative to wild-type BCL-X_L [16].

Altogether, the findings described above support the notion that insertion of the central hydrophobic helical hairpin in a transmembrane orientation is crucial for channel formation by BCL-2 type proteins, and that this activity is important for the cytoprotective function of BCL-2 and its close homologues. What remains to be elucidated, however, is whether a link exists between the channel-forming activity of BCL-2 type proteins and their ability to suppress the release of cytochrome *c* and other cytotoxic macromolecules from the mitochondrial intermembrane space. One possibility is that formation of ion-conducting channels by anti-apoptotic BCL-2 type proteins at the OMM serves to counteract inappropriate ion or metabolite imbalances required for breaching of this permeability barrier [8]. The OMM contains channels such as VDAC and TOM which display higher conductances than BCL-2/BCL-X_L channels in planar lipid bilayer membranes. Thus, it can be argued that it is unlikely that addition of a smaller conductance pathway to the OMM by BCL-2 type proteins could materially affect its permeability properties. However, VDAC and TOM channels do not exist permanently in an open configuration in the OMM *in vivo* [17], and they exhibit different properties with respect to selectivity and voltage dependence compared to BCL-2 and BCL-X_L channels. Since BCL-2 and BCL-X_L can reside in inner mitochondrial, endoplasmic reticulum, and nuclear membranes [5], it is also conceivable that formation of ion-conducting channels by BCL-2 type proteins at a subcellular membrane other than the OMM may indirectly inhibit the release of mitochondrial apoptogenic factors into the cytosol.

In addition to their similarity to pore-forming domains of bacterial toxins, another important aspect revealed by structural studies of BCL-2 type proteins is that their BH1, BH2, and BH3 domains create an elongated hydrophobic groove which can nestle an amphipathic helix formed by the BH3 domain of pro-apoptotic partners [3]. It is well established that the ability of anti-apoptotic BCL-2 family members to engage in this type of physical interaction with BAX-type proteins and/or with BH3-only proteins plays an important role in their cytoprotective function [1–4,18]. Within this framework, a widely accepted model holds that BCL-2 type proteins localized to the OMM block the release of mitochondrial intermembrane pro-death factors by directly binding to and neutralizing active pro-apoptotic family members which otherwise would breach this permeability barrier (see Section 3.2). Conversely, evidence obtained in liposome studies suggests that BH3-mediated binding of BAX-type proteins to BCL-2 type proteins may inhibit anti-apoptotic channel formation [19,20].

2.2. BAX-type proteins

In healthy cells the majority of pro-apoptotic BAX exists as a monomeric, inert cytosolic protein. Monomeric soluble BAX displays structural features similar to those of BCL-2 type proteins, namely a globular bundle of α-helices with a largely hydrophobic helical hairpin buried at the core of the molecule, and an elongated hydrophobic groove located at its surface [21]. Upon receipt of a death signal, BAX is activated through a multi-step process apparently driven by BH3-only proteins, which includes its translocation to the OMM, membrane insertion, and oligomerization [22]. NMR studies indicate that the structures of BAX in solution and in membrane-mimetic environments differ substantially [21]. The fact that BAX-type proteins tend to form high-order oligomers in the presence of lipids and detergents has precluded so far obtaining detailed structural information on the membrane conformation of BAX [3,22].

The exact role played by BAX oligomerization in the release of mitochondrial cytotoxic macromolecules during apoptosis remains controversial [1–4,22]. One suggestion is that upon oligomerization, BAX forms a proteinaceous channel at the OMM large enough to allow transit of cytochrome *c* into the cytosol. Pavlov *et al.* [23] reported a previously unknown increase in membrane conductance in proteoliposomes derived from mitochondria purified from an apoptotic cell line, which they termed mitochondria apoptosis-induced channel (MAC). MAC displayed electrophysiological characteristics akin those observed in planar lipid bilayers treated with a detergent-oligomerized, carboxy-terminally truncated form of BAX. On the basis of its maximal conductance, a diameter of about 4 nm was estimated for MAC, close to the diameter of cytochrome *c* (3.0–3.5 nm). In fact, addition of exogenous cytochrome *c* dramatically decreased MAC activity, probably due to plugging of the channel [24]. Based on these and other observations it was hypothesized that MAC is partially or totally composed of oligomeric BAX, and that MAC functions as an OMM channel allowing for cytoplasmic transport of cytochrome *c*, but not larger cytotoxic proteins trapped between the two mitochondrial membranes [25]. In an *in vivo* model of apoptosis, however, MAC activity was only detected in a late apoptotic mitochondrial population after substantial amounts of cytochrome *c* have been released, casting doubts about the contribution of MAC in OMM permeabilization [26]. In another study, the capacity of a carboxy-terminally truncated form of BAX (obtained by renaturation in the presence of detergents) to release differently sized molecules encapsulated within pure lipid vesicles was examined [27]. A Hill-plot analysis of the kinetics of vesicular contents release suggested that BAX creates channels of increasing molecularity in the liposome membrane, the largest one being constituted by four protein molecules. Exogenous cytochrome *c* decreased the efflux of intravesicular cytochrome *c* in this system as well, in agreement with the hypothesis that oligomeric BAX creates a channel large enough to transport cytochrome *c*, but not other mitochondrial intermembrane apoptogenic proteins. An important issue left unanswered by these studies, however, is the growing body of evidence indicating that activated BAX-type proteins create a permeability pathway in the OMM large enough to allow diffusion not only of cytochrome *c* (~12 kDa), but also of SMAC/DIABLO (~46 kDa for the native dimer), AIF (~57 kDa), OMI/HTRA2 (~111 kDa for the native trimer), and other even larger macromolecules [1,2,4,28–34].

3. LIPIDIC PORE FORMATION BY BCL-2 FAMILY PROTEINS

3.1. BAX-type proteins

In 1999, we reported that planar lipid bilayer membranes treated with full-length BAX present two characteristics unusual for typical proteinaceous channels: (i) a continuously variable increase in membrane conductance, and (ii) a notorious decrease of membrane lifetime [35]. These results are reminiscent of recordings of purely lipidic planar membranes subjected to high electrical fields [36], as well as recordings obtained in the presence of amphiphilic antimicrobial peptides including magainin [37] and HFIAP [38], or certain viral cytotoxic peptides, such as HIV-1 Vpr [39], and influenza A PB1-F2 [40]. In all these cases, membrane permeabilization is thought to occur through formation of pores with a surface covered completely or partially by hydrophilic lipid headgroups (i.e., lipidic pores) (Fig. 1, bottom). Theoretical treatments and experimental work indicate that evolution of both purely lipidic pores and peptide-induced lipidic pores is governed by two opposing physical forces: membrane surface tension, which acts as a driving force for pore opening, and membrane line tension, which tends to close the pore [36,41,42]. Electrophysiological measurements indicated that BAX and related pro-apoptotic molecules, but not BCL-2 type proteins, reduce the line tension of planar lipid bilayer membranes, supporting the view that BAX-type proteins permeabilize planar membranes by forming lipidic pores instead of proteinaceous channels [7,35].

Unlike in a proteinaceous channel, in a lipidic pore structure lipid monolayers bend sharply to line the edge of the transmembrane aqueous pathway (Fig. 1). From this perspective, subsequent investigations to try discriminating between the proteinaceous channel model and the lipidic pore model examined the effect of lipid monolayer curvature on BAX-induced membrane permeabilization. The bending propensity of a lipid monolayer is a function of the intrinsic curvature of its lipid components, which, in turn, depends on the relative cross-sectional areas of lipid hydrophilic and hydrophobic parts [43]. Lipids having similar cross-sectional areas at the level of the polar headgroup and at the level of the apolar hydrocarbon chains form bilayers composed of flat monolayers. Lipids possessing a hydrophilic part larger than the hydrophobic one spontaneously adopt non-bilayer morphologies with positive (normal) monolayer curvature. Finally, lipids having a hydrophobic part larger than the hydrophilic one organize themselves in monolayers having negative (inverted) curvature. Non-bilayer lipids having positive or negative spontaneous curvature can be incorporated in a planar lipid bilayer structure, resulting in a bilayer with two monolayers containing stored positive or negative curvature stress, respectively [44]. Membrane incorporation of non-bilayer lipids having positive spontaneous curvature, such as lysolipids, potently promoted the destabilization of planar lipid bilayers and the release of vesicular contents induced by BAX-type proteins, whereas membrane incorporation of non-bilayer lipids having negative spontaneous curvature, such as unsaturated species of diacylgycerol and phosphatidylethanolamine, had the opposite effect [45]. Importantly, changes in either the membrane affinity of BAX or its intramembranous oligomeric status could not account for the effects exerted by non-bilayer lipids on membrane permeabilization. Based on these results, we proposed that BAX-type proteins function by imposing curvature stress on the lipid bilayer of the OMM in such a manner as to form large lipidic pores having net positive curvature.

A further test of the lipidic pore model assessed the capacity of BAX-type proteins to induce lipid transbilayer redistribution. The rationale behind this approach is as follows: if BAX creates a lipidic pore in the membrane, its constituent monolayers would become continuous *via* the pore-lining lipids, thereby permitting movement of lipid molecules from one monolayer of the bilayer to the other (Fig. 1, bottom). Addition of BAX to liposomes triggered mobilization of lipid molecules between the two leaflets of the membrane concomitantly with vesicular contents release, strongly suggesting that both processes are mechanistically related [46,47]. On the contrary, neither BCL-2, nor *S. aureaus* α-toxin (a pore-forming toxin which creates proteinaceous channels by inserting a transmembrane β-barrel) induced substantial lipid transbilayer redistribution [46].

3.2. BH3-only proteins

BCL-2 family members belonging to the BH3-only sub-group cooperate with BAX-type proteins in the molecular pathway culminating with OMM permeabilization and release of intermembrane apoptogenic factors [1,4]. However, it remains a subject of debate exactly how BH3-only proteins assist BAX-type proteins to breach the OMM permeability barrier. Kuwana *et al.* [48] demonstrated that, in the presence of an apoptogenic form of BID, BAX induces the release of macromolecules much larger than cytochrome *c* from the interior of vesicular systems of decreasing complexity, mirroring the release of multiple mitochondrial intermembrane cytotoxic macromolecules triggered by pro-death BCL-2 family members upon apoptosis induction. Evidence was obtained suggesting that BID activates BAX through insertion of its BH3 domain into BAX's hydrophobic groove and that excess BCL-X_L inhibited this process by binding to and sequestering BID. Interestingly, despite its very large size, the membrane lesion caused by BAX together with BID could not be visualized by electron microscopy, and it was not clearly correlated with a specific oligomeric status of BAX. Moreover, membrane permeabilization required the presence of the mitochondrial lipid cardiolipin. These findings led the authors to hypothesize that BAX, BID, and cardiolipin cooperate to permeabilize the membrane through either an inverted-type non-bilayer lipid structure or a lipidic pore.

To advance in understanding how BAX, BH3-only proteins, and mitochondrial lipids cooperate to induce OMM permeabilization, we utilized liposomes emulating the lipid composition of sites of contact between the outer and inner mitochondrial membranes, where several BCL-2 family proteins and cardiolipin seem to co-localize upon apoptosis induction [46]. Our results lend further support to the notion that pro-apoptotic proteins form large, positively curved lipidic pores in the membrane which do not discriminate permeants according to their size. BCL-2 blocked the vesicular permeabilization induced by BAX and BID with a similar dose-dependence as it inhibited cytochrome *c* release from isolated mitochondria, adding further credit to the physiological relevance of this liposome system. An additional important highlight of our study was the finding that BID assists BAX in membrane permeabilization in a dual manner: (i) triggering BAX oligomerization, *via* its BH3 domain, and (ii) reducing the energetic cost of lipidic pore enlargement, *via* a lipid-interacting portion of the molecule which exerts positive curvature stress on the membrane. Notably, although soluble BID is a globular all α-helical protein with a compact fold, upon binding to lipid bilayer membranes BID adopts a conformation in which (i) the availability of its BH3 domain for interaction with BAX-type and/or BCL-2 type proteins is increased [49], and (ii) all its helices are localized on the membrane surface, as expected for generation of positive curvature stress on the membrane, due to the hydrophilic/hydrophobic mismatch in lipid bilayer packing around the protein [49,50]. Finally, studies with BIM using the same experimental system described above showed that BAX and BIM together did not cause efflux of large macromolecules from the interior of pure lipid vesicles [46]. BIM also lacked the capacity to exert positive curvature stress in the liposomal membrane. Remarkably, BIM reversed the inhibitory effect of BCL-2 on lipidic pore formation, probably due to the fact that BCL-2 possesses a greater affinity for BIM's BH3 domain than for BID's BH3 domain [51].

In summary, these results suggest that BH3-only proteins may utilize different mechanisms to assist BAX in forming lipidic pores at the OMM sufficiently large as to allow transit of differently-sized pro-death proteins from the intermembrane space into the cytoplasm.

4. CONCLUSIONS AND PERSPECTIVES

Studies using model membrane systems have provided valuable insights into the molecular mechanisms of membrane permeabilization induced by apoptosis regulatory proteins of the BCL-2 family. A model originated from investigations with planar lipid bilayers and liposomes that explains the release of multiple mitochondrial intermembrane cytotoxic molecules usually observed in the course of apoptotic is cell death holds that pro-apoptotic BAX-type proteins (aided by BH3-only proteins) form large lipid-containing pores in the OMM. However, the exact structure and composition of the OMM permeability pathway through which intermembrane apoptogenic factors exit mitochondria during apoptosis, and the putative role of mitochondrial lipids in this process remain important challenges for future research. On the other hand, although substantial evidence supports the view that anti-apoptotic BCL-2 and BCL-X_L can form ion-conducting channels in pure lipid membranes, important questions remain as to how this ion channel activity contributes to inhibition of release of mitochondrial intermembrane apoptogenic proteins and preservation of cellular viability. Hopefully, future studies with model membrane systems will serve to continue advancing in our understanding of the biological mechanisms of action of BCL-2 family proteins.

REFERENCES

[1] N.N. Danial, S.J. Korsmeyer, Cell death: critical control points, Cell 116 (2004) 205–219.

[2] D.D. Newmeyer, S. Ferguson-Miller, Mitochondria: releasing power for life and unleashing the machineries of death, Cell 112 (2003) 481–490.

[3] A.M. Petros, E.T. Olejniczak, S.W. Fesik, Structural biology of the BCL-2 family of proteins, Biochim. Biophys. Acta 1644 (2004) 83–94.
[4] D.R. Green, G. Kroemer, The pathophysiology of mitochondrial cell death, Science 305 (2004) 626–629.
[5] S.A. Oakes, J.T. Opferman, T. Pozzan, S.J. Korsmeyer, L. Scorrano, Regulation of endoplasmic reticulum Ca^{2+} dynamics by proapoptotic BCL-2 family members, Biochem. Pharmacol. 66 (2003) 1335–1340.
[6] A.J. Minn, P. Velez, S.L. Schendel, H. Liang, S.W. Muchmore, S.W. Fesik, M. Fill, C.B. Thompson, BCL-X_L forms an ion channel in synthetic lipid membranes, Nature 385 (1997) 353–357.
[7] G. Basañez, J. Zhang, B.N. Chau, G.I. Maksaev, V.A. Frolov, T.A. Brandt, J. Burch, J.M. Hardwick, J. Zimmerberg, Pro-apoptotic cleavage products of BCL-X_L form cytochrome *c*-conducting pores in pure lipid membranes, J. Biol. Chem. 276 (2001) 31083–31091.
[8] S.L. Schendel, M. Montal, J.C. Reed, BCL-2 family proteins as ion-channels, Cell Death Differ. 5 (1998) 372–380.
[9] K.M. Kim, C.D. Giedt, G. Basañez, J.W. O'Neill, J.J. Hill, Y'.H. Han, S.-P. Tzung, J. Zimmerberg, D.M. Hockenbery, K.Y.J. Zhang, Biophysical characterization of recombinant human BCL-2 and its interactions with an inhibitory ligand, antimycin A, Biochemistry 40 (2001) 4911–4922.
[10] W.A. Cramer, J.B. Heymann, S.L. Schendel, B.N. Deriy, F.S. Cohen, P.A. Elkins, C.V. Stauffacher, Structure–function of the channel-forming colicins, Annu. Rev. Biophys. Biomol. Struct. 24 (1995) 611–641.
[11] C.M. Franzin, J. Choi, Y. Zhai, J.C. Reed, F.C. Marassi, Structural studies of apoptosis and ion transport regulatory proteins in membranes, Magn. Reson. Chem. 42 (2004) 172–179.
[12] P.M. Kim, M.G. Annis, P.J. Dlugosz, B. Leber, D.W. Andrews, During apoptosis BCL-2 changes membrane topology at both the endoplasmic reticulum and mitochondria, Mol. Cell 14 (2004) 523–529.
[13] J.A. Silverman, J.A. Mindell, H. Zhan, A. Finkelstein, R.J. Collier, Structure–function relationships in diphtheria toxin channels: I. Determining a minimal channel-forming domain, J. Membr. Biol. 137 (1994) 17–28.
[14] S. Matsuyama, S.L. Schendel, Z. Xie, J.C. Reed, Cytoprotection by BCL-2 requires the pore-forming α5 and α6 helices, J. Biol. Chem. 273 (1998) 30995–31001.
[15] J.A. Silverman, J.A. Mindell, A. Finkelstein, W.H. Shen, R.J. Collier, Mutational analysis of the helical hairpin region of diphtheria toxin transmembrane domain, J. Biol. Chem. 169 (1994) 22524–22532.
[16] A.J. Minn, C.S. Kettlun, H. Liang, A. Kelekar, M.G. Vander Heiden, B.S. Chang, S.W. Fesik, M. Fill, C.B. Thompson, BCL-X_L regulates apoptosis by heterodimerization-dependent and -independent mechanisms, EMBO J. 18 (1999) 632–643.
[17] E.A. Jonas, D. Hoit, J.A. Hickman, T.A. Brandt, B.M. Polster, Y. Fannjiang, E. McCarthy, M.K. Montanez, J.M. Hardwick, L.K. Kaczmarek, Modulation of synaptic transmission by the BCL-2 family protein BCL-X_L, J. Neurosci. 23 (2003) 8423–8431.
[18] E.H.-Y.A. Cheng, M.C. Wei, S. Weiler, R.A. Flavell, T.W. Mak, T. Lindsten, S.J. Korsmeyer, BCL-2, BCL-X_L sequester BH3 domain-only molecules preventing BAX- and BAK-mediated mitochondrial apoptosis, Mol. Cell 8 (2001) 705–711.
[19] S.-P. Tzung, K.M. Kim, G. Basañez, C.D. Giedt, J. Simon, J. Zimmerberg, K.Y.J. Zhang, D.M. Hockenbery, Antimycin A mimics a cell death-inducing BCL-2 homology domain 3, Nat. Cell Biol. 3 (2001) 183–191.
[20] M.K. Manion, J.W. O'Neill, C.D. Giedt, K.M. Kim, K.Y.J. Zhang, D.M. Hockenbery, BCL-X_L mutations suppress cellular sensitivity to antimycin A, J. Biol. Chem. 279 (2004) 2159–2165.
[21] M. Suzuki, R.J. Youle, N. Tjandra, Structure of Bax. Co-regulation of dimer formation and intracellular localization, Cell 103 (2000) 645–654.

[22] B. Antonsson, Mitochondria and the BCL-2 family proteins in apoptosis signaling pathways, Mol. Cell. Biochem. 256–257 (2004) 141–155.
[23] E.V. Pavlov, M. Priault, D. Pietkiewicz, E.H. Cheng, B. Antonsson, S. Manon, S.J. Korsmeyer, C.A. Mannella, K.W. Kinnally, A novel, high conductance channel of mitochondria linked to apoptosis in mammalian cells and Bax expression in yeast, J. Cell Biol. 155 (2001) 725–731.
[24] L. Guo, D. Pietkiewicz, E.V. Pavlov, S.M. Grigoriev, J.J. Kasianowicz, L.M. Dejean, S.J. Korsmeyer, B. Antonsson, K.W. Kinnally, The effects of cytochrome *c* on the mitochondrial apoptosis-induced channel MAC, J. Am. J. Physiol. Cell Physiol. 286 (2004) C1109–C1117.
[25] M. Priault, N. Camougrand, K.W. Kinnally, F.M. Vallette, S. Manon, Yeast as a tool to study BAX/mitochondrial interactions in cell death, FEMS Yeast Res. 2003; 15–27.
[26] G. Guihard, G. Bellot, C. Moreau, G. Pradal, N. Ferry, R. Thomy, P. Fichet, K. Meflah, F.M. Vallette, The mitochondrial apoptosis-induced channel (MAC)corresponds to a late apoptotic event, J. Biol. Chem. 279 (45) (2004) 46542–46550.
[27] M. Saito, S.J. Korsmeyer, P.H. Schlesinger, BAX-dependent transport of cytochrome *c* reconstituted in pure liposomes, Nat. Cell Biol. 2 (2000) 553–555.
[28] G. Van Loo, H. Demol, M. van Gurp, B. Hoorelbeke, P. Schotte, R. Beyaert, B. Zhivotovsky, K. Gevaert, W. Declercq, J. Vandekerckhove, P. Vandenabeele, A matrix-assisted laser desorption ionization post-source decay (MALDI-PSD) analysis of proteins released from isolated liver mitochondria treated with recombinant truncated Bid, Cell Death Differ. 9 (2002) 301–308.
[29] L.Y. Li, X. Luo, X. Wang, M. Madesh, B. Antonsson, S.M. Srnivasula, E.S. Alnemri, G. Hajnoczky, Rapid kinetics of tBid-induced cytochrome *c* and Smac/DIABLO release and mitochondrial depolarization, J. Biol. Chem. 277 (2002) 5651–5659.
[30] M. Rehm, H. Dussmann, J.H. Prehn, Real-time single cell analysis of Smac/DIABLO release during apoptosis, J. Cell Biol. 162 (2003) 1031–1043.
[31] H. Yamaguchi, K. Bhalla, H.-G. Wang, Bax plays a pivotal role in thapsigargin-induced apoptosis in human colon cancer HCT116 cells by controlling Smac/DIABLO and Omi/HtrA2 release from mitochondria, Cancer Res. 63 (2003) 1483–1489.
[32] D. Arnoult, B. Gaume, M. Karbowski, J.C. Sharpe, F. Cecconi, R.J. Youle, Mitochondrial release of AIF and EndoG requires caspase activation downstream of Bax/Bak-mediated permeabilization, EMBO J. 22 (2003) 4385–4399.
[33] R.T. Uren, G. Dewson, C. Bonzon, T. Lithgow, D.D. Newmeyer, R.M. Kluck, Mitochondrial release of apoptogenic proteins: electrostatic interactions can hold cytochrome *c* but not Smac/DIABLO to mitochondrial membranes, J. Biol. Chem. (2005) in press.
[34] B.M. Polster, G. Basañez, A. Etxebarria, J.M. Hardwick, D.G. Nicholls, Calpain I induces cleavage and release of apoptosis inducing factor from isolated mitochondria, J. Biol. Chem. (2005) in press.
[35] G. Basañez, A. Nechushtan, O. Drozhinin, A. Chanturiya, E. Choe, S. Tutt, K.A. Wood, Y. Hsu, J. Zimmerberg, R.J. Youle, BAX, but not BCL-X_L, decreases the lifetime of planar phospholipid bilayer membranes at subnanomolar concentrations, Proc. Natl Acad. Sci. USA 96 (1999) 5492–5497.
[36] L.V. Chernomordik, G.B. Melikyan, Y.A. Chizmadzhev, Biomembrane fusion: a new concept derived from model studies using two interacting planar bilayers, Biochim. Biophys. Acta 906 (1987) 309–352.
[37] L. Yang, T.A. Harroun, T.M. Weiss, L. Ding, H.W. Huang, Barrel-stave model or toroidal model? A case study on melittin pores, Biophys. J. 81 (2001) 1475–1485.
[38] G. Basañez, A.E. Shinnar, J. Zimmerberg, Interaction of hagfish cathelicidin antimicrobial peptides with model lipid membranes, FEBS Lett. 532 (2002) 115–120.
[39] G. Basañez, J. Zimmerberg, HIV and apoptosis: death and the mitochondrion, J. Exp. Med. 193 (2001) F11–F14.

[40] A.N. Chanturiya, G. Basañez, U. Schubert, P. Henklein, J.W. Yewdell, J. Zimmerberg, PB1-F2, an influenza A virus-encoded proapoptotic mitochondrial protein, creates variably sized pores in planar lipid membranes, J. Virol. 78 (2004) 6304–6312.
[41] M.T. Lee, F.Y. Chen, H.W. Huang, Energetics of pore formation induced by membrane active peptides, Biochemistry 43 (2004) 3590–3599.
[42] H.W. Huang, F.-Y. Chen, M.T. Lee, Molecular mechanism of peptide-induced pores in membranes, Phys. Rev. Lett. 92 (2004) 198304-1–198304-4.
[43] G. Basañez, Membrane fusion: the process and its energy suppliers, Cell. Mol. Life Sci. 59 (2002) 1478–1490.
[44] N. Janes, Curvature stress and polymorphism in membranes, Chem. Phys. Lipids 81 (1996) 133–150.
[45] G. Basañez, J.C. Sharpe, J. Galanis, T.B. Brandt, J.M. Hardwick, J. Zimmerberg, BAX-type apoptotic proteins porate pure lipid bilayers through a mechanism sensitive to intrinsic monolayer curvature, J. Biol. Chem. 277 (2002) 49360–49365.
[46] O. Terrones, B. Antonsson, H. Yamaguchi, H.-G. Wang, J. Liu, R.M. Lee, A. Herrmann, G. Basañez, Lipidic pore formation by the concerted action of pro-apoptotic BAX and tBID, J. Biol. Chem. 279 (2004) 30081–30091.
[47] R.F. Epand, J.C. Martinou, S. Montessuit, R.M. Epand, Transbilayer lipid diffusion promoted by BAX: implications for apoptosis, Biochemistry 42 (2003) 14576–14582.
[48] T. Kuwana, M.R. Mackey, G. Perkins, M.H. Ellisman, M. Latterich, R. Schneiter, D.R. Green, D.D. Newmeyer, BID, BAX, and lipids cooperate to form supramolecular openings in the outer mitochondrial membrane, Cell 111 (2002) 331–342.
[49] K.J. Oh, S. Barbuto, N. Meyer, R.-S. Kim, R.J. Collier, S.J. Korsmeyer, Conformational changes in BID, a pro-apoptotic BCL-2 family member, upon membrane-binding: a site-directed spin labeling study, J. Biol. Chem. 280 (2005) 753–767.
[50] X.M. Gong, J. Choi, C.M. Franzin, D. Zhai, J.C. Reed, F.M. Marassi, Conformation of membrane-associated pro-apoptotic tBID, J. Biol. Chem. 279 (2004) 28954–28960.
[51] A. Letai, M.C. Bassik, L. Walensky, M.D. Sorcinelli, S. Weiler, S.J. Korsmeyer, Distinct BH3 domains either sensitize or activate mitochondrial apoptosis, serving as prototype cancer therapeutics, Cancer Cell 2 (2002) 183–193.

SUBJECT INDEX

D

E

F

G

H

I

K

L

N

O

P